Praise Fo

CONSCIOUSNESS IN

"Creative nonfiction at its finest!"

- the Narrative Self

"Everything you wanted to know about the subject of consciousness rendered in as few words as humanly possible."

- the Experiencing Self

"Easy to read, easy to follow....a masterful synthesis of philosophy, neuroscience, and evolutionary biology."

- the Awareness

"Never has the state of Nirvana been so eloquently explained."

- Social Media Persona

"Expect nothing less than a complete paradigm shift in our understanding about man's place amongst the stars."

- the Ego

"Not since *The Matrix* has a story delivered more vocabulary or insight into explaining our inner-worlds."

- the Shadow

"A powerhouse of information....The study of consciousness will never be the same."

- the Prospective Self

Consciousness in a Nutshell

Consciousness Management

A Psychonautical Odyssey

Written by **JAY NELSON**
Produced by **LINDY NELSON**

Illustrations by **KEVIN CUELLAR**

CATALYST PRODUCTIONS

psy-cho-naut \ sī-kō-nȯt \ n. [psych] from ancient Greek, *psyche*, "mind, soul" + naut from Greek, *naútēs*, "sailor, navigator" 1: Someone who likes to explore the mind either by using certain psychoactive substances, or by other means of inducing altered states. 2: an explorer of consciousness or sailor of the soul.

The information contained within this book is strictly for educational and entertainment purposes only. It is not intended to be a substitute for medical or legal advice. If you wish to engage in psychonautical explorations, you do so at your own risk. Additionally, while the topics presented have been thoroughly researched and cited, it is important to keep in mind that this is still a work of creative non-fiction, akin to a memoir; and that the 'know-it-all' neuropsychopharmacologists ("James" and "Ava") are actually just characters based on ourselves. Most of the names, locations, and incidents have been changed to protect personal privacy, although in some cases, a few characters have been collapsed into one to expedite the delivery of a certain point.

Cover art and illustrations by Kevin Cuellar

Printed in the United States of America
First Edition

ISBN: 979-8-9859802-0-2 - Hardcover
ISBN: 979-8-9859802-1-9 - Paperback
ISBN: 979-8-9859802-2-6 - E-Book

CATALYST PRODUCTIONS

For Eva

Contents

Introductions. xviii

PART 1 : THE STAGE | xxvi

Chapter 1: Trapped 1

    Methodology 6

Chapter 2: Source Code. 10

    Debunking Folk Intuition 15

Chapter 3: The Movie in your Mind. 19

    Filters 20

    Memory 24

Chapter 4: Consciousness as Software 30

    Compartmentalization 35

Chapter 5: Intro to ASCs 38

    Mental Cartography. 40

    Humans & Computers 41

    Intro to Self 48

Chapter 6: Three Cases of Brain Trauma. 54

    Metadata. 60

    Regarding Phrenology. 61

    A Parting Note on Plasticity 65

Chapter 7: Sensory Awareness. 66

    The Limits of CS5. 71

    The Dials of Perception. 74

Floodgates 77

Brief Recap 78

Chapter 8: Sensory Input = Consciousness Output* 81

Sensory Substitution 86

Chapter 9: What is Reality? 90

Eye Cameras and Vision as a Layered System 94

The Internal Model.. 96

Frame Rates of Reality & the Soap Opera Effect.............. 98

Chapter 10: Emotional Coding............................... 101

If This, Then That 102

PART II : ALTERED STATES | 111

Chapter 11: Cuckoos Part II................................ 112

The Age of an Arms Race 117

Back on the Couch 119

Chapter 12: Non-Ordinary States 126

Interactions with the Unconscious......................... 128

Lenses and Filters 130

Access Points 134

Chapter 13: Unlearning.................................... 137

Flow States.. 139

The Deep Now 142

Chapter 14: Pinning it Down 145

Localization & Lateralization............................. 147

Observing the Mind 150

Chapter 15: Disentangling the Self 154

The Developing Brain 157

Experience vs. Memory 166

Practical Examples 167

Final Thoughts 170

Chapter 16: Mental Disorders 171

Dysfunction and the Default Mode Network 172

The Problem with Time 176

Chapter 17: MDMA, Psilocybin and LSD...................180

    LSD (lysergic acid diethylamide): 181

    Psilocybin: 183

    Mind Manifesting.. 187

Chapter 18: Piecing it all together............................191

    DMN 193

    LSD 194

Chapter 19: Experiences. 198

    Dissolvability of Self 199

    Identity 201

    Blackouts and Light Pollution 203

    Self-Image & the Belief Box 204

Chapter 20: DEMP Cont'd 210

    Malleability 211

    Possibility 213

    Conclusions about the Four Rites of Passage 216

Chapter 21: The 'Chair' Incident.............................218

    Koreatown Cont'd 222

    Fear and Time Dilation.................................... 223

Chapter 22: Koreatown Revisited 227

    Patterns of Thought 230

    The Oddball Effect.. 234

Chapter 23: Synesthesia 237

    Koreatown Cont'd.. 241

Chapter 24: Synesthesia Cont'd..............................245

Chapter 25: Off the Rails....................................249

    Major Synesthesia 250

    Koreatown Cont'd.. 251

Chapter 26: What Really Happened 257

    Debriefing & Story B 259

    The Birthplace of Story B 262

    The Magic Camera 265

PART III : ORIGINS | 271

Chapter 27: Gold, Diamonds and CS5 272
(*The Origin of Everything*)...................................272
    A Crash Course in Cosmology 273
    How Did We Get Here? 276
    Sex 101... 279
    Brief Recap 284
Chapter 28: UNITY..288
    Fundamental Assumptions 291
Chapter 29: Concepts of Infinity 296
    The World of the Dream 299
    Infinity Cont'd... 301
    Philosophers Cont'd.. 303
    Science Cont'd... 304

PART IV : INTERPRETATIONS | 309

Chapter 30: Piecing it all Together II 310
    Three Worldviews Cont'd 312
    Tracing the Idea through Time.............................. 316
    The Monk's Final Lesson 319
Chapter 31: Speciation Events................................322
    Darwin's Finches 324
    Islands... 325
    Mosquitos in the Underground 327
    Artificial Selection 332
Chapter 32: Interdependence.................................335
    The Problem with Emergence 336
    Eastern Philosophy 339
    Redefining the Environment 341
    Time-lapse 344
Chapter 33: Cuckoos III & Story B............................347

The Void 350

Piecing It All Together III 351

Chapter 34: A Parting Note on Opposition. 359

Infinity. 362

Tao Cont'd 365

Brief Recap 366

Chapter 35: The Ava Ending 373

Acknowledgements 389

Endnotes/Bibliography 391

Image Attributions 435

Glossary. 439

About the Authors 449

(Scan QR Code for color images of all photos and figures)

www.consciousnessinanutshell.com/gallery

"To be is to be perceived."

- George Berkeley
(philosopher)

Editor's Note: This book is written in the style of creative non-fiction, or narrative non-fiction. In this genre, authors tend to use the techniques and literary elements of poets, playwrights, and fiction writers to present a non-fiction narrative.

A good example of this might be legendary acting teacher Konstantin Stanislavski's book *An Actor Prepares*. After realizing that much of the subtle nuances and emotionality of his method were sure to be lost in a strict transcription of his lectures, Stanislavski decided to have his lessons live on a different form: a diary. Written from the perspective of a first year acting student attending a prestigious acting school. Reading from this mock diary, it becomes increasingly clear that the highly revered acting teacher, Tortsov, is really just Stanislavski under a different guise. Yet, by unpacking his teachings through the psyche of a young student, this great teacher allowed his readers to revel in the craft of acting versus being told to. Making the lessons presented not only more memorable but a little more personable as well.

Somewhere between an extended "Thought Experiment" and the style of *An Actor Prepares* lies *Consciousness in a Nutshell*.

Consciousness Unleashed

A Psychonautical Odyssey

Introductions

First his eyes widen, then he creeps down low, watching. Inching closer and closer to his prey, he flattens himself out and raises one paw. Eyes narrowing, he starts to wiggle his butt and sway his head from head side to side. You can tell he's moments from pouncing, when you start to wonder...*Why is he moving his head and body like that?*

Or how about when you're at a restaurant. You're sitting in the main dining room, just waiting for your friends to arrive, when somewhere behind you a clumsy waiter lets a champagne flute topple off his tray and come crashing to the floor. Hearing this, you instinctively swivel around to your five o'clock with enough time to see a piece of the stem break off from the bowl and come skirting across the dining room floor.

But how did you know where to look?

Or what about when you're running late to a movie theater across town. You thought you knew the way, but after a couple minutes of driving, you start to get the sneaking suspicion you might be lost. Panicking, you fire up your phone's GPS, and within seconds your phone's receiving pings from at least three or four satellites (each orbiting thousands of miles away). And yet, somehow, based on the time-lag differences between these pings, your device is able to work out your exact location inside of ten feet.

How is this possible?

In all three of these scenarios—the champagne flute, the cat, the GPS satellites—the concept of triangulation is what's saving the day. Technically, with GPS it's called "trilateration," but the same basic principle still applies. If you know where satellites A, B, and C are, then based on these three known data points, you can determine an unknown data point (where you are).

With each sway of the cat's head, not only is he able to capture a different image on the backs of his retinas, but he's also helping his brain to compare the difference between what his left eye sees and what his right eye sees—that way, when he finally does make the jump, he does so with pinpoint precision.

With the champagne flute, the sound of the breaking glass hits your right ear about eight millionths of a second before hitting your left ear, and without even having to know what *interaural intensity difference* is, or what *interaural time difference* even means, your unconscious mind is able to localize the sound, orient your attention, and stop a piece of the flute sliding across the floor with your foot.

Three cases of triangulation; three scenarios in which we harness the power of known data points to solve unknown ones. So now, the question is, when it comes to tracking down the elusive definition of consciousness, might we be able to use a similar strategy?

However, in lieu of answering that question directly, we've opted to write this book instead. A narrative non-fiction novel that seeks to inform as it entertains, calling forth only the best arguments from neuroscience, biology, and philosophy to answer the age-old question: What is consciousness?

In many ways, we already know what consciousness is. We feel it every morning when we open our eyes. We experience it waxing and waning as the day marches on, until sometime after sundown when it all starts to slip away. Descending into an altered state of consciousness we call *sleeping*; here we're not exactly awake, but we're not exactly shut down either. Rather, we exist between worlds, as our brains consolidate memories and clean house.[1]

This is consciousness. Fleeting, but persistent. Inseparable, yet mysterious. You could say it's the very thing that makes life worth living, that it's the very thing that defines our capital "B" Being. Hell, you could even argue that consciousness is the only reason that something like *meaning* even exists, for without some sort of experiencer or conscious agent, how could there be meaning?

Meaning necessitates an interaction, a subject. And as far as we know, the only thing that's capable of producing meaning in the entire universe is us.

Too often, at lectures and events, I'll hear physicists and cosmologists purposefully trying to trivialize our existence. Standing on stage, lecturing about the immense size of the cosmos, they'll say things like, "We're just an average-sized planet, orbiting an average-sized star in a relatively small-sized galaxy, that has somewhere between 100 and 400 billion stars in it." Then, holding their hands like they were bragging about some big marlin they just caught, they'll boast about how our galaxy is just one of trillions.[2] After that, they'll either mention the fact that there's a supermassive black hole at the center of each galaxy, or they'll start comparing the age of our universe (about 13.82 billion years) against the fact that our species has only been around for the last 300,000.[3]

That we've only been recording history for the last 5,000 years.

Then, as if they were actually going to say something profound, they point at the ground and slow their speech. Like a commander leading their troops into battle, they milk that last sentence for all its worth, capping off their little diatribe with something to the effect of, "So in the grand scheme of things...*we* don't matter."

And right out the gate I have to vehemently disagree with this position. Not with the facts but with the conclusion.

Sure, we might be living on a small mote of dust amidst the grand sea of the cosmos but the fact that we have conscious experiences at all is no trifling matter. For a moment, just consider the idea that it took nearly 14 billion years for the universe to produce something that definitively has consciousness, and that the oldest fossils on record indicate that life has been existing on this planet for at least the last 3.5 billion years.[4] Conservatively speaking, that means it's taken at least 3.5 billion years for life on this planet to be able to ask the question "What am I?" and only about a couple thousand years to generate a reasonable response.

To quote Sir Julian Huxley, "Evolution has become conscious of itself."[5] And if we only took an extra second to really think about what that quote implies, I seriously don't understand how anyone could ever utter the words, "*We* don't matter."

Put it this way, between your ears sits this three-pound lump of flesh—the most complicated object we've ever discovered, mind you—and yet, somehow, this bundle of neurons can ask questions about itself like: Where do thoughts come from? Are other animals conscious? Or, why do we even have brains in the first place?

Surely in the age of fMRIs, EEGs, and Positron Emission Tomography we've made some headway with this, haven't we?

For nearly a decade, this has been the subject of our research, collecting and collating all of the greatest minds and philosophical insights under one roof, and then breaking them down into more easily digestible chunks. If we've succeeded, then I think you'll find what follows to read a little less like a medical journal and a little more like a one-on-one chat with an old friend.

Like most books in the creative non-fiction genre, there's a narrative to help guide you through the topic. In Stanislavski's landmark acting book, *An Actor Prepares*, it's a mock diary you're made to learn from, whereas for *Consciousness in a Nutshell*, it's a letter. Told through the eyes of an up-and-coming neuropsychopharmacologist.

You see, by many people's accounts, Konstantin Stanislavski was the greatest acting teacher who ever lived. He knew he could never recreate the emotionality and weight of the acting conservatory experience simply by publishing his lectures verbatim, so what he did was publish three "fictional" volumes entitled: *An Actor Prepares*, *Building a Character*, and *Creating a Role*—all written from the perspective of an observant student named "Kotsya."

Stanislavski could've theorized abstractly, or listed his favorite exercises plainly, but he opted to have a dialogue instead. And so, in 1928, he began writing his teachings down, in a diary, as if he were one of the students himself.

Pulling bits from his life, his lectures, and then weaving it all together with a bit of fiction, Stanislavski found that he could expose more of the psychological components to acting than any textbook ever could, helping to secure his three-part series into the top five best books on acting ever written. Granted, this book is not about acting per se, but seeing as, teaching someone 'what' a psyche is can be just as problematic as teaching someone how to recreate one on stage, this book cannot be presented in your usual run-of-the-mill non-fiction form. Instead, it'll be presented more like a dialogue.

Like light, consciousness is still one of those subjects that's hard to wrap your mind around. We so desperately crave the kind of definition that could fit on a T-shirt like: "Light is a particle" or "Light is a wave," when the truth is actually much stranger. *Light is both a particle and a wave*. And understanding this key fact about light is step one in understanding the complexity of consciousness.

Consciousness isn't just the contents of your conscious experience because it's also the unconscious processes that give rise to your conscious experience.

To show you what I mean, just pretend we were trying to describe what light is instead. Supposing that were the task of this book, the first thing we'd probably do is have you close your eyes and picture the rainbow. After that, we'd ask you to picture all the colors you've ever seen (every sunset, every full moon, every shade of green). Then, once it felt we had some confidence about what light was, we'd pull out some game-changing fact like, *All the light that our eyes can even detect is less than one-half of one percent of the entire electromagnetic spectrum.*[6]

Of course, whenever we usually start thinking about light, the tendency is to only picture visible light, like sunlight or fluorescent lights, because more often than not, we forget about x-rays, gamma rays, ultraviolet, and infrared. Simply because we can't 'see' them, we forget about microwaves and radio waves too, and yet, the fact still remains, they're all forms of light; we just don't tend to think of them that way. Truth is, they all exist on the same spectrum—inside the very room you're sitting in—it's just that human beings only ever evolved the sensory modalities necessary to perceive one ten-trillionth of the entire electromagnetic spectrum.[7]

Against this backdrop, if we truly were trying to have a discussion about light, it would certainly follow that we couldn't just talk about the teeny tiny band of light humans can detect because we'd also have to talk rather extensively about all the light we can't. Effectively, we'd be contrasting visible forms of light against "invisible" forms of light; and through the process of selective comparison, we would eventually come to understand the transmutable nature of light (i.e., its ability to be X-rays, or gamma rays, or infrared, etc.). And in a similar way, that's exactly what we've done here—only it isn't light, it's consciousness. And instead of focusing all our attention on the baseline, wakeful conscious state we have to talk about when it bends and breaks. When it's...*altered*.

Much like a cat swaying its head from side to side capturing different images of what its prey looks like, we too are going to have to consistently shift our perspective so we can tackle this subject from many different angles. Call it a multidisciplinary approach, call it a systematic disentanglement of a complex issue through various fields of study...*this* is the only sensible way to talk about consciousness. A multimodal conversation that spans the fields of psychology, cosmology, ethology, astrophysics, evolutionary biology, filmmaking, coding, cognitive neuroscience, philosophy, mythology, ecology, psychonautics, and neuropsychopharmacology.

To be fair, not a single one of these fields can give you the answer to 'what' consciousness is, or 'how' it came about, but the combined effort of all of them can. And to be perfectly honest, you don't even have to be an expert in any of these fields to understand this topic. All you really need is to have a working understanding behind these various fields of study, which is where "James" comes in (our faithful narrator). By following the story of his research and how he came to understand the subject of consciousness, you will be catapulted into and out of all of the aforementioned topics.

Presented in four unequal parts, it should probably be stated at the outset, that everything about this book was constructed with momentum in mind. By carefully leading the reader through some of the biggest breakthroughs in science and philosophy we're actually going to solve this problem once and for all.

In Part I, we'll mainly be discussing the brain and a few key characters as we seek to gain knowledge around some of the neural correlates of consciousness and begin to understand how the structure of the brain informs that experience. Then, in Part II, we'll provide the reader with a story that contrasts normal, unaltered brain states to non-ordinary states of consciousness like anesthesia, dreaming, sensory deprivation, the psychedelic state, and so on, in an effort to really solidify our vocabulary. In this regard, the first two parts of the book are really designed to help prime our brains for how to talk about consciousness correctly, while Part III exists to explain where it all came from. Here, we'll finally be able to map our understanding of what consciousness is onto an actual timeline. Showing just how and why something like an embodied, conscious observer could have even evolved in the first place. Just after that, we'll be discussing the philosophical implications of conscious minds before whipping everything we've learned into a frenzy of comprehension in Part IV.

Make no mistake, there's a great deal of information one needs to be familiar with in order to conceptualize consciousness properly, but by the end of Part III, we will have already answered the question "What is consciousness?" (and from multiple different angles). Currently, there is no smoking-gun type evidence anyone could ever point to as the definitive origin of consciousness, however with as many polymaths, great thinkers, and philosophers that have come before us, we make the claim that there is an answer to the "hard problem" of consciousness—and that's what makes up the bulk of Part IV.

This final portion is appropriately titled "Interpretations" as there are many different ways one could interpret the facts. For instance, some people look at the size of our universe only to conclude that its empty, void, and meaningless. While others, such as ourselves, see just the opposite: a rich, grandiose tapestry of life which we're not only a part of but deeply connected to. Matter of fact, that's one of the biggest advantages to having a brain like ours in the first place. It's this ability to shift and change perspectives. To ingest long-form arguments, in the form of a lecture or a book, and then come to see our place in the world in a new and novel way. Just like time, consciousness is still one of the great mysteries of science, but this story should get you much closer to understanding what it means to be alive because what you're actually holding in your hands isn't just one person's take on the matter, it's humanity's greatest crack at the age-old mystery. A celebration of sorts, not only for how far we've come but for that curious thing we all seem to share as a species: *consciousness*.

Chapter 1

Trapped

WE WERE WATCHING the nature channel when it happened. Sitting on that gaudy leather couch when the ground beneath my thoughts just completely drops away, plunging me into the world of thoughts and dreams; ideas and connections. The place where sheer chaos meets practical order.

From somewhere beyond the screen stands everyone's favorite wildlife enthusiast, holed up in some recording booth, sipping on some herbal tea, as he masterfully narrates the scene. This drone shot of some wildlife preserve suddenly gives way to a couple aerial shots of this beautiful, grey bird gliding along the tree line. With long grey feathers and piercing gold eyes, at first we think it might be a hawk—that is, until the narrator starts in with his famous whisper narration.

"The common cuckoo," the narrator says, "is technically, a parasite." Parasitic to a species of birds known as the reed warbler, the guy everyone secretly wishes was their long-lost grandfather goes onto explain how *Cuculus canorus* might be Nature's most ingenious con-artist.

Before some clocksmith decided to incorporate its mating call into one of his clocks, before the word came to be associated with someone crazy or "off their rocker," ornithologists have long been fascinated by this odd species of bird. Namely, because cuckoos are one of the few avian species that don't bring up any of their young. Instead, they trick other species of birds into doing all the work for them.

Ava sitting next to me, clutching that big blue bowl we only use for popcorn, she pulls up the plush blanket coiled around her feet and shoots me a look.

"Watching from afar, the cuckoo must wait patiently for the mother to leave her nest, for when she finally does, the cuckoo will only have seconds."

Completely transfixed by what's on the TV, Ava and I watch as this rather innocent-looking bird, about the size of a sparrowhawk, swoops down to an itty bitty nest perched above some flooded reeds and stands on its edge. Head looking both ways, we then watch as this golden-eyed trickster dips her head down into the nest only to resurface holding a spotted green egg seconds later.

Then, she tilts her head back.

Like a pelican, she jerks her head forward a few times so she can swallow the egg all in one go, and if that weren't bad enough, she even lays one of her own in its place.

"The whole endeavor takes less than ten seconds," boasts the narrator, "and within that time, the cuckoo has just destroyed this entire family's hopes of reproductive success."[1] Knowing where this is going, I reach for a fistful of popcorn and tell Ava, "Watch this."

The next shot is of some poor unsuspecting mother, about the size of a sparrow, flying back to her nest completely unaware that one of the eggs is not her own. What's worse, this impostor egg she's about to help incubate will be the first to hatch, and when it does, the first thing it'll do is slaughter all its would-be brothers and sisters.

Just then Ava turns to me with arched eyebrows and says, "This is crazy." The way she's sitting, it's the same way you'd sit if you were watching a slasher flick. Legs tucked underneath her. Legs tucked off to the side. She's sitting all the way up, blindly grabbing for popcorn, when the baby cuckoo on TV finally busts out of its shell. Naked and featherless, it then walks on limbs it's never used before over to one of the four remaining eggs and starts to do the unthinkable. Using its back and its legs, it rolls one of the eggs up the side of the nest, hoisting its foster-brother up onto its shoulders, like it was Atlas about to move some furniture, when suddenly, the cuckoo staggers back. She's standing at the edge of the nest now, the very place her mother stood eleven days before, when the baby cuckoo finally flicks its "wings" and the egg goes flying.

"The cuckoo will repeat the process until there's nothing left," whispers the narrator. But even in that sultry British tone, these words send a shiver down my spine. Not even ten minutes spent on the other side of that shell and this chick has already committed a quadruple homicide.

Halfway around the world, sitting on that tattered leather couch, you almost can't help but feel anger towards this little monster and nothing but sympathy for the reed warblers duped into feeding it.

But, of course, this isn't really the right way of looking at it.

First and foremost, we need to remind ourselves that both creatures are only ever trying to do what we're all trying to do: *survive and reproduce*. And second of all, we actually have no reason to suspect that either party actually knows what they're doing. Which isn't to say that birds don't have some semblance of awareness. More like, they might be better conceptualized as being unconscious robots acting largely on instinct. "Clockwork robots" is the way the world's leading evolutionary biologist, Richard Dawkins, would describe it.[2] But sitting at home, munching on popcorn, it's just way more convenient to start making up stories. To start picking sides and assigning agency.

To think, that somehow, this blind bird (no older than a few minutes) had some kind of diabolical plan etched on the inside of that shell, the first item of which read: *Kill all your would-be brothers and sisters*.

The baby cuckoo almost certainly doesn't think that.

It doesn't toss out all of its would-be brothers and sisters because it 'wants' to, it does this because it was 'programmed' to do this. Just as a newborn baby has instincts for sucking, a newborn cuckoo has instincts for bucking. An instinct for balancing an egg up atop the hollow bone in its back and then ejecting it off the side just as easy. In actual truth, the poor cuckoo never meets its real mother. It never builds its own nest or rears any of its own young. It's never explicitly shown all the tricks behind successful brood parasitism—of which there are several—and yet, in less than one year's time, she too will make the return journey all the way from Africa to this same marshy area of Wicken Fen to play the trick on another host family.[3]

You want to talk about learned behavior vs. instinctual behavior? About the ingenious tricks of survival crafted by eons of time and that righteous caretaker we call Mother Nature?

Then let me introduce you to *Cuculus canorus*.

By the time the nature doc was finished, Ava turns to me with a face that didn't know what it wanted to be—anger, fear, curiosity, respect—just a mishmash of conflicting emotions. And even though there weren't any words exchanged, knowing Ava, I still knew what she was thinking. She was thinking something to the effect of, "This is *life*?"

Then, me looking back at her, feeling like an alien or some kind of robot trapped in my own skin, I force a half-smile. My eyes looking at the floor, then back up to find hers, I reply in kind.

Yes, this is life.

But for some reason this nonverbal response stirs up something dark in her, and in that moment Ava grabs the remote. Mashing the button with her fingers, she hits the button labeled "OFF."

—

Here's the thing, in six hours…I will be dead. In less than six hours, I will be nothing more than a memory in the minds of the few people that knew me—and you're going to have your whole life ahead of you—so don't waste your time investing emotions into someone like me. I'm no hero. I'm no saint. I just happened to be in the right place at the right time when an idea just fell into my lap.

Correction: *our laps.*

And then, well…life happened. I don't want to bore you with the details, so let's just leave it by saying that, through no fault of my own, I wound up getting trapped in a room without doors or windows. Trapped in a place where if I'm not dead in six to eight hours, I'll wish I was.

The story I want to tell, it's only the greatest story of all time, quite possibly because this is the story of how we got here. And by "we" I don't mean Ava and me, I mean it in the grandest sense of the word. We meaning *all life*.

You know, in a perfect world, this might be the kind of story I'd tell you over the course of a few semesters. Slowly but surely building our arguments from the bottom up, but sadly, we just don't have that kind of time. On account of me dying and all, we simply aren't going to have enough time for me to answer every question, address every outlier, or go on every tangent. However, I will be able to offer you the abridged version of how we got here.

Right about now, I'm sure you're probably wondering why I'm even writing to you in the first place. I mean, you don't know me. I don't know you. But truth be told, that's probably a good thing. Sometimes, you can be more honest with a total stranger than you can be with a friend, and for this story to work, I'm going to have to be brutally honest about something I regret.

Once upon a time, the way I used to tell this story was through a giant lecture series. Starting with a firm understanding about what evolution is, and then building upon that foundation with ever-increasing complexity. In fact, this whole thing used to be more like a laundry list of facts than a story. It was this gigantic, monstrous, detail-ridden manuscript dog-eared with hundreds of those little Post-it notes and factoids. Until one day, I caught a glimpse of how to make it shorter. The moment after that, the love of my life disappears and somehow I wind up here. Trapped in the kind of place you wouldn't wish upon your worst enemy.

I apologize for having to put the matter so bluntly, but when you can literally calculate the number of seconds you have left in your head you tend to stop sugar-coating the inevitable...*I'm dying.*

Dying I can live with.

What I can't live with is the idea that I'd be breaking a promise I made to an old friend.

You see, a long time ago, I made a promise that whatever happened I'd figure out a way to share our research.

"No matter what," I said.

So if I can't somehow figure out a way to distill a decade's worth of research into this one letter to you, then my entire life's work will have been for nothing.

Where I'm trapped, Ava isn't here, and neither is that manuscript. Although, in order to tell this story correctly, I'm going to have to imagine she is. In all fairness, I'm going to have to imagine that all of my mentors, teachers, and colleagues are standing behind me if I'm to make it through this thing. So if you ever catch me using the royal "we" as I try and explain something, just realize that, for me, everything about this story is a collective endeavor. And if I'm being perfectly honest, the only reason I'm even able to explain some of the finer points about consciousness or evolution is because I'm standing on the shoulders of giants, who were once standing on the shoulders of giants, and I simply cannot stress that point enough.

Who I am is of no importance; what I have to tell you is.

That said, for the remainder of this thing I'm going to have to rename you to Tom, if that's alright. I realize that's not your real name, but calling you Tom will make some of our thought experiments go a lot easier, and in the interest of all fairness I'll take on a pseudonym too. From here on out, you can call me James.

Methodology

Back in the mid 1930s, the father of modern ethology, Konrad Lorenz, was studying the imprinting behavior exhibited by greylag geese when he made a discovery of sorts. Just for context, Konrad Lorenz wasn't the first person to discover that little gosling's will imprint on the first moving object they see, like a pair of Wellington boots or a model train, but he was the first to really popularize this finding, helping to secure ethology (the study of animal behavior) as a legitimate sub-discipline of biology.[4,5] Although, if you asked Lorenz, he'd have probably said that his most significant contribution to science wasn't the discovery of a new animal or a behavior, but rather a new way of interpreting the facts. More specifically, it was Konrad Lorenz who first got hold of the idea that instincts, or "fixed action patterns" as he called them, could be better thought of as being like anatomical organs.[6,7]

Of course, nowadays, we take this sort of observation for granted.

Because this idea came to us before you and I were even conceived, it seems rather easy for us to accept the idea that we might be able to identify a certain species of animal solely by watching their behavior, but at the time, this idea was revolutionary. Like most great ideas, it seems rather obvious to us in hindsight. However, when it first came out (the idea that certain behaviors are *innate*, while others are *learned*) certainly caused some ripples in the intellectual community. But then, by employing some clever experimentation techniques, Lorenz was able to show the limitations of these instincts.

How so?

Basically, it breaks down like this: Genes build bodies (we have tons of genes that contribute to the building and shaping of bodies through the processes of embryology), but when a gene exhibits a particular effect on a body, say slightly longer tail feathers or piercing yellow eyes, we call those effects *phenotypes*. And what Konrad Lorenz bravely conjectured nearly a century ago was that these fixed action patterns weren't being consciously worked out by the individual; they were inborn. Or, put in another way, these rather stereotypical patterns of behavior weren't being consciously passed down or imitated, because really, they were just phenotypic expressions of a certain gene.

So, for example, the way a bird constructs a nest isn't by watching its bird parents or by going to some bird primary school. Rather, a bird builds a nest in much the same way it grows a liver or it grows a stomach; it just does it. A bird doesn't have to know how or why it follows these highly stereotypical patterns of behavior when it comes to building nests or bringing up chicks, and it doesn't have to.

Thing is, these fixed action patterns once helped its ancestors to survive; these patterns were subsequently selected for, implying that in our times, a bird can now exhibit wildly complex patterns of behavior, like brood parasitism or egg-smashery, without ever being shown how or why.

Why'd we start off talking about cuckoos and not human behavior?

In part, it's because instinctual behavior is a lot easier to spot in birds than it is in humans, but more importantly, it's because the cuckoo story just happens to be the perfect vehicle for describing some of the more nuanced points about evolution.

With most birds, even the most severe skeptic could always make the claim that somehow these birds were teaching their offspring how to hunt or how to find mates—but not with cuckoos. With cuckoos, we can be certain the parents aren't passing down any life lessons to their kids because by the time their chicks hatch, the real mother is already halfway back to Africa. Presumably, on her way to accept some kind of *Mother of the Year* award, seeing as she just laid between eight and twenty-five eggs that season while suffering none of the economic costs involved in bringing up a family.[8]

No nests. No child support. No visits on weekends.

What's more, you know the foster parents never taught these cuckoos how to parasitize another host family, so how else do we account for this behavior?

This right here—this is just one of the many intellectual hurdles we're going to have to leap over in order to fully understand what consciousness is because, believe it or not, there are unconscious forces operating within us that drive the majority of what we do. In all honesty, we're not even aware of most of the learning that shapes our behavior.

———

Over the course of this conversation, I hope to be able to explain some of the differences between instinctual behavior and learned behavior, between unconscious cognition and conscious cognition. And being that this is the big elephant in the room when it comes to this subject, I thought you two should meet. The real problem is, our brains are so biased in the way that we learn, think, and perceive that we often have a hard time comprehending unconscious learning and unconscious behavior. Which, when you think about it, is kind of a big deal, given that 95 to 99% of our cognition is unconscious.[9]

For the sake of clarity, we should probably mention that most experts fall into the 99% camp. However, because there are some who've low-balled the estimates of what the subconscious is capable of, I suppose we'll grant the conscious mind a little footing before we systematically start pulling the rug. Also for clarity, please note that we'll be using the words unconscious and subconscious rather interchangeably, but please don't let prefixes like "un" or "sub" mislead you into thinking that these processes are somehow beneath you in some way. Because really, these unconscious processes give rise to you and your thinking. And now that we've covered that, we may now draw our first line in the sand together.

On one side, I want you to imagine all the conscious aspects of consciousness (that which you're aware of and can see). On the other, a buzzing flurry of electrical activity (that which you have no awareness of and cannot see). Arguably, this is the side of consciousness that's of most interest to someone like me, for without these sorts of unconscious processes I wouldn't be able to walk in a straight line, pick up a glass of water, or even write a letter to a stranger, such as yourself. As amazing as the conscious mind is, it really does fail to compare against what the subconscious is capable of. And to explain this one massive, massive point, we're going to have to flashback to the first time you could say I really "met" Ava.

Back when I thought I knew everything, but really, didn't know much of anything.

For the record, I still don't think I know anything, but what little I do know, I'll tell you in exchange for a favor. A question, really. So there you have it, everything I know about the subject of consciousness in exchange for one little favor.

If that's not a deal of a lifetime, I don't know what is.

Chapter 2
Source Code

THE FIRST TIME I really met Ava, she was designing websites to help pay for graduate school. She'd build them for college kids and local businesses in the bay area. Watching her write thousands of lines of code just so a local delicatessen could have their logo animate across the page and explode into a sandwich never seemed like a good time to me, but she liked it. It was creation. Her way of painting and I dug that.

Call me crazy but I thought it was attractive. Watching her turn nothing into something. Watching her take a blank page and turn it into someone else's dream. The front-end of their start-up; the web domain for their brain-child. She'd talk to me about Style Sheets and Search Engine Optimization, and the whole time, I'd just stare at her face. Watching her lips curl up when she got really excited about learning Java or designing her own CMS.

The first time she told me she knew three languages, I said, "What, like German?"

She thought I was being funny.

We never saw eye to eye when it came to web design, mainly because Ava was hypercritical of every site I ever logged on. If she saw something she'd never seen before, she'd take my mouse, right-click, select "View Source," and suddenly, we'd be taken to a land of confusing shit. A land she called "the source code." Ava would later explain to me how there's a front-end of a website (what a consumer might see whenever they log onto amazon.com or whatever), and then, there was a back-end (what a programmer might see).

Watching her work, I got to see first-hand that the front-end of any website is really just an interactive illusion put on by the source code. Every animated cat GIF, every desired font, every backlink all has to be accounted for and coded into the source code. Which also means that some poor schmuck has to sit behind a screen for hours just so they can turn a blank page into something like this:

```
<html lang="en" class="focus-within">
▶ <head>…</head>
▼ <body class="focus-within">
  ▶ <script type="text/javascript">…</script>
    <script>window.clientSideRender = false;</script>
    <!--pageHtmlEmbeds.bodyStart start-->
    <script type="wix/htmlEmbeds" id="pageHtmlEmbeds.bodyStart start"></script>
    <script type="wix/htmlEmbeds" id="pageHtmlEmbeds.bodyStart end"></script>
    <!--pageHtmlEmbeds.bodyStart end-->
  ▶ <script id="wix-first-paint">…</script>
  ▶ <pages-css id="pages-css">…</pages-css>
  ▼ <div id="SITE_CONTAINER" class="focus-within">
    ▼ <div id="main_MF" class="focus-within">
        <div id="SCROLL_TO_TOP" class="qhwIj ignore-focus" tabindex="-1" role="region" aria-label="top of page">
        <button id="SKIP_TO_CONTENT_BTN" class="_2SeL5 has-custom-focus" tabindex="0">Skip to Main Content</butt
      ▶ <div id="BACKGROUND_GROUP">…</div>
      ▼ <div id="site-root" class="focus-within">
        ▼ <div id="masterPage" class="mesh-layout focus-within">  grid
          ▶ <header tabindex="-1" id="SITE_HEADER_WRAPPER">…</header>
          ▶ <main id="PAGES_CONTAINER" tabindex="-1" class>…</main>
          ▶ <div id="soapAfterPagesContainer" class="page-without-sosp">…</div>
          ▼ <footer tabindex="-1" id="SITE_FOOTER_WRAPPER" class="focus-within">
            ▼ <div id="SITE_FOOTER" class="_3Fgqs">
```

Little lines of code she'd charge you 1500 bucks for.

Sure, they might look like useless strings of characters right now, but feed them into an internet browser and you got a beautiful landing page replete with an adaptive layout, navigation bar, and a content slider.

Again, not exactly how I like to spend my free time, but there was at least one thing I latched onto during all those lessons and it was that lovely pairing of words "source code." A beautiful concept, I thought, since years before she and I actually linked up, I was already loosely writing a book about consciousness.

Which, incidentally, is kind of how Ava and I met.

I suppose our paths would've crossed eventually, I mean ██████ ██ not that big of a campus. But she was one of the first people to speak at our writers' group. When it was time for feedback, she got up and told me that I didn't really have a voice. Not that I couldn't speak, but that my writing lacked originality. "And in a field already saturated with droll papers," she said, "mine was never going to stand out." She even went so far as to reread one of my passages in front of the whole group, capping it all off with, "This could've been written by anybody." Her style of feedback, the complete opposite of a compliment-sandwich. My face got hot when she said it, and sure I got defensive, but sitting in that circle, in a group I started, my ego was crushed. It took me a couple days...

Correction: *A couple weeks*

To realize her critique was spot on, and that I probably owed her an apology. Funny how the people you like the least end up being the people you respect the most.

Flash forward to about three months later when Ava practically lives with me, and now she's trying to teach me how to write source code. Although, I wouldn't really call it teaching as much as I was just watching her talk and letting her think out loud.

Nice couples do that for each other.

Staring at her mouth, admiring her passion, I could feel my eyes moving from her lips to her eyes, her strawberry-blonde hair and then back to her mouth again, all the while thinking, "I'm probably gonna marry this girl." I liked the way she thought. I loved the way her mind worked. I actually liked the fact that she challenged me—I just had to hate it first. The first few weeks of our relationship, it drove me nuts. I wasn't used to being told to keep up on a hike or being told I needed to get out more. Looking back, it really was just the Adult Version of shoving the person you like on the playground. Here we are thinking we're all grown up, and yet we're still playing the same games.

Movies are really just the Adult Version of *Make-Believe*.

Love, the Adult Version of *Hide and Seek*.

The way I see it, life is an endless series of games.

Ava was playing the Coding Game; I was playing the Philosophy/Writing Game; we were both playing the Neuroscience Game. I think it's incredibly useful to know which games you want to play, but perhaps, more importantly, which games you don't.

Coding was never going to be my game, and I knew that. Which is precisely why I never ended up taking those lessons of hers all that seriously. At any rate, even half-distracted, I still ended up walking away with a few key principles about web design, like SSL encryption, user experience, and user-friendly. Concepts I would later find to be indispensable when trying to talk about consciousness abstractly. Because, as you may have already gathered, we don't see the back-end side of consciousness, we see the *user-friendly version*. The beautifully animated front-end with pretty bubbly fonts and custom-tailored pop-ups that tell us things like:

> *Your laundry's still in the dryer.*
> *Tomorrow's street sweeping day.*
> *...and don't forget to book that flight!*

That's the side of consciousness we're privy to. That's the side of consciousness we're 'allowed' to see. What we don't see are the billions of neurons, popping off hundreds of times per second, that give rise to these thoughts, instead, we just experience the thoughts. And while there certainly has be some kind of behind-the-scenes to all this magic, that's definitely not what you and I see. We don't perceive the source code, we see the user-illusion, and really, that's the idea I want you to play with here. That consciousness, at the level of experience, is like an illusion.

———

You see Tom, consciousness is an illusion in the same way that movies are illusions. Chris Hemsworth isn't actually Thor; he can't really fly through a brick wall unscathed. And, sorry to say, but that's not his real accent. Layer upon layer, these cinematic tricks stack up to create something greater than the sum of their parts (a fantastical movie about an arrogant god named Thor), but piece by piece these tricks can be dismantled and explained. And consciousness isn't all that different, just complicated. I'm not saying this movie in your mind is a complete fabrication. All I'm saying is, some movies give you the courtesy of showing you a title card that reads: "This story is based on true events," and consciousness does not.

***Notes from the Editor:**
Instead of bogging the reader down with too many qualifying statements and asides, the narrator has chosen to streamline the discussion, distilling every case study and philosophical point down to its quintessence. If you ever get lost,

just keep in mind that there's a glossary of terms near the back, in addition to the endnotes on each chapter. Traditionally, these sorts of extraneous details and asides might appear at the bottom of the page, but at the request of the authors, we've decided to place them inside the endnotes section to make the arguments even more streamlined.

—Don't get me wrong, websites are illusions too, it's just that whenever you log onto a website there's this implicit understanding that what you see isn't all there is to be seen. Whether there be a back-end or a source code, whenever people log onto a website, on some level, they already know they're only seeing one layer of the website's reality. And yet, when it comes to the subject of consciousness, most people don't assume they're seeing an illusion at all.

Matter of fact, most people assume they're having a direct perception of reality.

To help explain what I mean, let's momentarily pretend you were just a website instead. Supposing that were your entire reality, could you imagine how difficult it might be to convince you that you're part of a layered system? That you might exist as thing we call a front-end, but you also exist as a back-end, a source code, and once we get down to the level of the computer's processor—a string of 1s and 0s we call binary. I mean, we could talk for days about the idea of a source code, but if you never saw your literal "webpage-self" transition from webpage to source code and then back again, would you even believe me, or would you cast me off as "just another fanatic" talking about something that can't be seen?

Do you see how this is precisely the kind of situation we're in?

There is a source code to your conscious experience (something your unconscious mind owns and operates) and Line 1, Rule 1 of that source code goes something like:

Deny all knowledge that consciousness is an illusion.

And really, this is what we're up against here—the fact that we have biases in our own thinking. Cognitive traps we can't always think our way out of. Most of us tend to think we just open our eyes and "there's reality" when the truth is, all we've ever seen is a slightly delayed reconstruction. Our brain's best guess. Granted, it isn't particularly easy trying to describe some of the unconscious mechanisms that give rise to this illusion, but if I haven't bitten off more than I can chew, I don't quite know how to live with myself.

That being said, my hope is, that if I tell you enough stories, facts, and studies about consciousness being a layered system (a reconstruction of reality that takes around 500 ms to synthesize)[1] that eventually, this idea will start become second nature, and by the end, you may even be trying to pick up where I left off.

Debunking Folk Intuition

You might not realize it, but optical illusions are some of the best ways we have for studying vision. Because when optical illusions work, they show us where the visual system breaks down; and when optical illusions don't work, they show us where human perception triumphs.

See *Hollow-Mask Illusion*

See *Necker Cube*

See *Checker Shadow Illusion*

If we saw the way folk intuition says we do, 'looking out' behind these little holes in our head, then these illusions wouldn't work at all. But because most of vision is actually an active process taking place inside our heads, what we actually see is just a low-resolution reconstruction synthesized by our brain.[2]

An internal model or representation of the world, not reality itself.[3]

In this way, consciousness is really just a complex and layered illusion. Kind of like a movie. The only difference is, with movies, there are tens, sometimes even tens of thousands of people working together to pull this trick off. And yet, when it comes to the proverbial movie in your mind there's just you. Well, you and your subconscious.

—

Next time you're at the cinema, and you're watching a movie on the big silver screen, I'm sure it's going to seem as if those big 40 ft tall faces are actually saying words and running from the cops. When in reality, you and

I both know, it's just an illusion. We both know the sound was recorded separately—that it was either sweetened in post, or replaced altogether by some foley artist—and is now being played off multiple speakers, the majority of which aren't even located behind the screen! Even still, even knowing this is all a trick, the brain still makes it *feel* like the sound is still coming out of their faces because this is the brain's best approximation of what's happening.

This is an example of an illusion gone right.

And most of the time, this illusion actually works. We watch people be charming and say catchy things, and we actually believe this dialogue is coming out of their face. But every once in a while, an editor, a projectionist, or a sound mixer will make a horrible error that only reveals itself after one of the characters opens their mouth to say something...but nothing comes out. Then, as it usually goes, about a second or two behind the moment, you hear the words "Back off" rattle out of the speakers, letting you know, without a shadow of a doubt, the audio is out of sync.

Why does this happen?

Admittedly, there can be a million reasons as to why this happens, but at the base of all of them is the idea that the footage you're watching isn't actually moving. Forgive me if this is elementary knowledge, but the reason we call them motion pictures is because each second of a film is actually comprised of 24 frames or images. And it's the rapid succession of these images that produces the illusion of motion (in your mind).

Just like those little flip books we used to play with as kids where you flipped the pages between your thumb and your forefinger really fast. Your brain actually "paints" on motion, when technically speaking, there isn't any. Flip them at a slow enough pace and it's just drawing after drawing, but as you increase the speed of your flips, eventually you come to a point where the brain starts to interpret motion and causality between the individual drawings. And this is precisely what's going on with movies. You may think you're sitting in a theater watching the actors on the screen move. When the reality is, you're sitting in a chair being shown a series of pictures, which when flashed at a fast-enough rate (24 frames per second) causes your brain to paint on motion and interpret causality.

This is another example of an illusion gone right.

However, even this can all go off the rails if, let's say, an editor edits the footage in a 23.98fps timebase while the sound mixer edits the audio in a 24fps timebase. You wouldn't think so but those fractions of a frame add up rather quickly, causing what my old roommate Brian used to call *audio drift*—industry talk for, "Someone just lost their job."

This is an example of an illusion gone wrong.

—

At the end of our coding lesson, Ava throws her head back and starts running her fingers through long, semi-tangled hair. To the ceiling, she says, "You know I usually charge for this sorta thing," and without missing a beat, I immediately get up from my dining room chair and make a beeline towards the kitchen. While she's distracted, I start fumbling through our kitchen's miscellany drawer, pushing aside dead batteries and old manuals until I finally find something small and metallic. Some kind of token for my appreciation. I'm making so much noise trying to find this thing Ava has to dramatically pause what she's doing and turn back in my direction. With her head framed between her hands, she twists around and pulls her hair out of that scrunchy. Then, with eyes trained on me, she moves the scrunchy to her wrist, gathers up all the hair between her two hands and then does some kind of magic trick only girls know.

Moving in front of the drawer, I tell her to close her eyes, and point in the other direction. I wait just long enough to watch the slightest of smiles wash over her face, and when it does I place something small and brass in the middle of her palm.

"You can open your eyes now," I whisper in her ear.

Except when Ava looks down, she's still wearing a slightly confused look on her face. It only takes her a second to realize it's a key.

But not just any key. Because this is a key that means something, and immediately she starts beaming.

"Can you get the mail for me while I'm gone?" I ask, trying to hold a stupid look on my face. That's when Ava punches me in the chest and kisses me on the mouth.

I'd never given a girl a key to my apartment before, but this was a girl I never wanted to watch leave. I think it was her bravery and her honesty that first made me fall in love with her. And basically, she and I have been inseparable ever since.

Chapter 3
The Movie in your Mind

RIGHT NOW, THERE'S a movie running. A movie of your life. Except in this movie, you're not only the star but you're also the writer. You can move the story forward or you can push the story back, but more than that, this isn't just the type of movie you just sit back and enjoy because it's also interactive. It's playful! Consciousness is an immersive story of which you're not only the lead actor, but the director, the camera operator, and the editor. Naturally, you may not think of yourself as being the editor or the DIT guy, simply because it doesn't *feel* like you're doing any of these jobs—although, technically speaking, you've just outsourced this labor to a part of the brain you don't consciously control. A part we call the subconscious.

Whether you realize it or not, all five of your sense organs are streaming in an overabundance of data into your subconscious mind every single second. So much so, that 99% of it just ends up getting tossed by the wayside.[1] What's probably even less obvious is that all these senses inform one another too. Take away your sense of smell and your sense of taste is never going to be the same. Take away your sense of sight or your inner-ear, and your sense of balance is going to be affected as well. You can even test this for yourself too, all you have to do is stand up and pick a point on the wall. Then, once you've found that, pretend you're a flamingo and raise one leg. Your mission is to just try and hold this stance for about sixty seconds and notice how you feel.

Hell, even in my helpless state I could easily hold this position for a minute or two, so let's up the ante. Now, let's try it with both eyes closed.

You wouldn't intuitively think that the visual stream would dramatically affect your sense of balance, but I bet you couldn't even hold that stance for more than fifteen seconds. Of course, it doesn't stop there either because your sense of sight is also helping you hear. You probably wouldn't identify yourself as being a lip reader, but I'll wager a guess and say that this is a trick you unconsciously pull out whenever you're in an extremely loud environment, like a noisy cafe. Even though you may not be aware of it, your eyes are watching the person's mouth, helping to inform your brain about which sounds to focus on, and more importantly, which ones to screen out.

***Notes from the Editor:*

"Screen out" is a term we'll come across often throughout this book, largely because that's a huge part of what the brain is doing all the time: filtering and channeling information. Kind of like my job. The only reason editors have work to begin with is because authors tend to write too much and too hastily. The aftermath of which being first drafts that drone on for hundreds and hundreds of pages. To circumvent this, authors tend to hire an outside source to "kill their babies" as it were, and snip out any information that is deemed non-essential to the theme of the book. At first glance, this may seem like an unnecessary digression, but there exists a large-scale network in the brain, the so-called *Default Mode Network*, which performs a similar task. In fact, you can almost think of the default mode network as being like the "Editor of the mind" as it is continuously policing how much sensory information actually makes it up into the conscious realm.[2]

Filters

We live inside an algorithmic bubble. Every time you go on Facebook, every time you search on Google, and every time you make a purchase, you're handing over information about yourself. And not just what you're searching for, but your location, your likes, your dislikes, your screen size, and of course, all those geotagged photos.

All this data, that you knowingly and unknowingly fork over every day, contributes to the size and shape of bubble in which you live. In this way, the next time you're on Facebook, participating in the Adult Version of *Show and Tell*, try and keep in mind that you're not seeing an objective Facebook reality. Rather, you're seeing a subjective Facebook reality.

The version of Facebook that you see isn't the same one that I see, and this goes way beyond the fact that we have differing friend groups. The version of "Facebook reality" you get is specifically tuned to your settings, your likes, your purchases—even *your friend's* purchases. There's an alliance of algorithms that dictate what you see when you log onto a site like Facebook, and in a similar way, there's an alliance of algorithms that dictate what you notice when you walk into a cocktail party. You've probably heard the expression:

"We see the world, not as it is, but as we are."

- origin unknown

And what this is essentially saying, in just a few words, is that consciousness is the most subjective experience of all time. We don't have an objective view of reality, we don't see the world raw and unfiltered, we see the world based on our previous experiences.

To show you what I mean, let's consider a girl who's a self-described "fashionista." Someone who takes an hour to pick out their outfit, an hour putting on their make-up, a half-hour on their hair, and whose room is covered in glossy magazines and chic multi-canvas wall art. Then, on the opposite side of the spectrum, we can picture a guy who couldn't tell you what he wore two days ago. A guy who might only know a handful of designer names, but is a total cut-up at parties. Sure, he might not remember what you wore or how much it cost, but he's one of those guys who can really spin a yarn.

If we were to then imagine these two people going to the same party, could you imagine just how different their experiences might be? They could even be side by side, "working the room together," as they say, and still walk away with a completely different experience.

How is this possible?

It's possible because our experience of reality is very much like that custom-tailored version of Facebook. Although, this is probably the only scenario in the world where Facebook has *less* data. In any case, the subconscious mind is very much like Facebook in that it uses every single datapoint possible to give you the second-by-second newsfeed of consciousness.

With Facebook, it's relatively easy to explain. You go about your day leaving digital breadcrumbs about where you've been and what you're up to, while a team of algorithms harvest and process that data. From there, the algorithms then decide which ads to push to your feed, which stories to feature, and which ones to shove towards the bottom. I say this is easy to explain because we already know how this came about. A gang of coders built and refined these algorithms with the intent of keeping you on the site for as long as possible. And yet, with the subject of consciousness, things are a bit more complex. In the case of consciousness, the algorithms that dictate your reality weren't designed by a crew of coders; they were designed by you. Albeit, a subconscious you. Making your subconscious something like the ultimate firewall.

What this means is, before you and I consciously become aware of a sensation, that sense data has already been filtered by the subconscious. Meaning, everything you've ever seen first went through an advisory committee you never approved of, and everything you've ever heard had to pass through a censoring system you never green-lit. Even the most intimate memories of your life (the ones you wouldn't trade the world for) well, even they all went through an editing process "you" never authorized.

Nothing about consciousness is raw and unfiltered; everything is processed. Everything has a filter on it, including yourself.

Correction: *Especially yourself.*

Our conscious experience, as amazing as it is, as incredible and convincing as it can be, is the end result of an illusion gone right. Like a movie, it's an artificial recreation of events that may or may not have happened.

You've heard people say, "Don't trust everything you see on TV." You've heard people say, "Don't trust everything you read." And now, take it from someone who looks at brains all day, "Don't trust everything your brain tells you."

See *Optical Illusions*

See *Phantom Limb*

See *Memories*

There's actually a well-documented phenomenon called the McGurk effect which perfectly demonstrates how the visual stream can 'edit' what it is we hear. In this particular study, we see participants being shown a video of a man saying "Ba" over and over. Just a guy saying "Ba...Ba... Ba" again and again. Now, when participants are first shown this clip they hear "Ba" because they can clearly *see* and *hear* the guy saying "Ba," which is exactly what you'd expect. But when researchers layer a video of a man saying "Fa...Fa...Fa" over the original audio "Ba...Ba...Ba" the participants then hear "Fa." Please note: the audio track hasn't been altered in the slightest, all that's really changed is now the visual track is of a guy saying "Fa" instead of "Ba." If you close your eyes and just listen, you can still hear that it's a guy saying "Ba," but the second you open your eyes and register the lip-curling movements associated with a "Fa" sound, your brain *autocorrects* what you hear to match what it sees.[3]

In other words, it changes the sound from "Ba" to "Fa."

What's more, you could even go watch this video right now, knowing full-well that your brain is going to try and perform this correction, and still not be able to stop it.

Bearing all these facts in mind, if the McGurk effect teaches us anything, it's that according to the brain, not all senses are created equal. And given the option, the brain will actually correct *what it hears* to match *what it sees*.[4] Furthermore, if the brain is confronted with an audio-visual stream that's out of sync, it will actually correct the sync up to about 80 milliseconds. So, to the average person, a movie that's out of sync by 80 ms or less won't be detectable, but as our audio-visual cues approach being out of sync by 80 ms or more, the brain can longer perform this correction.[5]

Notes from the Editor:
One of the reasons why the brain has a built-in "auto-sync function" is because light and sound travel at different speeds (186,000 miles per second and 767 miles per hour, respectively). Combine that with the fact that these sensory inputs are then processed by the brain at different speeds and it becomes quite apparent that syncing data streams is a very important function of the brain.[6]

—Still don't believe me about your brain editing reality, then here's another little experiment you can try. Something I originally heard from neuroscientist and professor at Stanford University, Dr. David Eagleman. However, to make things a little more ostentatious, we'll add one slight modification. To do this experiment properly, you'll need to go to the bathroom and bring a video camera with you, angled towards the mirror—that way you can fact check what you see against what the camera sees.

Now, without looking at the camera, I want you to stare into the mirror and focus entirely on your right eye. Then your left. Then your right. Then back to your left. Keep alternating the focus between your left eye and your right eye for about fifteen seconds, and what you should notice is that according to your conscious experience, your eyes never even moved. Looking in the mirror, you might be able to feel the tiny muscles in your eyes start to move as you shift focus from eye to eye, but of course, the brain won't show you these little saccades. Truth be told, if you want to see these tiny eye movements at all, you'll have to consult that camera. And, provided you've hit your angle right, you'll finally be able to see all those little saccades your brain has deemed unnecessary for you to see.

Memory

Memory and vision are both extremely complex parts to the conscious experience, so instead of spending five sections in a row trying to explain them, we'll be parceling it out as we go. In all honesty, that's mainly what we'll be doing the whole time (parceling out information). This is a technique I picked up from an acting teacher back when I was just an undergrad. Back then, I used to get the most terrible stage fright whenever I had to present a project or speak in front of a group. I stuttered. Whatever paper I was holding shook. It was debilitatingly horrible. So one day my creative writing teacher pulls me aside and says I should consider taking up an acting class.

"To help get rid of the nerves," she says.

Following her advice, I take up *Scene Study 101*, and in that class ended up learning a valuable insight about how our brains hold onto information.

Each Friday, a horseshoe of seats would form around our make-shift stage, which also meant that each Friday somebody new was crying. Just not for the reasons you'd expect. These students weren't crying because their scene required them to, they were crying because our teacher was tougher than nails. Most scenes never made it past the first twenty seconds before he'd hold up his hand and tear them to shreds. And after witnessing this process happen to a few unlucky souls that first week, my scene partner and I decided we weren't going to let that happen to us. If nothing else, we were going to make damn sure we knew our lines—*inside and out.*

Me personally, I went home and recited the lines to myself about thirty times over. In the mirror, on the drive to work, I quietly repeated the lines to myself over and over.

This method of learning, what we like to call "rote memorization" can be highly effective for memorizing things like multiplication tables or a phone number, but it's basically poison for an actor. Of course, I had no way of knowing this. So when the following Friday rolled around and my partner and I got up to perform our scene, I started saying the lines exactly as I had rehearsed them at home. Right down to the pauses and everything. And it went pretty well. I hardly stuttered. We didn't forget anything. For me, it couldn't have been a more perfect scene—that is, until the teacher tried to give us a redirect.

Repeating the lines over and over meant even the inflections had gotten burned into my memory, and once that happened, there was no changing them. I tried taking the redirect, but after just a few lines I found myself slipping back into my original delivery.

Seeing an opportunity to help me, and one to lecture, our teacher got up and said, "This is a perfect example," then paused for dramatic effect, "of what *not* to do."

I won't go into detail about how many times my brain's delivered up that soul-crushing sound bite over the years, but just know it was a lot. My second week in acting and this guy already had me experience the two things I feared most in life: performing in front of people, and being told I did it wrong.

From a psychological perspective, he got me to overcome my fears through a form of exposure therapy, but don't give him that kind of credit because I'm pretty sure he was just being a dick. Later on in life, once I started getting into psychology and neuroscience, I finally realized the full scope of what he was doing. He was creating an emotionally charged event for me to learn something.

Me, in front of the class all vulnerable and exposed, him detailing all the ways I failed at my execution, I'm pretty sure night terrors aren't this bad. If it were a dream, this would've been the part where most people would've woken up soaked in a pool of their own embarrassment, but seeing as this was real life, that meant I just had to squint my eyes and choke back what wanted to be tears.

Inside my brain, a structure called the amygdala was pumping out a memory enhancing cocktail of neurotransmitters, like dopamine and noradrenaline, in hopes that I'd never forget the moment. And this is just one of the things the brain does when it's in a highly fearful situation. The big idea being, if you can remember in vivid detail about the time you were in physical or emotional danger, then maybe you'll be less likely to end up there again.[7,8] This cocktail is the reason why you tend to remember the really bad moments in life with such clarity, and it's also the reason why I'll never be able to forget the lesson he was trying to impart.

"You need to learn the emotional life of the scene first," he says, pacing around the room. Why you're saying what you're saying? What are the reasons your character is there? What's he trying to accomplish? "As you practice the scene," he urges, "don't try to learn the lines; that'll come later." You need to rehearse the *emotional life* of the scene first. Once you nail down all the whys: Why you move from this side of the stage to that, why your character is choosing this tactic over another—then and only then, should you begin focusing on the lines. That way you lock in the emotions of the character, first. The subtext.

The "current behind the words," he calls it.

The way I had tried (top-down) was essentially flawed from the get-go because I was trying to intellectualize the scene instead of just living it. Rote memorization might be a useful trick for memorizing something like your license plate or the ABCs but try reciting any of those backwards and you'll soon see the one-directionality of that program's recall.

And yet, with the bottom-up method of learning lines, we were working with the subtext first, letting the emotions inform the words, not the other way around. Our teacher used to always say, "You give Gary Oldman five takes and he can deliver the lines in six different ways," and he was right. Because not only did this method of rehearsal give us more flexibility, but it made learning lines about a billion times easier.

"It's not that you did it wrong," he says, "It's just that you did it *backwards*…which was wrong." British people truly have mastered the art of condescension. Cutting you down in a way that if you're not fully paying attention, you could miss.

—

When it was all said and done, I never ended up having to take another semester of Scene Study, mainly because, after that first week, I didn't really need to. Oddly enough, my fear of public speaking pretty much dissipated after being ridiculed like that in front of a jury of my peers. And for some reason, him being British, made the whole thing feel more official. Life certainly has a way of teaching us the lessons we might want but would never dare ask for. And that acting teacher, as much as I cursed him and told the other kids he was just pissed because he messed up his audition to be on *American Idol*, ended up becoming one of my favorite teachers of all time. Which, upon further investigation, is actually an interesting little aside.

My *Experiencing Self*, the traumatized little kid getting schooled in front of fourteen other students, hated this guy, while my *Remembering*

Self, the one that's speaking to you now, likes him. Apart from him showing us a new way to learn, I discovered a new way to teach: bottom-up. In other words, starting with a simple outline or picture and then slowly building on top.

From the familiar to the unfamiliar, memorization works best when you can daisy chain new information to past information, and what we're doing right now is laying the groundwork for what consciousness even is.

Much like when a painter takes on a new project, they rarely try to nail it all in one go. Rather, they first start with a series of sketches (each one more complicated than the last). And, piece by piece, they *discover* their masterpiece. Growing up with two artists in my household, I got to see first-hand the graduated progressions of thought one has to go through to produce something that stops you in your tracks. And in case you hadn't guessed, you and I are going to be following a similar workflow.

As you might expect, the first few models are bound to be rudimentary. In all honesty, they might only last us a few sections before we trade them in for a model that's a bit more sophisticated. However, as we carry on, I promise our descriptions about consciousness will start to become more dialed in. Not to mention, by parceling out information in this way, instead of blocking it all together, each time we revisit an idea or a concept, you'll effectively be forced into strengthening the connections to that idea.

This method of memorization is what learning researchers like to call "interleaved practice." And, crazy as it sounds, but it actually goes a lot farther than just studying one subject for an extended period of time.[9,10]

Notes from the Editor:
Of all the study tips available to students, spaced repetition and interleaving have proven to be the most effective for long-term recall. Instead of studying the topics A, B, and C in the usual blocked form "AAABBBCCC," which is the way most schools still teach. Studies going back as far as 1986 suggest that interleaving studies like "ABCABCABC" produces a better recall than merely chunking it all together. Not only is this true for academia, but mixed practice can also enhance the acquisition of a new skill in sports or in music.[11]

—Outside looking in, something like consciousness seems so stupefyingly complex that many philosophers and neuroscientists still believe it can never be explained. But that's only because they're not in a position where their life literally depends on it. You'd be surprised at what you're capable of when your foot's caught in a trap. Because when you literally have no other option, you end up finding a way of making the impossible possible. A way of turning that limitation into a wellspring of creativity.

With this in mind, I want to offer you the first abstract picture of consciousness that we have: *That consciousness is layered, and that it's layered like a totem pole.*

At the base of the totem pole sits Tom, the conscious agent of your brain. Stacked on top, all the layers of your conscious experience (your memories, your language, your gender, race, nationality, job, age, sub-personalities, physiological state, relationship status, and so forth).

Tomtt, *the real Tom,* is that single point of awareness that sits at the bottom of that totem pole and is forced to 'look out' through all the layers piled on top.

If you think this clouds the way you see the world—you're right. If you think this tarnishes the image you have of yourself—you're also right. We don't see the world as it is, we see the world *as we are.* And, while we may not be given the keys to the algorithmic bubble we're all trapped inside of, the more we can learn about how the brain constricts the mind, the less likely we are to be prisoners of it.

Chapter 4
Consciousness as Software

\<div class="fl-rich-text"\>

\<h3\>\<strong\>Consciousness as Software\</strong\>\</h3\>

\<p\>If the brain is the hardware, then consciousness is the software. This metaphor can take us quite far, so let's really take a moment to hammer out the details of what we will now call \<em\>Consciousness 5.0\</em\>\</p\>

01110111 01101000 01110100 00100000 01110010 00100000 01110101 00100000 01101000
01110000 01101110 01100111 00100000 00110010 00100000 01100110 01101110 01100100
00100000 01101000 01100100 01101110 00100000 01101001 01101110 00100000 01100010
01101110 01110010 01111001 00100000 01100011 01100100 00111111 00100000 01101101
01101110 01100111 00111111 00100000 01110000 01110010 01110000 01110011 01100101
00111111 00100000 01101111 01110010 00100000 01110000 01110010 01101000 01110000
01110011 00100000 01100001 01101110 00100000 01100001 01100100 01110010 01110011
00100000 01100110 00100000 01110011 01101101 00100000 01101011 01101110 01100100
00111111 00100000 01101001 00100000 01110100 01101100 01100100 00100000 01110101
00100000 01111001 00100000 01101001 01101101 00100000 01101000 01110010 01100101
00101100 00100000 01101110 01110111 00100000 01110100 01101100 01101100 00100000
01110101 01110011 00100000 01111001 00100000 01110101 01110010 00100000 01101000
01110010 01100101 00111010 00001010 00001010 00110001 00110001 00110101 00110101
00100000 01010011 00100000 01001000 01100001 01110110 01100001 01101110 01100001
00100000 01010011 01110100 01110010 01100101 01100101 01110100 00001010 01010011
01110100 01100101 00100000 00110001 00110001 00100000 01010000 01001101 01000010
00100000 00110001 00110000 00110101 00111001 00001010 01000001 01110101 01110010
01101111 01110010 01100001 00101100 00100000 01000011 01001111 00100000 00111000
00110000 00110000 00110001 00110010

"If the brain is the hardware, then consciousness is the software. This metaphor can take us quite far, so let's really take a moment to hammer out the details of what we will now call Consciousness 5.0"

Your ability to make sense of that first paragraph depends on your ability to read source code. The second, binary. The third, English. I'm sure there's someone on the planet who can read all three of these, but surely, that is not me. If you're a coder, the first paragraph is child's play. "<h3>" denotes the heading style. "" denotes that the following words are going to be in bold, and "</h3>" denotes that all words before this command will be bolded and in "Heading Style 3." Luckily for me, there are tools at my disposal that can read all three of these, but I can really only make sense of that third one. Main reason being, I'm not a computer.

Which one of these paragraphs is true?

Technically speaking, they're all true, it just depends on your ability to decode them. And your ability to decode them is dependent upon your experiences. If you never studied how to convert binary into ASCII text, then the second paragraph probably just looks like noise, but once we get down to the level of the computer's processor, it should be known that all information exists this way. Every picture, every song, and every video eventually gets crunched down to a string of 0s and 1s—and at the level of the processor—that's all that even matters.

Are you a 0, or are you a 1? On or off?

Not surprisingly, computers share a lot in common with their creators, so in an effort to explain how our minds work, first let's explain theirs.

Now, it used to be that Windows ran on top of a background operating system called MS-DOS, and for the sake of this next analogy, let's just assume that all computers still work this way. In that sense, every computer has three levels of processing. There's the user-friendly operating system called Windows complete with a start menu, desktop, and icons. Then there's MS-DOS (the background operating system) that's actually carrying out commands like:

C:\GAMES\solitaire>solitaire

And then there's the level of the processor, the CPU, which is actually executing the game as a series of 1s and 0s. Before Windows existed, you used to have to know commands like "defrag C:" in order to defragment your C drive, but then Microsoft developed a user-illusion software called Windows that ran on top of MS-DOS providing the user with that classic desktop experience we all know and love.

Once this happened, you no longer needed to type in:

C:\GAMES\SOLITAIRE>solitaire into the MS-DOS command prompt. Instead, you could just locate the icon on your desktop, double-click it with your mouse and as if by some force of magic, you're playing solitaire.

Double-clicking the icon that looks like a pack of cards, Windows tells MS-DOS to run the Solitaire program; MS-DOS tells the computer's processor to initiate a hefty chunk of 1s and 0s; and then somehow, on the other side of that screen, you're playing a game with yourself. In many ways, this is very similar to the flow of consciousness we experience. Although, before we continue, we should probably establish a good baseline definition for this synthesis of sensory information that's going on in between our ears.

If we're likening the brain to being like the hardware—or "wetware" to borrow Dennis Bray's term—then consciousness is like the software. And I've personally taken to calling our particular brand of human consciousness, *Consciousness 5.0* or *CS5*.

Just as Windows used to run on top of MS-DOS, Consciousness 5.0 runs on top of a background operating system that we can now call the Subconscious OS.

So, for computers, the flow of information is:

Binary—> MS-DOS—> Windows

While for us, it's:

Neuron spikes—> Subconscious OS—> Consciousness 5.0

As the metaphor suggests, we can't 'see' the background operating system of our minds because, like MS-DOS, the Subconscious OS isn't meant to be seen. It's meant to be subterranean. To play a game of solitaire you don't need to see the CPU burning through 0s and 1s, nor do you need to see MS-DOS turning every mouse-click into an executable pathway. All you really need is for the ace of spades to drop on the first stack when you 'ask' it to. And, in a similar fashion, we as conscious agents don't need to see which networks of neurons are popping off, hundreds of times per second, in order for us to ride a bike. In fact, the more we consciously start to question the mechanics behind shooting a free throw or riding a bicycle, the more likely we are to just get in our own way.

—

Back when I was first learning to shoot free throws at about eight or nine years old, I needed my Consciousness 5.0 (my conscious attention and awareness) to learn how to aim and where to place my hands. My dad showed me what a good arc looked like, what a "swish" sounded like. Watching him, I had a decent idea about how much power I'd need. Then, after modeling the shot a few times, he tosses me the ball and says, "Don't stop 'till you hit ten in a row."

Of course, for the first few minutes all I could hit was the rim, but eventually, I started to make more shots than I missed, and once that happened, my brain really started taking statistics on what was working and what wasn't.

That whole first hour, I needed my Consciousness 5.0 firing on all cylinders to make sure I had the right angle and the right release. But after that day, the angles and arc needed to sink a shot were no longer things I had to consciously attend to; they had become automatized, background processes that just activated whenever I had my hands on the ball and my toes at the free throw line. In a way, it was almost as if my Consciousness 5.0, which we can now abbreviate to CS5, had written a program for my Subconscious OS to own and execute: the free throw program. Something that seems rather unique to our brand of consciousness.

Most animals are born with programs pre-loaded on their mind's OS (we call this pre-bundled software *instincts*). But humans have an ability to make updates to their software in a way that no other species can. We have the ability to develop new instincts, write new programs, or even override existing ones. Obviously, we're not the only species that can learn or acquire new skills, but no other species on the planet can imitate or pass on information quite like we can.

***Notes from the Editor:*

Procedural memory includes tasks like: driving, snowboarding, skating, swimming, playing an instrument, learning a dance routine, etc. All these activities require complex sequenced movements that need to happen in rapid succession one right after the other. Stored in an area of the brain called the cerebellum (Latin for "little brain"), once learned, this type of memory is rarely forgotten. Even in patients suffering from neurodegenerative conditions like Alzheimer's, the cerebellum is one of the last regions to deteriorate.[1]

—At one point, walking, talking, and reading all took tremendous amounts of mental resources (blood, O2, glucose, wattage), but as we began to practice these tasks more and more, these processes started to shift from the foreground of focal attention towards the backdrop of conscious awareness.

If you had tried engaging me in conversation while I was learning to shoot free throws, you wouldn't have seen the 'spinning wheel of death' indicating that my attentional resources were maxed out—although you probably would have inferred as much as I fumbled to make words. Just as installing new applications and running multiple programs at once weighs heavily on a computer's Random Access Memory, learning a new skill weighs heavily on our CS5. Even if we don't have an exact "mental RAM" equivalent, we do have short-term memory, working memory, and only a finite amount of variables we can keep track of at any given moment.[2]

Case in point, next time you have two friends around you, have one stand at your left ear and the other stand at your right. Then, with you standing in the middle, have each one tell you a story about sixty seconds in length. Your mission is to try and listen to both stories at the same time and then repeat them back once the minute is up. Without question, you'll be able to *hear* both stories at the same time, but I'd bet my life on the fact that you won't be able to recall both of them with at least 90% fidelity. Because while we might be able to listen to one person's story and do another task like driving or unloading the dishwasher, the amount of bandwidth it requires to listen, comprehend, and decode two stories at once, is beyond us. Which is also why people tend to turn down the radio in their car when they're trying to think or parallel park.

Where I'm from, they'd look you up and down and say, "Can't walk and chew gum at the same time, huh?" A derisive saying, nonetheless, but it speaks to a very important point about all this, which is, that we have an attentional-based consciousness.

A consciousness that runs on attention and awareness.

In other words, thinking and perception both cost something, and that something, we call attention. Thinking costs attention; performing motor acts costs attention—and you simply cannot be aware of everything at once. For this reason, it appears we've evolved a brain that pays attention in at least two different ways: One way, which has been likened to that of flashlight or "spotlight of attention," and the other way, which is more akin to a floodlight.[3] Generally speaking, the spotlight way of looking at the things is brought to us by the left hemisphere of the brain, while the floodlight view comes courtesy of the right, providing us with a more global, broad, and sustained form of attention.[4,5]

Compartmentalization

Typing on a keyboard used to be something I had to watch myself do. Index fingers firmly planted on F and J, I used to have to peer over my hands to spot which vowels were where. Getting my pinky to pull its weight, making sure my fingers stayed in their individualized jurisdictions, typing on a keyboard used to be this painful, laborious thing. That is, until the summer of seventh grade came around and it finally started to become an unconscious process.

The change didn't happen overnight, but somewhere over the course of that summer, I watched my need to correct my hands become less and less frequent. And because this conscious form of attention had been diverted so often, it's like my brain had created a new piece of software to help expedite the process: the typing program. Now, years later, I'm proud to say I can type 65 to 70 words per minute with minimal errors. Just don't put an empty diagram of a keyboard in front of my face and ask me to fill in the blanks. Because if you did that, I probably couldn't even get ten right.

Which, by the way, is actually an interesting little fact: Just because one part of your brain knows something, that doesn't mean all the rest of it does too. For example, my cerebellum might know where all the keys are, and be able to tell each of my ten fingers where to move, but that doesn't necessarily mean that I (the conscious agent of my vessel) *know* where they are too. So, even though, we might speak about the brain being this unified thing—or "you" being this unified thing—we're actually quite complex and compartmentalized little creatures.

When I say Tom, I am of course referring to you and all that you contain. But once we really start trying to tease apart what and who you are, it becomes rather unfair to call you by just one name, because realistically, there's just more than one self. There's the Experiencing Self, the Remembering Self, the Prospective Self, the Bodily Self, the Unconscious Self—all living and cohabitating inside the same brain, mind you.

We can talk with ourselves, fight with ourselves, bargain with ourselves. Hell, we can even reminisce with our *selves*. Think about that. Sometimes, you'll be driving to work or in the shower, and your brain will press play on a memory you didn't even know still existed, effectively entertaining itself with itself.

Who pressed play? And who hid the memory from whom?

Consciousness is the tangled web we perceive.

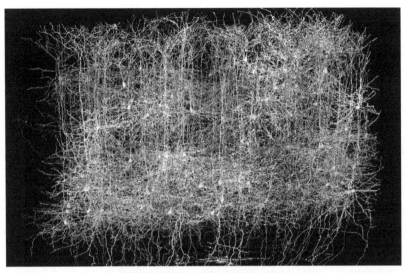

Image Credit: www.microns-explorer.org

(Inside your skull, each one of your 86 billion neurons is making around 10,000 connections to its neighboring neurons, creating a vast intricate neural network like this.[6,7] However, this rendering actually comes from a mouse, and is only showing 200 neurons.[8] The sort of processing power and imaging techniques required to capture a human connectome with more than half a quadrillion connections is far beyond our current capabilities. How do we conceptualize half a quadrillion? Well, we could take a trillion and multiply it by 500, or we could write the number "5" on a sheet of paper and then follow it up with fourteen 0's, because that's half a quadrillion).

Chapter 5
Intro to ASCs

HERE'S THE THING, if you're drinking a cup of coffee right now, you're in an altered state of consciousness. If you're enjoying a glass of wine, or it's been twelve hours since your last cigarette, same story. In fact, it doesn't take too much to knock us out of our *baseline wakeful conscious state* (what we're calling Consciousness 5.0 or C5). And whenever this happens, we can measure the quantitative and qualitative effects of that transformation. You do this enough times—across enough patients or volunteers—and what starts to happen is, you build up a repository of reliable effects generated from a given compound, technology, or practice. This idea is at the heart of how psychonautics and neuropsychopharmacology play into one another, as it's only after we've gotten knocked out of our baseline state of awareness that we begin to understand what the word "baseline" even means.

Think back to the first time you sipped a cup of coffee or drank a beer. The first time you dreamt a dream so real you had to phone a friend to make sure you were really out of it. Or how about the first time you were in the presence of something so magnificent, so awe-inspiring, that you lost your train of thought.

So dumbstruck by a piece of art, or a ball of fire descending into the night sky, that all you could do was just stand back and admire.

The question we need to be asking ourselves here is: *Did these experiences teach us anything?*

If they did, then that's all you really need to keep in mind before moving forward (that altered states of consciousness are capable of showing us different ways of experiencing a moment); but also that altered states of consciousness (or ASCs) unpack at different levels and at different times in our lives. Just like falling asleep, you can't always force them into happening. However, you can set up the preconditions necessary for these ASCs to come about, which is where Ava and I come in.

You might wonder how it is that we're even able to study something as complex and difficult to access as consciousness, and the short answer is roundaboutly.

Being neuropsychopharmacologically-minded (combining the tools of psychopharmacology with the knowledge of neuroscience), teams of researchers tend to study consciousness by finding and isolating the sorts of compounds and techniques needed to induce a whole range of altered states. From elation to awe, caffeinated to sedated, we study the hypnotized, the anesthetized, the inebriated, various forms of intoxication...

***Notes from the Editor:*

Altered states of consciousness (ASCs) are sometimes also referred to as non-ordinary states of consciousness (or NOSC).

—"There's flow states, mystical states, psychedelic states, trance states, meditative states, out-of-body experiences, runner's high, states of delirium, high fever, shock, panic, various stages of coma, vegetative states, locked-in syndrome, sleeping, dreaming, lucid dreaming...the list goes on. In any case, the most important thing to keep in mind here is that (a) these are all just *transformations in consciousness*, and (b) these transformations can be brought about in at least one of three ways: (i) physically, which would include yoga, movement meditation, fasting, holotropic breathwork, stroke, sickness, lack of sleep, lack of food, lack of oxygen, head trauma, and so on (ii) technologically, which would include VR, binaural beats, transcranial magnetic stimulation, transcranial direct stimulation, sensory deprivation, alpha and theta biofeedback, and last but not least, (iii) pharmacologically, which would include ingesting some sort of compound that induces changes in thought and perception. In other words, psychotropic drugs like caffeine, alcohol, sleeping pills, mood stabilizers, anxiolytics, antidepressants, psychedelics, and the like.

Mental Cartography

Before all our magnificent imaging techniques like MRIs, fMRIs, CT Scans, PET Scans, and Single-Photon Emission Computed Tomography, doctors and anatomists were only able to make maps of the brain either by studying cadavers, or patients who'd suffered lesions or stroke. You see lesions to the back of the head in nineteen or twenty different soldiers, all of whom subsequently went blind, and it's reasonable to assume that this area of the brain probably had something to do with visual processing.[1,2] So back in those early days, when anatomists were still just toying with the idea of a visual cortex, I'm sure they welcomed patients with damage to the back of their heads with open arms. Their accident, how- ever unfortunate, was a step forward for science. Which might sound harsh, but honestly, it's how we learn. Almost everything we have is built on the backs of deaths, failures, and accidents.

See *Microwaves*
See *America*
See *Evolution by Natural Selection*

Needless to say, but all these logical deductions were later confirmed using all our modern technology. But in the same way that we've had maps of the world centuries before satellites, we've had maps of the brain long before magnetoencephalography and diffusion tensor imaging. If we then take those same principles of accidental discovery and curiosity-driven science and apply them towards the subject of consciousness…well, you just can't not study altered states.

———

Just consider the lengths you'd have to go to if you were trying to make a map of the world before satellites. You'd have to send out numerous ships headed every which way and ask them report back on what they've discovered and where. You'd have to compare those reports against all your previously collected information, carefully contrasting estimated distances with measured distances, while, at the very same time, reviewing tons of personal accounts and hastily drawn sketches, just praying for those accounts which featured some kind of overlap. Effectively, you'd be drawing faraway lands you'd probably never get to see. And then—

only after an extremely long and arduous process—might you be able to produce a map of some utility.

Looking at all our maps from the 1600s, you'd almost expect them to be more primitive than they actually are (bearing all sorts of glaring and obvious mistakes). But, as you well know, much much can still be achieved using this older method of approach. Incidentally, this process isn't too far off from the way we're having to study the uncharted waters of consciousness right now because, truth be told, we have no direct access. Sure, we might have some fancy new imaging techniques, but the brain is still very much a black box. We have no way of 'porting' into someone's brain and knowing something about their internal experience, so in a way, it's kind of like we're still operating in an age before satellites.

Much like a cartographer who's trying to make a map of all the conscious states available, we still have to send out exploratory missions and wait to hear something back. And just like any good mapmaker that came before us, we should take an extra second to acknowledge those first brave explorers who made the ultimate sacrifice.

Before they left the harbor, many of them had to make peace with the idea that they'd probably never see their families again. Hell, for all they knew, once they crossed that horizon, they might have been about to sail right off the edge of the map.

Humans & Computers

Look, eventually, all our computational metaphors are going to break down. It's not really a question of if, it's a question of when. And when this happens, we're going to have to trade in these ideas for a model that's a bit more sophisticated. Be that as it may, there's still a lot we can accomplish by comparing the brain to a computer, so long as we keep in mind the glaring and obvious differences. Like the fact that we're alive and they aren't, or the fact that we have stomachs and they don't. I realize this may sound rather elementary, but this one fact about being alive has many downwind implications that aren't exactly self-evident. For instance, because we have an instinct to survive, our bodies and our brains have motivations, desires, and capabilities for adaptation far beyond any computer you've ever sat in front of. As "smart" as your laptop is, it doesn't have the ability to make changes to itself, like structural hardware changes, while we, on the other hand, have the ability to repair ourselves, rewire ourselves, and continuously adapt our hardware/software to the ever-changing environment.

On the contrary, computers receive a steady-state of power—and save for over-clocking a computer's processor—the processing power you have is pretty much the processing power you have. Furthermore, when a computer first boots up, you can ask it to do almost any calculation you desire. You don't have to wait until it's been turned on for two and a half hours to get a straight answer out of it, whereas with humans, sometimes, this is the case. Time of day is extremely important to the quality of one's brain function, as time of day generally correlates with one's feeding regimen/sleep-wake cycle. And if you don't think that has any effect on your state of mind, then I invite you to go ask anyone who's recently been denied parole, "What time of day their hearing was," because it's highly likely they'll either say, "Just before lunch," or "Right at the end of the day."

In a study of over 1100 judicial rulings from 8 different judges, across a 10-month period, researchers found that judges tended to be more lenient just after a scheduled break or lunch. More specifically, according to this study, your chances of being paroled go from about 65% (at its highest), to as low as 0% as the day marches on. And yet, after each of the 2 scheduled breaks, the rate of a favorable ruling returned to that original 65%.[3]

So what's going on here?

Perhaps it's another reminder that we're not mindless automatons, but rather, emotional, embodied creatures, who need to take breaks from time to time. Because apparently, making decisions about the fate of someone else's life weighs heavily upon mental resources. And if our goal is to stay more objective, then perhaps we need to start being more mindful of our blood-glucose levels.

I believe this point to be extremely instructive because while we can make allusions to the brain being like hardware, and consciousness being like software, the hardware/software divide isn't really tenable. The term "wetware" really does seem to fit the bill since it continuously forces us to remember that the brain isn't just some machine, but that it's composed of living, organic material that is constantly re-organizing itself based on the inputs it receives from the environment.[4] And, to keep this living material in an optimal state of awareness/functionality requires the proper amount of rest tempered with the proper amount of quality nutrition. Too much food and all our blood rushes to our GI tract to aid in digestion, leaving less blood flow for everything else—like thinking. Too little food and we start to get short and snippy, or as my old roommate Brian used to call it "hangry."

When it comes to food, there certainly is a Goldilocks zone for optimal brain function. And I'm sure we can all agree that just after a Thanksgiving meal, or on day two of a Master Cleanse clearly isn't one of them. At any rate, there are still tons of different ways one could get their body/mind into a peak state of performance from biohacking to breathing exercises, supplements to nootropics. There's juicing regimens, mindfulness meditation, movement meditation—the list could literally fill the rest of our time here, and even then it wouldn't be exhaustive. So instead of potentially telling you something you already know, let's just leave it by saying the following: Almost anything you put on or in your body likely has an effect on you and your brain state, and that effect can range from slight to significant.

What happens when you stack up a whole bunch of these tools and tricks?

Well, you get flow states, Olympians, and top performers in every field.[5]

—

Think about it, everything you've ever heard, every poem, every piece of music, every "I love you" all happens within the confines of your skull. It's certainly not intuitive to think this way, but the brain doesn't see or hear anything. The brain doesn't feel; it doesn't smell or taste; the brain *processes*. It's an information-processing machine. Not a piece of hardware, but rather, a piece of wetware that sits inside your skull and is fed streams of data taken in by the sense organs.

If we're comparing the brain to a computer, you don't type on the motherboard, you can't 'pinch to zoom' on the motherboard, nor can you 'see' you. Of course, things might plug into the motherboard. Things like your speakers, your keyboard, a USB webcam, etc. But your motherboard just sitting on a desk couldn't do any of those things, and likewise, neither can your brain. Point of fact, I could be touching your brain right now and you'd never even know…Why?

Because the brain doesn't have pain receptors; it feels nothing.

I could be placing a spoonful of creme brûlée on your parietal lobe right now but you'd never end up tasting it…Why?

Because the brain doesn't have taste buds; it tastes nothing.

The experience of sweetness might originate inside the parietal lobe, but the lobe itself only deals with neurological 1s and 0s (neurons either firing or not firing). In this regard, when you place a forkful of cheesecake in your mouth, the taste of the cheesecake gets converted into electrical signals *at the tongue*. And because the sensory cells that detect sweetness, sourness, saltiness, savoriness, and bitterness, exist on the tongue—and not in the brain—we can liken this conundrum to our experience of pain.[6]

Slam your hand in a car door and you're likely going to be in a lot of pain, but before doctors ever start operating on your hand, first they're going to give you an infraclavicular brachial plexus nerve block that essentially blocks the signals of pain coming from your hand from ever reaching the brain. In this sense, the pain hasn't really gone anywhere. Instead, it's more like we've unplugged one of your peripheral devices and your arm is no longer able to transmit the signal of pain.

Against this backdrop, what we really need to draw our attention to is the fact that the world of the brain is very much like the world of the motherboard, because eventually, it all comes down to binary. On or off. Plus or minus. Neurons either fire or they don't fire.

have to keep in mind that everything that either has happened to you, or will happen to you takes place inside the Virtual Reality simulation put on by your brain. Accordingly, if a paranoid schizophrenic sees a man in his room with a gun, he actually sees a man with a gun. The fact that no one else sees him hardly makes a difference; his perception is his reality.

—You see Tom, although you have about 86 billion neurons in your brain, and although they make a vast intricate spider-web of connections with each neuron making around 10,000 connections to its neighboring neurons. The neurons in our brains don't actually touch. Instead, the way they communicate is two-fold: through electrical excitation (aka neuron spikes) and through a chemical language (aka neurotransmitters).

Neuron spikes trigger the release of certain chemical messengers, these neurotransmitters subsequently float across the synaptic gap (the space between the transmitting and receiving neuron), and in turn either have an excitatory effect or an inhibitory effect on the receiving neuron. Some of the more common neurotransmitters, I'm sure you've probably heard of, like serotonin, oxytocin, dopamine, acetylcholine, and histamine. But to simplify things just a bit, let's knock that ginormous number of neurons in our head down to just two.

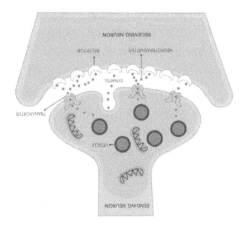

In this scenario, how it works is, Neuron A will send a message to Neuron B using a neurotransmitter like acetylcholine. This chemical messenger then floats across the synaptic gap until it finds an acetylcholine receptor site on Neuron B, and voilà!

The message is received: *Contract your bicep.*

Now obviously, this is a massive over-simplification as contracting your bicep involves more than just two neurons, but we can take this rather simple explanation and begin to scale it up in our minds. While we're at it, we can also start to envision that the way these neurotransmitters operate is akin to a lock-and-key system. Neurotransmitters being like the little **keys** which only fit into their corresponding **locks** on the receiving neuron. So dopamine only fits into dopamine receptor sites, serotonin only fits into serotonin receptor sites, histamine only fits into histamine. And it just happens to be the case that these "keys" can get hijacked by chemicals that look similar enough in structure.

These chemicals, we pejoratively call *drugs.*

For example, America's number-one drug of choice, caffeine, looks so much like the neurotransmitter adenosine that when it slips into the adenosine receptor site on Neuron B, it blocks it; the end result of which being: You don't feel tired anymore. To be clear, Neuron A is still sending the neurotransmitter associated with tiredness, but seeing how the adenosine receptor site is being blocked, the message is never received.

So what's the result?

Increased neuron firing—something adenosine is supposed to help down-regulate. But because the caffeine molecule is sitting inside the adenosine receptor site (literally blocking the message) your brain senses panic, starts to release the fight-or-flight hormone we call adrenaline, and the next thing you know, you start feeling more alert.

In this situation, I'd like to point out that the coffee isn't actually giving you any energy, it's just tricking your brain into releasing dopamine and adrenaline.[7] In a similar vein, the only reason opioids like morphine, codeine, vicodin, oxycontin, and heroin even work on us at all is because they resemble a certain endogenous pain-blocking neurotransmitter we produce called endorphins. In actual fact, that's even where endorphins got its name (endogenous morphine).[8]

***Notes from the Editor:*

Plants are extraordinary chemists. Being immobilized, plants can't readily ward off predators in the same way a human might. So instead, they produce chemicals that either draw animals in, or keep them at bay.

—Likewise, if you're a follower of the medical marijuana movement, then you'd already know the only way marijuana is able to generate its effects in humans is because we already have an endogenous cannabinoid system (or endocannabinoid system) that responds to the hundred-something cannabinoids in marijuana, like THC.

The endogenous version we produce is called anandamide, and it's the main chemical responsible for the feeling of "runner's high."[9]

All this is to say that we have a lock-and-key system for chemical messengers, and it just so happens that Mother Nature tends to produce endless variants on these molecules and then scatter them about in plants. To be fair, the plants are generating these chemicals for their own ends either to lure animals in, or to deter predators from ever coming back. Nectar, pollen, and fruit would all be examples of a plant luring another animal in to help spread its seed. Then, on the opposite end, you have plants which generate bitter-tasting alkaloids to avoid being eaten. Some plants, like the pitcher plant *Nepenthes pervillei*, even produce a lovely aromatic perfume designed to temp insects into stopping by for a visit. Once there, the insects then fall into their perfectly designed pitcher pot deathtrap, where they eventually drown and are devoured by the plant.[10]

With this in mind, if there were ever any doubts about who the best pharmacologist was, just ask any chemist worth their salt and they'll tell you...*it's plants*.

Aspirin, quinine, penicillin, cocaine, novocaine, alcohol, Sudafed, and any opiate derived pain-reliever all have origin stories involving plants. The problem is, we have seven major neurotransmitters (each with their own laundry list of responsibilities), and as it turns out, the one that's intricately wound up with thought, perception, and memory has been made extremely difficult to study because of some bullshit that happened during the Nixon administration.

See *Counterculture*

See Hippies
See *Timothy Leary*

Don't even get me started on that guy. People still talk about him like he was some kind of psychedelic evangelist, but not in my circle of friends. People like him, in conjunction with the overzealous fear-reports of the time, are the reason why some of our best tools for self-discovery wound up in the most restrictive drug-use category on the planet.

Schedule I.

At some point, we're going to have to discuss some of the experiments surrounding some of these serotonin agonists like mescaline (aka peyote), "magic mushrooms" (aka psilocybin), and a fungus derivative called lysergic acid diethylamide (aka LSD). But before we get into any of that, first we need to jump backwards in time.

***Notes from the Editor:*

The real difference between this book and most other books on the topic is that traditional non-fiction doesn't tend to feature intricate subplots with archetypal characters and interconnected themes. Where the studies and findings consistently seek to push the frontiers of our knowledge forward, the interwoven stories seem to keep the emotional, embodied nature of consciousness present in our minds while, at the same time, providing much needed moments for reflection and review.

Intro to Self

Flashback to when I was an undergrad. Flashback to before the writers' group, before I took up psychology, before I ever even considered a career in neuroscience. Back then, I was just a stupid kid writing for the university newspaper. Each week having to interview at least one new professor and publish a 400-word blurb about them. What they're working on, what classes they teach, what makes them worth studying under, all that jazz.

This particular interview, I remember rather vividly, because I was supremely hung over and definitely did not want to be there. My clothes still smelled like a bar. My skin still smelled like alcohol. I was sweating. This was clearly not my best day.

So I start the interview with Professor Roberts, a guy who's been teaching Art History since before the Common Era, when I realize this is probably going to be another one of those disaster interviews my senior editor takes her big purple pen through.

Huge purple X marks, the size of each page, staring back at you from your inbox letting you know your article'd been cut. But, seeing as that wasn't my call to make, I still had to show up at Taper Hall that morning and say, "No, no, the pleasure is all mine."

The way Professor Roberts looks, it's the same as one of those guys you see playing chess on a park bench with a parrot on his shoulder. One look at this guy and you just know he has tenure or mites, but probably, he has both. So there we are sitting in his office, I'm leaning in trying to give off the impression that I actually want to be there when I go to ask my first question. Holding my pen against my mouth, I start by asking, "Professor Roberts, if you could..." but before I even have the chance to finish, he fires back with "Doctor," correcting me.

Yeah, one of those people.

Annoyed but trying not to show it, I start again, "Dr. Professor Roberts, if you could meet anyone—alive or dead—who would you want to meet?"

Now, normally, I wouldn't even dream of publishing the warm-up questions, but when a guy with two doctorate degrees hanging on the wall, hiding behind reading glasses the thickness of *Great Gatsby* goes, "Myself," I have to drop my pen.

In this moment, I just have to drop my pen so I can lift my head to make sure he's not smoking a wooden pipe and wearing a tweed York-shire cap when he says it. In my head, I picture him taking a long drag of his pipe to accentuate his prestige, staring vaguely in my direction and moving his mouth around a lot, but perhaps, I'm getting ahead of myself.

My first thought was, "You could meet Charles Darwin or Nikola Tesla and you want to meet yourself?" Shaking my head, I purse my lips together and blink at the floor. My blinks feeling punctuated and loud. In my mouth, I can feel my tongue tapping against the bottom row of my teeth as I try to resist the urge to say something. When all of a sudden, the words "Why yourself!" fall out of my face, and I just roll with it. Twisting part of his mustache like he were winding up a watch, Professor Roberts shifts himself in his chair. "Because we don't see ourselves like other people see us." Pushing those thick glasses tight against his face, he goes, "We only know of ourselves through indirect measures—looking at our actions and behaviors *after the fact*—so I'd like to meet myself." Then leans back in that big burgundy chair, "See myself through a different pair of eyes."

Now, this question has been a part of my getting-to-know-you process for years. It must've been asked at least fifty times before this day, and yet this reply is the only one I ever bothered remembering not to forget. Because as Professor Roberts went on to explain: *We don't quite know ourselves as intimately as we like to think we do.*

We don't always know why we're mad or why we're sad. We rarely know what the cause of our depression is.

"When we get into it with our partners," he says, "sometimes we don't even know what the fight is actually about."

We don't know what we're truly interested in, that we have to stumble into.

We don't always get to choose what we have motivation for, which is why caffeine, Adderall, and Tony Robbins are so sought after in the first place.

When it comes to what we're into sexually...or what we find funny...even that we don't get to choose. We've all laughed at something we probably shouldn't have; tried to love someone we thought was "good for us"; or found ourselves attracted to someone, who at first, we couldn't even stand. Making his point, Roberts goes, "We have patterns, but that doesn't mean we always get to see them." And right as he says this, I'm immediately transported back to the first time I ever saw the movie *Liar Liar* with Jim Carrey.

I mean, I didn't see the movie with Jim Carrey, but he was in it, and in this movie he played a guy who couldn't tell a lie for 24 hours.

Anyway, Jim Carrey's character, Fletcher, is talking to his ex-wife and she's trying to explain to him how much his actions and promises matter to his little boy. So Fletcher starts to defend himself.

"Now let me tell you something," he says to his ex-wife, but seeing as he can't lie, he ends his sentence with, "I'm a bad father!"

A realization that shocks even him.

Up until that point, you could tell Fletcher couldn't see (or had been willing *not* to see) all the pain and disappointment he was causing his little boy. But after that painful revelation, the movie takes an unexpected turn, and he actually does start acting like a better father. I share this memory with Professor Roberts who only nods his head and opens up his hand.

"We have ideas about ourselves, and who we think we are," he says, "but that doesn't always make it true."

For example, the first time I heard myself recorded on an answering machine, I couldn't believe it was me. I didn't want to believe that was my ugly prepubescent voice on my friend's recording, so I backed away from the machine and said, "That's not me," as if distancing myself from the machine was going to make it any less me. Later, I would come to find out this phenomenon actually happens to a lot of people (they see themselves or hear themselves recorded, and then they start to cringe). To save face, they say things like, "Well," the camera adds ten pounds."

But I gotta ask you, Tom, what's your intuition about all that? That a recording device magically adds ten pounds to whomever it's focused on, or that maybe there's a mismatch between how we look and how we think we look? Between who we are and who we *think* we are.[11]

No lie, but I've actually heard a few actors tell me they don't like to watch themselves on camera, and to that I always say, "How do you think we feel?"

Not out loud, obviously. But the voice in my head, he says it, which, oddly enough, doesn't sound anything like my real voice—so what the hell's going on here? And who's the real "me" anyway? Why does the voice in my head sound so much younger than the face I see when I look in the mirror? Why does the body I use to move through space always feel more defined than the photos I see when I look at my phone?

As Professor Roberts went on to suggest, one of the reasons we're so hypocritical is because we have a view of ourselves that's incomplete. There's a wallet-sized picture of ourselves we like to carry around, an Ego Self, he calls it. But that picture is really just a low-resolution snapshot of who we are as people.

Professor Roberts's other PhD was in developmental psychology.

"The Ego Self isn't complete and authoritative," he adds, "it's air-brushed and sanitized."

After jotting this little gem down in my notebook, I'm quick to realize that Professor Roberts is a lot wiser than I judged him to be. And that maybe...just maybe...I shouldn't be so quick to cast out judgment based on looks.

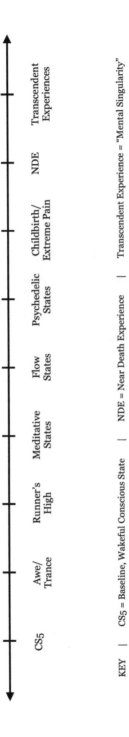

An Incomplete Spectrum of
Altered States

CS5 | Awe/Trance | Runner's High | Meditative States | Flow States | Psychedelic States | Childbirth/Extreme Pain | NDE | Transcendent Experiences

KEY | CS5 = Baseline, Wakeful Conscious State | NDE = Near Death Experience | Transcendent Experience = "Mental Singularity"

Chapter 6

Three Cases of Brain Trauma

WAKING UP IN someone else's body changes you. Don't believe me, go ask Harvard-trained neuroanatomist Dr. Jill Bolte Taylor about the morning she woke up with a blood clot the size of a golf ball and she'll tell you the same thing. The person who went into the stroke and the person who came out the other end were two very different people. Of course, on paper, she was the same Dr. Jill Bolte Taylor she'd always been. But go pick up her book or check out her TED talk and you'll hear Dr. Taylor speak about life before the stroke as if she was a completely different person. And why? Quite simply, because in many ways, she was a different person. Her neurocircuitry was different. The way her brain harvested and processed reality was different. And after suffering a stroke that lasted just a shade over four hours, that neurocircuitry was forever changed.

It took eight years for Dr. Taylor to make a full recovery. But on the day it happened, on some level, Dr. Taylor knew she was never going to be the same person again. So instead of trying to cling to her old identity, she became Dr. Jill instead.

There's very few people who ever suffer a stroke of this magnitude and live to talk about it. But perhaps even fewer who go on to give a TED talk that's in the top ten most watched TED talks of all time, who go on to write a *New York Times* bestseller, and who go on to make *TIME Magazine's* "100 Most Influential People in the World."

In her non-fiction novel *My Stroke of Insight,* Dr. Jill describes the whole experience as being "one of the greatest blessings" she's ever received.[1] And if you're anything like me you gotta wonder how that's even possible. I mean, how could a genetically predisposed hemorrhagic stroke—that thing we all fear in life—be a good thing?

Well, to summarize her account in just a few sentences, she experienced a non-ordinary state of consciousness (a near-death experience) gifted by an arteriovenous malformation. Or, in simpler terms, a congenital brain disorder she didn't know she had, gave way to a golf-ball-sized blood clot that wound up bursting inside the left hemisphere of her brain, off-lining a good chunk of her brain's functionality. Pressing on her language centers, which are typically located within the left hemisphere, this clot was spewing out more blood onto her neurons every single second (something that's actually toxic to brain cells).[2] Meaning that with each passing moment, this congenital brain disorder was killing more and more of her brain structures, and in turn, more and more of her.

Waking up that morning, Dr. Jill says she could sense something was off, but as strokes often go, she didn't know she was having one. Being that she was only 37 at the time, having a stroke was hardly on her radar for potential morning ailments, so by the time her right arm went limp and fell dead against her side, calling 911 wasn't even an option. The part of her brain associated with numbers was already completely off-lined and in that moment, her first thought was, "Wow! This is *so* cool!"[3]

Her language centers drowning in a pool of blood and she's thinking, "This is *so cool!* How many brain scientists have the opportunity to study their own brain from the inside out?"

Reading her account of what happened, her fascination for the human brain oozing off every page, you really get the sense that for all the people to have a stroke, she really was the perfect person. Because in addition to having the right skillset, she was basically gifted front-row tickets to see just what happens when a patient suffers one of these 'brain attacks.' Then, on her eight-year road to recovery, she got to experience what it felt like to be wounded and in rehabilitation. Experiencing the successes of modern medicine, and the pitfalls. The treatment, and the mistreatment. The good, the bad...Dr. Jill saw it all.

Her recommendations for recovery ought to be a must-read for any medical professional or rehab specialist.[4] But perhaps even more incredible was her ability to articulate this experience in such a way that people could finally appreciate, at least from a neurological perspective, that they are not their "brain chatter." Sure, their brain chatter might be a part of them, but it isn't the only part.

Because after that section of her brain essentially died that day, the "narrator of her experience" might've fallen silent, but she, herself, did not die. Rather, it was more like she was reborn inside her own body. Due to the fact that this hemorrhagic stroke had laid waste to the majority of her left hemisphere, Dr. Jill could no longer walk, talk, move, or recognize her own mom. She ended up having to relearn everything from what colors were, to how to pick out a person's voice amidst background noise; she even had to relearn her own name. But despite losing all of this, Dr. Jill still calls this experience "a tremendous gift," since it was only by losing her left hemisphere that she ended up finding a bigger piece of herself to begin with (her right hemisphere consciousness and her right hemisphere voice). A voice, which always seems to get drowned out by the domineering left side. Consequently, when the left hemisphere of her brain finally fell silent, what Dr. Jill really found was peace. *Nirvana,* she calls it.

And because she had found nirvana, she knew everyone was capable of finding it. And that's what motivated her to recover, she says.[5]

So what happened here?

To answer this, it must first be acknowledged that in many ways we are our brain structures, and that the way our brains are wired up is uniquely tailored to our experiences.

Those 86 billion neurons in your head are making something like 800 trillion connections to one another, and it's that wiring diagram (what we call your connectome) that makes you...*you.* Supposing you had an identical twin brother, it would be the differences between your connectomes that would make you different, as it certainly wouldn't be your DNA. As you'll recall, identical twins share 100% the exact same DNA. Therefore, the real difference between you and your hypothetical twin brother would lie in the connections your brains make, not the genome itself.

Different preferences, different ways of thinking, different experiences—all this gets coded into the brain and causes dissimilar connections. Armed with this knowledge, it should be much easier for us to understand how head trauma could cause subtle or even profound changes in personality, given that, in many ways, the brain is us.

It almost feels foolish to point this out, but I think it's important to highlight the fact that this doesn't happen with any other part of the body. Tearing your ACL or suffering a compound fracture is likely going to put a dent in your day, but it's not going to have long-lasting effects on the way you reconstruct reality.

And yet, when it comes to the brain, most often this is the case.

One of the most frequently cited examples of this comes from a guy named Phineas Gage. A railroad worker who had a 43-inch-long tampering iron blow through his left cheek and then exit out the top of his skull.[6] The amazing part is, the guy pretty much remained conscious, and was even walking around within minutes! So the doctors patched him up, and then he went on to live another 11 years. The only problem was, after the injury, everyone started to notice that he was exhibiting some markedly different personality traits.

Apparently, a 13.5 lbs steel rod passing through your dome can do that.

The physician who attended to Phineas, Dr. Harlow, writes: "the balance between his intellectual faculties and animal propensities seems to have been destroyed." Friends of his said he was "no longer Gage," as he had become increasingly difficult to be around, when, at one time, he was "a great favorite."[7] Surprisingly, Phineas was able to return to work less than four months after the accident, although when the day did finally come, the railroad wouldn't even take him back. Which, if you know anything about their checkered past, is really saying something. Nevertheless, the crucial point under consideration is what the loss of these lobes did to him. Because while Phineas might have physically survived the accident, mentally he would never be the same.

———

Or consider the case of Clive Wearing: a renowned British conductor and musician working for the BBC, until the day he contracted herpes encephalitis. Usually, the virus doesn't cross the blood-brain barrier, but in this one rare case it did, causing massive brain swelling and an almost complete destruction of an area which is heavily involved in learning and memory (the hippocampus). The damage was actually so profound that Clive became notoriously known as "the man with the seven-second memory."[8]

***Notes from the Editor:*

In London, where Taxi-drivers are forbidden to use any form of GPS and are forced to learn every street, landmark, and route possible in order to pass an exam they call "the Knowledge"; their hippocampi have been shown to be significantly larger than average. Passing this test is among one of the hardest examinations in the world and typically takes anywhere between two to four years of training.[9]

—One of the greatest filmmakers of our time, Christopher Nolan, once made a movie about a guy suffering from a rather peculiar brain disorder. In the movie, the main character suffers a blow to the head during a robbery, and subsequently develops a condition where he can no longer form new memories. The movie was of course *Memento*, and believe it or not, but this condition actually exists. In the film, Guy Pearce's character could still remember everything that happened to him up until the accident, just not after. Although, in the case of Clive Wearing, the damage to his hippocampi was so severe, not only does he suffer from anterograde amnesia (the form of amnesia displayed in *Memento*), but he also suffers from retrograde amnesia as well. Meaning, he can't remember anything that's ever happened to him *in his entire life*.

Paradoxically, he can still read, write, talk or even play the piano. But every seven to twenty seconds, it's as if someone comes by and shakes the etch-a-sketch of his mind, completely wiping his memory slate clean.

His diary, is riddled with entries like:

7:46 am—I wake for the first time.
8:07 am—I am now totally, perfectly awake. 1st time.
8:31 am—Now I am Really Completely AWAKE (1st time).

In the documentaries about Clive, you see a man milling about his room discovering consciousness for the very first time, again and again. Seeing another human for the very first time, hearing someone's voice for the very first time, and then rushing over to his diary to document the historic event.

This is a momentous occasion...I AM ALIVE!
Hurray! X Hurrah!
I am now COMPLETELY awake for the first time.

—But once he notices a similar entry made only a few minutes prior, he carves them out. Rather violently, I might add.[10] Assuming he'd either written them unconsciously or perhaps the diary itself was an elaborate forgery. With no memories he can call to mind, and no ability to plan for the future, Clive is forever trapped inside a "now moment." Watching the way he is with his notebook, you can tell Clive wants to have more than one moment again, but with the severity of his condition, he no longer can.

In one of the documentaries about him, aptly titled *Prisoner of Consciousness* we're shown just how strange and compartmentalized consciousness can really be. For example, his wife Deborah (about the only person he can recognize) walks him over to a piano and starts whispering to him that he can play. That he plays beautifully. But Clive just shakes his head and says, "I have no memory of that."

Sitting him down on the stool, he's still swearing up and down that he doesn't know any music, that this is the first time he's even been conscious since he got sick. But then, just a few seconds later, you'd be hard-pressed to know anything was wrong.

Clive can still sing, sight-read, even conduct a concert. You can sit him down at an organ, inside one of the most prestigious churches in England where he used to conduct, and watch him come alive as he improvises a beautiful piece of music. He can move you to tears, instinctually playing all the right chords to make those little hairs on your arm stand on end, but after a couple blinks time, has zero recollection.

Echoes from the song he just played still bouncing around the cavernous halls of the church, and he'll say, "I've never heard a note of music in my life."[11]

———

Clive's case is a sad one, but it offers neuroscientists and psychologists a window into just how modular the brain—and the memory system—really is. Clive may have lost his ability to recall memories or make new ones, but oddly enough, he still knows certain details. For instance, if you had a career at the BBC, or got married, I'm sure you'd have a plethora of memories you could call to mind surrounding either event. The difference with Clive is, he no longer can. He knows he's married; he just can't recall getting down on one knee. He knows he worked for the BBC—and that he got really sick—yet, anytime he goes searching for a memory it's:

Error 404: File not found

Metadata

Years ago, back when I first met Ava. Before she moved in, but after I got used to her correcting me. She and I were back at my place making some home-made lasagna when the topic of Clive came up. Trying to describe what she thought was going on, Ava started quoting some psychology class she took eons ago about that phenomenon that happens whenever you go searching for a word but can't seem to find it.

"You know that you know it," she says, "but when you go looking for it…it isn't there." I remember laying down the first layer of noodles when Ava fell silent. Paralyzed. Her face looking like it was about to sneeze.

"There's a name for it," she says, but ironically, she can't say what it is. Eyes darting left-to-right, you could literally watch her trying to find the words to describe the exact condition she was experiencing. Frustration soon gives way to anger as she slaps her hand down on the counter and goes, "Come on, you know it!"

Pressing my knife sideways on some garlic, breaking its shell, I raise my shoulders and shake my head.

"When it's on the tip of your tongue? Come on, what's it called?"

But, for some reason, I'm blank too.

Grabbing the ricotta, I can distinctly remember being in the classroom where we learned about it, but for whatever reason, I can't picture it either. Finally fed up with the frustration, Ava storms out of the room while I go back to laying down the second layer. About a minute after that, Ava walks back into the kitchen, tosses her phone on the counter and says, "Well, *that* was anti-climactic."

Then, cradling her temples between her thumb and middle finger she goes, "It's called the tip-of-the-tongue phenomenon." Most times, scientists like to use words with Greek or Latin-sounding names, but every once in a while, they just call it like they see it:

Screaming Hairy Armadillo
Alice in Wonderland Syndrome
Tip-of-the-Tongue Phenomenon

No explanation necessary, reading the title is all you need to know.

Regarding Phrenology

Take your hands and start feeling around your skull. Start with the back of your neck and move upwards and what you should notice are little bumps and valleys. Spongy spots. Spots small enough for your thumb to get caught in, and other places that are just hard bone, a quarter of an inch thick.[12] Hands on both sides of your head, you're probably going to notice places where your skull changes shape, or where the angles start to shift, and it's these tiny differences—coupled with the growing evidence for the brain's modularity—that ended up seeding this idea called phrenology.

Back in the 1800s, when this idea first became popular, there was considerable evidence for specific brain regions being tied to specific brain functions, which inadvertently ended up seeding a runaway train of thought regarding skull shape and personality. This idea (that you could predict a person's character, intelligence, or personality based on the relative sizes of these bumps) is complete pseudoscience, but like some ideas, it's a hard one to kill. Even now, hundreds of years later, people are still trying to dust off the old phrenology meme and give it a face lift, when in reality, there's just no evidence to support it. It is the case that there are a lot of commonalities across all brains, including specific areas for vision, hearing, facial recognition, somatosensory cortex, and more. But at the same time, it also bears mentioning that no two brains are *exactly* alike. Moreover, just like any real map, once you're on a plane five miles up, you may start to realize that thick black line separating the two countries isn't really there.

Motor and Sensory Regions of the Cerebral Cortex

On a map, it's pretty straight-forward to see that Germany sits right next to Poland, but if I were to say to you that all Germans live in Germany, you'd instinctively know that isn't the case.

In actuality, the closer you get to the border, the more likely you are to see this free flowing of Germans heading into Poland and Poles heading into Germany.

Staring at a map for too long, we can trick ourselves into believing that these thick black lines have to exist, when really, if there wasn't a fence, mountain range, or river separating the two countries, we might not know where we are.

Am I in Poland or am I in Germany?

The point is, just as with maps of the world, our maps of the brain aren't always so cut and dried. And to say that all memory happens in the hippocampus—and only in the hippocampus—is on par with saying all Germans live in Germany.

Unfortunately, cases like Clive Wearing are not as rare as you might expect, and upon closer investigation, we can actually start to understand just how complex and layered the memory system really is. Because, as we just saw, even though Clive may have suffered bilateral hippocampal damage, not everything about the memory system was smeared out. This, and many other citable cases involving lesions or stroke, clue us into the fact that memory isn't localized to one particular area of the brain. But is instead a distributed phenomenon, occurring with pieces and parts in the cerebellum, the prefrontal cortex, the amygdala, the hippocampi, etc.[13] More on this topic later.

Notes from the Editor:
In patients suffering from Alzheimer's disease and some forms of dementia, the hippocampi are among the earliest and most severely affected structures. Hence their inability to form new memories.

—Back when we were making lasagna and we were unable to remember that term "tip-of-the-tongue phenomenon," we should probably go ahead and mention that a part of our brains *knew* that we knew the answer, while another part did not. When Ava went, "You know what I'm talking about," she wasn't lying; I did actually know what she was talking about. Sure, we might've forgotten the specific name of the term, but we both remembered the concept. And had we not Googled the condition, one of us probably would have remembered the name, eventually. Yet, at the exact moment she and I both wanted to know—it was *Error 404*.

This type of thinking, "meta-cognition" (thinking about thinking), is an interesting one because a part of our brains knew that we knew the information while the part that actually knew the name (that cluster of neurons) had been used so infrequently, it was going to take another moment or two for our subconscious to actually localize the pathway.

Dissecting this memory flub for a second, I think it's rather safe to say that the human memory system works nothing like a computer. You can't just hit a search button and expect nearly instantaneous results. However, there is a sense in which a part of the brain knows what the brain knows without having to access the memories directly. Or, to put the matter in another way, it seems as if a part of the brain has already indexed the memory system.

Take, as an example, the feeling of seeing someone you know but in an entirely different context. Like spotting a co-worker from three jobs ago in a foreign city. Of course, you may notice her immediately and think to yourself, "Wow, she looks familiar," but still not make the connection until hours, sometimes even days later.

This one time, when this exact situation was happening to Ava, she goes, "It feels like my brain is *buffering*," and it made me laugh. It was the perfect line to encapsulate the emotional agony we all experience when a part of our brain knows something that the conscious part does not. Similar to the search results you might get from Google, sometimes the memory system only spits back what it's indexed (the headlines and the tags, not the actual sites themselves). To see the full site, you still have to click on the appropriate link. And just as it can sometimes take a minute or two for the webpage to load, sometimes it can take a moment for our memories to load too.

And yet, for someone like Clive Wearing, these pages never seem to load. For patients like Clive Wearing, the only type of memory they seem to have left is a pithy amount of metadata.

***Notes from the Editor:*
Technically, Clive has a semantic memory that's mostly intact (general world knowledge and facts) and procedural memory. It's his episodic memory that's been the most severely affected.

—At age 37, Dr. Jill had to come to terms with the fact that a part of her (the Harvard-trained brain expert) had died that day and that who she was going to become next was ultimately going to be up to her.

Did she actually die that day?

That depends on who you ask. According to the state of Massachusetts, the answer would've been, "No." Yet, if you asked Dr. Jill, the answer would almost certainly be, "Yes, and no," because while her heart never stopped beating, she did lose a good chunk of her left hemisphere. Not just bits of her personality, but her ability to read, write, or even move. Too often, neuroscientists will cite a case like this and talk about a loss of function, like language. But what's often swept under the carpet is how an injury like this might off-line a component of experience as well, like her ability to define herself from that which is other. To know where she ended and the world began.[14]

Of course, we could say that pieces of her died that day, but in all the ways that matter, "No, she did not."

A Parting Note on Plasticity

From just three cases of brain trauma, we're beginning to uncover just how *modular* and specific the brain can be in regards to the localization and lateralization of certain brain functions. However, we mustn't let this idea be taken to its extreme.

The extreme is, every gyri and lobe is designed for task X and only task X. The extreme is phrenology.

Without question, the brain does have some hyper-specialized components, akin to a graphics card or a sound card. But what's been understated thus far is the amount of redundancies, feedback loops, and back-up systems Mother Nature has hardwired into us. Matter of fact, the more you read about brain trauma and strokes, the more you start to realize just how plastic and resilient the brain can be. For example, some stroke survivors might have lost their ability to speak but could still sing, and by using melodic intonation therapy, these patients were actually able to recover their speaking voice by first learning how to sing their messages.[15]

Resiliency, it seems, isn't just a personality trait—it's something that's baked into the fabric of who we are as human beings.

Chapter 7
Sensory Awareness

EVERY SINGLE SECOND, it's estimated that 11 million bits of information come streaming into your body through all of your sense organs. All that information (with the exception of olfaction) then travels through an area of the brain we call the thalamus, where it is subsequently pared down to around 200 bits* for your *Experiencing Self* to perceive and attend to.[1] Now, I'm sure you're wondering how it's even possible to pare down that much information into a workable user-interface like Consciousness 5.0, and the answer is, the brain takes shortcuts. **Lots** of them.

Take vision for example. To explain the complexities of vision, the first thing we really need to do is gloss over how digital cameras work. And the way we can accomplish this is by imagining we were trying to livestream our favorite college professor giving a lecture.

So there they are, standing in the front of the auditorium, lecturing about Eastern Philosophy or whatever, while you're off somewhere towards the back, standing next to your tripod. Mounted on top, a jet black DSLR camera that has an HDMI cable running to a laptop that's connected to internet.

In this scenario, light's flowing into the camera's lens, it's hitting an image sensor back where the film strip used to be, whereupon the light gets converted by the camera's computer into millions of tiny picture elements (aka pixels), which then get assigned a certain light value... which then get coded into a distinct pattern of 1s and 0s (aka binary). After that, this information then travels *out* the camera, through the HDMI cable and into the laptop it's connected to. Upon arrival, these 1s and 0s then pass into the computer's CPU, where it gets translated and reassembled back into video by some fancy computer software, so it can ultimately be displayed on the screen of your laptop.

Now, if that sounds like a clunky flow of information, then buckle up, because the way our brains reconstruct reality (*keyword being *recon-struct*) is far more complex. We'll be rounding out the picture of how vision works over the course of our time here, but for now, it'll suffice to say that we don't see like our intuition tells us we do. I realize vision feels like you just open your eyes and there's reality, but that just isn't the case. Far from it, actually, because technically, we don't even see with our eyes; we see with our brains.

More precisely, we see with our *internal model.*[2]

———

When light first comes into our eyes, it hits the back of our retinas where it's broken down into a series of neuro-electrical impulses. Then, it's sent back along some cables (nerves) to the visual processing center of the brain where it can be stitched and reconstructed into a three-dimensional representation of reality that you eventually get to perceive.[3] With digital cameras, the fidelity in which this reconstruction happens is pretty spot on. Despite some pixelation here and there, the camera will essentially capture as much reality as its tech allows. A better image sensor equates to more pixels; more pixels results in a higher resolution photo, which ultimately amounts to a higher fidelity representation of reality. However, when it comes to brains, we employ a lot of guess work, a great deal of filling in the gaps, and a ton of prediction based on expectation. Motion, edges, color, contour, shape, shadows, objects, these are all internal processes that are going on inside your brain right now.[4-7]

To be even more precise, there are teams of neurons whose sole purpose it is to track and identify objects, just as there are teams of neurons whose sole job it is to detect edges.[8] Just like that flip book we all used to play with as kids, sometimes your brain "paints" on motion when it believes there to be motion (see Fig 1.1); and sometimes, it paints on shadows when it thinks there ought to be some too (see Fig. 1.2 and 1.3). If we actually perceived the world the way our common sense says we do, then optical illusions wouldn't work at all, change blindness wouldn't be a thing, and magicians would be out of a job.

Fig 1.1

For the Checker-Shadow illusion, take a close look at the squares marked A and B and tell me if they're the same color or not.

Edward H. Adelson

Fig 1.2

Now look again.

Edward H. Adelson

Fig 1.3

Most people tend to say that Square A appears darker than Square B—that is, until they look a second time. We'll be returning to some finer points about vision soon enough, but for now, just realize that everything you've ever seen is a reconstruction.

Your brain's best guess.

Just like that livestreamed lecture, everything you've ever seen had to be deconstructed into two-dimensional retinal data before it could then be *re*-constructed inside your mind's eye, leading us to the inevitable conclusion about the way vision really works.

Namely, that you don't see, but that you learned *how to see*. And nowhere is this more self-evident than in the studies from humanitarian Dr. Pawan Sinha.[9]

Back in 2002, this neuroscientist was visiting his father in New Delhi, when he began to notice just how prevalent childhood blindness had become. In India, and other developing nations, many children are born with treatable forms of blindness, like congenital cataracts. And yet, because these kids have such limited access to healthcare, the majority of them never end up getting the sight-restoring surgeries they need. What's worse is, in the past, the parents were told these kids would never be able to see.

Long story short, but none of this sat well with Dr. Sinha, so in 2005 he launched an initiative to help restore vision to these kids, free of charge. Project Prakash, he called it, which in Sanskrit means "light." The only problem was, even Dr. Sinha wasn't too sure about how much light he was going to be able to restore, given that the majority of these kids had already passed the so-called "critical window" for visual development. At the time, it was generally believed that the critical window for visual development might slam shut around the eighth year of life.[10,11] But still, Dr. Sinha felt he had to try.

Best case scenario, he and his team might be able to restore vision to some impoverished kids; worst case, they would finally be able to study how the "newly sighted" come to see the world. Of course, folk intuition would have you believe that if you just removed the opaque cloudy lenses blocking their vision, these kids would finally be able to see reality, right? But, as we're about to find out, "seeing" reality is a much more difficult task than anyone had previously assumed. Sure, vision might feel effortless to you and me right now, but that's only because our brains have been doing R & D on how to see ever since the moment we took our first breath and cried.

Just take the following image for instance (Fig 1.4).

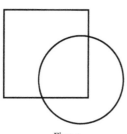

Fig 1.4

If I were to ask you how many objects there are in this photo, without even thinking about it, you'd probably just say "Two," and then roll your eyes at me for even asking you a question that simple. On the other hand, if we were to ask someone from Project Prakash the exact same question, it would take them a few months to perceive that there is, in fact, a circle and a square here, and not three separate objects.[12]

What appears to be so effortless for you and me (perceiving a circle overlapping a square), turns out to be extremely problematic for a brain just learning how to see. Even identifying a human face—that thing which appears to happen so automatically we take it for granted—even that takes around four weeks to pick up.[13]

Against this background, if you ever thought of your eyes as being like these little tiny windows, out of which *reality* just comes pouring in, then think again, because evidently, learning to see is a skill that takes years to master. And sadly, if the brain is deprived of this ability, even for just the first couple months of life, the literature is quite suggestive that there will always be impairments to the visual system that no amount of training seems to shake.[14,15]

Notes from the Editor:
Manoj Yadav, one of the patients of Project Prakash, was actually eighteen years of age before he was able to get his congenital cataracts removed. Because of this, it would take him another eighteen months before he could actually see things more clearly.[16] For example, if you were to show him a picture of a cricket ball set against a white background, and then asked him, "How many objects do you see in this picture?" Yadav would say three, because to someone that's never seen anything before, the world appears to be over-fragmented. Their brain hasn't yet figured out where one object ends and another begins, so to them, even a ball's shadow appears to be its own object.[17]

—Even though Dr. Sinha's studies might've finally put to bed the long-held assumption about a critical window for learning how to see, the results of Project Prakash seem to indicate that there is a critical window for visual acuity, which does seem to close some time before the age of eight.[18]

The Limits of CS5

When you're on an airplane and you're staring down at a huge body of water like the Pacific Ocean, it truly feels like your field of vision is capturing hundreds of square miles in stunningly crisp high-definition. The way consciousness presents itself, it almost feels like we can see all the intricate details of the waves at any given moment. And that if we could only *pause* our internal experience, we could zoom in on any one wave and the image would never pixelate.

At least, that's the way it feels.

But take a good long look at Ninio's Extinction Illusion (Fig. 1.5), and tell me if you still feel the same way.

Basically, I'm curious to know how many dots you can perceive at once.

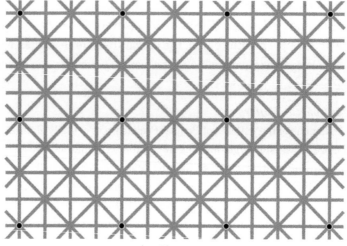

Fig 1.5

Most people can only see four dots at once, even though there are clearly twelve featured. If visual illusions of this type are suggestive of anything, it's that (a) our visual system is not as good as we like to think it is, and (b) that if you're not looking directly at something, your brain's liable to just fill in the blanks with whatever it 'thinks' is there.

So where do those other dots go?

Good question. Although, the real answer seems to lie in the way our visual system operates. Or doesn't operate, rather, because while it may subjectively feel like your vision is being captured in 8K or higher, the real truth is, the bulk of it's closer to 240p.[19] Make no mistake, the center part of our vision has clarity, thanks to an extremely dense concentration of photocells located there. But once you start moving away from that tiny circle, there's a tremendous drop-off in terms of clarity and contrast sensitivity. From something like 1080p at the bullseye, down to something like 240p towards the periphery. And this is precisely why you couldn't see all twelve of those dots at once.[20]

According to photoreceptors on the periphery of your retinas (of which there are fewer) those black dots didn't even exist. But once you shifted your gaze, effectively re-registering the visual data with the parts of your eye that can actually discern fine-level detail, a few more dots magically started to appear.[21]

The first time I showed this to my old roommate Brian, a gamer/video editor who's always bragging about how good his vision is, he told me it felt like his brain was playing visual Whac-a-Mole. Dots disappearing and reappearing just as fast; he even went so far as trying to memorize where the other dots were, hoping his brain would hold onto the memory. But once the photoreceptors on his periphery saw that intricate, grey, cross-hatched pattern, his subconscious just filled in the blanks about what it 'thought' must've been there, effectively covering up those dots.

***Notes from the Editor:*

The fovea centralis is a little depression or dimple on the back of the retina where cone concentration is at its highest. Due to the densely packed concentration of photoreceptors there, it's no surprise that visual acuity is also at its highest. Conversely, just below the fovea sits the optic nerve, a place where there are zero photoreceptors. As a result of this, both our eyes have a substantially sized "blind spot" that our brain is continuously working to fill in.[22]

—Looking at a page of a book, it truly does seem like I can see all 300 something words printed on each page, but in actuality, I don't. I can't even read more than a few words at a time before I have to shift my gaze to capture a few more words.

And when I want to finish that sentence...

I'll probably have to shift my gaze...at least once more...because that's how small the fovea actually is. Only about 1% of our retinas can perceive fine-level detail; the remaining 99% is just low-res peripheral vision roughly analogous to peering through frosted glass.[23]

How important is this foveal data?

Well, when you consider that about half the entire brain is either directly or indirectly tied up with vision in the first place.[24] And then you take into account that about half the visual cortex is devoted solely to processing the information it receives from the fovea—I'd say it's pretty damn important.[25]

The Dials of Perception

Slowly but surely, as we continue to tear through more and more of these studies, you may begin to notice the outline or *shape* of this user-illusion. Comparable to placing a VR headset over your eyes and then doing your best impression of a dog trying to dry itself, all the facts, findings, and experiments are designed to help us accomplish something similar. With VR headsets, it's easy to shake your head from side to side thus exposing the latency between what's seen and what's supposed to be seen, yet with CS5, the experiments necessary to expose the illusion are a little more complex.

So far, we've already shown how the brain edits reality and screens out saccades; that it continuously works to fill in those blind spots and places where visual acuity is rather poor (the remaining 99% of your retina). And then, not too long ago, we even talked about how an injury to the brain might off-line certain components to that experience, like Dr. Jill's ability to tell where her body ended and the rest of the world began, or Clive's ability to extract himself from the present moment. After that, we contended with the fact that CS5 actually lets us feel like we have better vision than we do. And the reason it does this, we submit, is because vision is a bit more complicated than previously anticipated.

We don't capture the entire visual field in stunningly crisp detail, we only think we do; and while CS5 may continuously give us the illusion of having more detail, in reality, we have to acknowledge that's just a sensation. Bearing all this in mind, we might then say: It's as if our Subconscious OS has tricked our user-illusion software into thinking it has more information than it actually does. And this is just one of the ways in which the Subconscious OS is able to pare down that 11 million bits into something more consciously manageable.

To help get a visual grasp on this, let's imagine that your subconscious mind has a tunable dial on it. Five dials to be exact. And that each one of these dials is directly tied to one and only one of your sense organs. So a dial for the amount of visual information you can see, a dial for the amount of auditory information you can hear, so on and so forth. At any given moment, when you add the values of these dials together, you should get

200 bits. Implying that the value on each one these dials is directly tied to that sense organ's information allotment. From this perspective, we could begin to imagine that, right now, your brain is giving you 20 bits out of that 200 for smelling, 90 bits for seeing, 15 bits for your sense of touch, another 15 for your sense of taste. And since you're listening to some music right now, let's say your brain has allotted you 60 bits for listening to the melody, beat, and lyrics.

Add all those values together and you get 200 bits—but let's keep in mind that's just for 1 second!

$$
\begin{array}{r}
20 \text{ bits (olfaction)} \\
90 \text{ bits (vision)} \\
15 \text{ bits (touch)} \\
15 \text{ bits (taste)} \\
+ \ 60 \text{ bits (audition)} \\
\hline
200 \text{ bits}
\end{array}
$$

Each second your brain is having to figure out, on the fly, what's the appropriate amount of information to allot for each sense organ; this implies that the *dials of perception* are always in flux, sometimes upping your visual acuity, sometimes dampening it. Sometimes boosting your sense of touch, while other times, like when you've been sitting for twenty minutes straight, straight up ignoring it.

An even more clear-cut example of this might be when someone whose just showered in perfume first walks into a room. At first, the smell is offensive, as you can literally taste their smell. But then, after about five minutes, the smell seems to go away. This isn't because the body spray magically lost its potency, it's because your Subconscious OS 'decided' that focusing on it was no longer a good use of resources. Therefore, we can imagine that after about five minutes, it's like the tune-able dial for olfaction got the volume turned down, and that particular smell, has effectively been screened out from conscious awareness.

***Notes from the Editor:*
The technical term for this is called habituation.[26]

—Now, am I suggesting that those nice, neatly rounded numbers are spot on for your particular brand of consciousness? Of course not. First off, it's extremely difficult to measure unconscious processing capabilities and then assign a certain value to them. However, the estimates for what our conscious selves can process and attend to range from 40 bits per second up to around 200, while the estimates for what our subconscious is capable of ranges from 11 million bps well into the *billions*.[27]

Either way you slice it, conservative or not, it seems rather self-evident that the subconscious can process and attend to much more information than the conscious self ever could. In any case, however, while we might have some degree of control as to where we point our spotlight of attention, there's not a person on the planet who can will their subconscious into giving them 180 bits for seeing, any more than there's somebody who can make their fingernails grow faster. To put it plainly, your subconscious owns and operates these ratios of sensory awareness, and it does so largely without your consent or control. For now though, the general idea is this: We are constantly being bombarded by a barrage of noisy, conflicting sense data, and it's literally too much for our conscious selves to manage or attend to. As a consequence of this, the Subconscious OS has to select out *an interpretation* about what's just happened, and then send that up to the executive self in the form of a 200-bit story.

And again, that goes for every second of your waking life.

Every second, your subconscious is spinning you a story about what's just happened, and in order to funnel that 11 million bits into something more useable, the brain is constantly connecting new information to past information. In this regard, I can imagine a hypothetical, albeit impossible inner monologue of the subconscious to go something like this:

Okay, so Tom's looking out the window of this plane down at the ocean below and he's seeing miles and miles of choppy waves, except he's only really only focusing on that little white sailboat, so let's let him see that in vivid detail while we 'phone in' the waves on his periphery. Being that we've already seen choppy ocean water a thousand times before, I'll just keep sending up stock footage of ocean water—that way I can keep the dials of visual perception at a tight 90 bits per second. I'm gonna give him 20 bits to smell that coffee that's being brewed in the galley, and another 20 bits into feeling his butt-cheeks because he's been sitting for about 3 hours straight and I need him to be aware of that fact. It sounds like someone just said "Tom" or "bomb" so let's up his hearing allotment from 40 to 85 because it might be urgent and I need him to pay close attention to what's about to happen next...

To be clear, I don't actually believe the subconscious has an inner monologue like the one featured above, but sometimes, it can be rather instructive to imbue the Subconscious OS with the kinds of rational decision-making powers we enjoy to better understand a point. Truth is, we still have yet to work out how the brain makes decisions about what *you* get to focus on, as often times we hear things we wish we didn't; we see things we wish we wouldn't; and we feel things we wish we hadn't. Like it or not, but what our brains habituate to (the feeling of the seat underneath your butt, the tension in the back of your neck, the soreness in your calves...) is mostly out of our control.[28]

The reason I say *mostly* out of control and not completely out of control, is on account of the fact that there is a massive, growing body of evidence surrounding the topic of meditation and its ability to grow your pre-frontal cortex, enhance your meta-awareness, and sharpen your attentional focus.[29-31] In short, if attention is like a muscle, then meditation is the practice that helps you define that muscle.

In any event, it's still the case that your brain is making decisions about the ratios of sensory awareness you get to perceive, but it's definitely not something "Tom" or "James" gets to decide. Rather, it's an unconscious decision mediated by an area of the brain we call the thalamus which we can think of as being like the brain's relay station. Then, wrapped around that, you have this thin band of cells that are involved in sensory gating and attentional modulation: the thalamic reticular nucleus.[32]

Although, we're likely to continue anthropomorphizing the subconscious with decision-making capabilities, it's actually more likely that the Subconscious OS arrives at decisions in much the same way that computers do. Algorithms, a source code—basically, just a long list of *If this, then that* commands.

Floodgates

I'm a big picture person, so here's another one of those place-holder sketches I'd like you to consider in regards to the dials of perception. For this next one, I want you to close your eyes and really try and imagine five tunable dials sitting on a control panel. Same as we did before. Only now, I want you to visualize that these five tunable dials are connected to five corresponding floodgates. So, with each turn of the dial labeled "sight," the floodgate for vision would start to open a degree or two; and with each turn of the dial marked "touch," another floodgate would begin to open, allowing more tactile-based sensations through. Moment by moment, each of these five floodgates are either opening a few degrees or closing a few degrees, carefully regulating the amount of sensory information that enters our sphere of awareness.

The following image isn't meant to be taken literally, as there is no official subconscious section of the brain. However, even despite its simplicity, this visual aid can still help us appreciate some of the complexity behind what we're actually talking about.

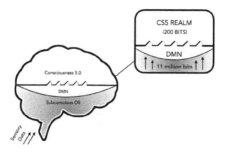

Here, we can clearly see sense data flowing up from the body and into the realm of conscious awareness, although before it even makes it there, this information has to be sorted and processed first. So, the Subconscious OS does all the processing; the reticular nucleus, the thalamus, and the default mode network do all the regulation and integration (always restricting the amount of sense data to a tight 200 bps). That way, Tom, sitting up in the proverbial "theater of consciousness" gets to do all the feeling and experiencing.

Brief Recap

To drive all this home, let's revisit the McGurk effect one more time. In this particular study, we got to observe just what happens when the brain is presented with two conflicting streams of information. Not having the cognitive bandwidth to support both, this meant the brain had to choose. The stream from the ears was, of course, reporting back that the guy was just repeating the same words over and over again ("ba...ba...ba"). While, at the very same time, the eyes were vividly registering a guy biting his bottom lip with his two front teeth and then flicking it down in a way that could only mean an "F word" was coming.

So what did your brain do?

It chose the visual stream. Not half the time. Not some of the time. *Consistently.* Even one of the lead researchers in the field, Lawrence Rosenblum, says he's still affected by it and it's been over twenty years.[33] If the persistency of this illusion reveals anything to us, it's that the brain likes to play favorites with the visual stream. And for this reason, we can imagine that the floodgate associated with vision would be a little more open than the one tied to hearing, causing a tilt in perception. The end result of which being, you hear "Fa" instead of "Ba."

To summarize, your ears still heard it correctly (i.e., you still had the sensation of hearing "Ba" falling on your eardrums), but, at the level of experience, where senses are unified and integrated, what you actually *perceived* was a man saying "Fa."

It's a subtle point to make, but not all sensations graduate to the level of perception. And in all sincerity, most don't. Again, the reason for this is because we have an attentional-based consciousness, and you simply cannot be aware of everything at once.

***Notes from the Editor:**
It has been argued by Dr. Iain McGilchrist that the conscious reality we perceive is like a map version of reality, not reality itself. Dr. McGilchrist then goes on to point out that maps are actually useful in what they leave out, arguing that a map with too many details distracts more than it helps.[34] In keeping with this, consciousness never seeks to provide the organism with every single available detail, just the most relevant information for each and every moment.

—Now, here comes the tricky part. Make sure you're sitting down for this one because the first time I learned it, I was a little unnerved to say the least. Those streams of data...they don't come labeled.

The information streaming in through your eyes doesn't have a little note attached to it saying this is visual data, nor will you find any note that reads, *This is hearing*. The body just takes in all these patterns of data, sends it up to the thalamus, and then figures out which is which based on memory, prediction, and expectation.

How does it accomplish this?

This is a question we'll be returning to again once we've uncovered a bit more about how the brain operates. Although, one of the best ideas about this comes from one of the biggest names in neuroscience: Dr. David Eagleman. His hypothesis, which I whole-heartedly agree with, is that "it's all about the structure of the data coming in." With vision, you have "two two-dimensional sheets coming in. With hearing, you have a one-dimensional signal through time."

"Your fingers," he says, "are picking up on this high multi-dimensional signal," and all this must look different on the inside (i.e., the pattern of the data must be different).[35] What also can't be ignored is the sheer amount of data captured by each sense organ in the first place. With 11 million bits steaming into your brain every second, and about 10 million of that coming in from just your eyes, does it really surprise you to learn that the brain would favor the visual stimuli?

I mean, there's just way more of it.[36] With this in mind, we can begin to imagine that with each passing moment the eyes are sending back a story they think is the most accurate. The ears, the tongue...all those skin receptors, they're each sending back a story they swear is the truth, the whole truth, and nothing but the truth—but sometimes they make mistakes. Other times, they disagree, and the subconscious just has to wade through that.

This is why you hear "fa" instead of "ba."

This is why you see a face in the Hollow-Mask illusion.

This is why your brain corrects audio and visual cues that are out of sync (up to 80 ms).

"Our perception of the world is a fantasy that coincides with reality."

- Chris Frith (psychologist)[37]

Chapter 8

Sensory Input = Consciousness Output*

A POLAR BEAR can smell a seal from twenty miles away, I tell him. Beat that!

"A duck-billed platypus," Terry says, "doesn't need its eyes or ears or nose to hunt for prey; all it needs is its bill."

That's not true, I fire back.

"Ask your phone," he says, folding his arms across his chest. "They hunt by electroreception…tiny sensors in their bill." "That and they're poisonous," he adds.

What he means to say is venomous, but since I'm too busy Googling: *duck-billed platypus electroreception*, I don't bother correcting him.

I don't want to burst Terry's bubble twice.

But the second I start the "e" in *electroreception*, Google finishes my thought and what do you know, Terry's right. Article after article about the duck-billed platypus's sixth sense: electroreception.[1] As if they didn't already take the cake in weirdest animal of all time.

That's when Terry starts to laugh. Raising my head to look at him, I ask, what's so funny? And Terry goes, "You got quiet," and then puts the tips of his fingers together like some kind of evil super-villain.

"Are you reading about how wrong you are?"

Flipping the switch for silent mode, I slide the phone back into the front of my jeans and turn to look at him, only Terry's still looking off in the distance.

"You're on your phone, right?"

Terry hates it when people get lost in their phone, but being blind from birth, he doesn't actually get to see how often it happens; he only feels it. Too often, he'll tell me, he'll be in conversation with someone—and everything will be going fine—when suddenly, the other person just starts to trail off. Their attention being diverted by a tiny black rectangle slipped from a pocket.

It happens so often, Terry says, that nowadays, he can actually tell the difference between when someone gets lost in a memory vs. an email. Apparently, Terry can still hear the soft taps on the glass when someone is typing feverishly. And what I can't believe is, how that even makes a noise.

You know back when I had a life. Before I wound up stuck in the kind of place you wouldn't wish upon your worst enemy, I was interviewing anyone and everyone I could find about their thoughts on consciousness. The blind, the sighted, the deaf—I was genuinely trying to get to the bottom of what life was like for them, and how this internal experience could morph and change. On account of being born blind, Terry taught me a lot about how the conscious experience could stretch. Just hearing the stories about what childhood was like for him made me slink down in my chair and mull over all the ways I took vision for granted. Take dreams for instance. One day, we're sitting in Terry's office talking about blind dreaming, when he finally just says, "Look, they're just as weird as your dreams—it's just I can't see anything."

Evidently, dreaming for the blind is more about the feelings of things. The sounds and the smells. But since I can be rather thick sometimes, Terry makes me close my eyes.

Rubbing his hands together like he's about to make a fire, he goes, "Now I want you to imagine that you're at a restaurant."

Lifting himself out of his chair, I picture Terry talking with his hands. "You know you're at a restaurant because it *sounds* like a restaurant. It *smells* like a restaurant. It *feels* like a restaurant. It doesn't feel like it feels when you're at home, or when you're in a car. It feels like you're seated at an open-air restaurant, okay?"

Then, for added effect, Terry starts pacing around in circles. "Across the room," he says, "you can hear a screeching espresso machine." The sound of pants shuffling from the left side of the room pulls my focus as Terry attempts to make this dream a Dolby surround-sound experience.

"There's a table talking really loudly over here," he says, "and one of the girls laughing sounds kinda hot." Those beige umbros shuffling once again, Terry takes a few steps and then throws his voice across the room, "All around you is a sea of conversations and smells. You hear forks scratching on plates, ice clinking in glasses, laughter coming from the patio, when suddenly, you hear the familiar sound of your waiter who says 'Hot plate!' So you pull your elbows off the table, you go to sit back in your chair and *BAM!* The next thing you know, you're floating in a pool."

Six inches from my ear, stands the new narrator of my experience.

"You know you're in a swimming pool, because it's not like the bathtub. It's not like the spa. It's not like the ocean. It *feels* like a swimming pool. It smells like a swimming pool. And there you are paddling around, feeling the weight of the water move through your hair…you're smiling. You can taste the chlorine seeping into your mouth when you think to yourself, 'Hey, where's that pasta I ordered?'"

Clapping his hands together really loudly, Terry snaps me out of the dream and goes, "That's what blind dreaming is like," and collapses back into his chair.

—

At a later date, Terry would go on to tell me all the ways in which being born blind had given him an entirely different internal model than someone like me. Seeing as Terry never saw anything—never saw a single thing ever—his brain never learned *how to see.* He never got to find out what depth perception was, or what colors were, so when we tried to talk about them, Terry had no frame of reference. In much the same way that you and I will never know how it feels to hunt via electroreception, Terry had a hard time understanding how I could work out something was twenty feet away using just my eyes. He and I actually had to talk in circles before I finally had enough sense to fish out a finished Rubik's cube from my bag and plop it into his hands. Then, placing his fingers on the edge of the blue side, I tell him to, "Walk 'three squares distance' and then stop."

"That's the other side of the cube," I tell him. It's three squares of distance away.

"Depth perception is like being able to count the squares of distance between two or more objects with your eyes," I explain. "Only there doesn't need to be any squares painted on the floor because I can just 'feel' the distance with my eyes...you know?"

But the truth is, Terry doesn't know.

Concepts like color and depth-perception aren't so self-evident to someone who's never seen anything before. Case in point, when I tried telling Terry I could draw a three-dimensional object, like that cube, on a two-dimensional sheet of paper, he said it felt like his mind was caving in. Contrast that against someone who's gone blind later on in life, and you soon find that their internal experience is still largely similar to yours or mine, it's just completely devoid of any visual input.

To explain what I mean, let's just say a guy suffers an accident and goes blind at age forty. Being that his brain has already constructed an internal model about the world that's largely based on vision, whenever he picks up a familiar object and holds it in his hands, his brain can still recall all the imagery associated with it. Likewise, because he already knows what colors are, he can still imagine whatever you hand him to be burnt sienna or cerulean blue; he can even still dream about the things he's already seen.

Most importantly, however, is the idea that his brain has already constructed an OS, and an internal model that's largely based on vision. And all this has an effect on the way the brain physically looks on the inside, because when your eyes are open and functioning properly, they're in communication with your occipital cortex. But once a person goes blind, this cortical real estate is swiftly taken over by other senses.[2]

How long does it take before this cortical reorganization starts to occur?

Amazingly, there are parts of the visual cortex that will start to get reassigned in less time than it takes to watch that show *60 Minutes*.

That's right, within as little as forty to sixty minutes of being blindfolded, parts of your visual cortex will start to process another sensory modality, like tactile processing.[3] And this is just one reason why we'll truly never know what it's like to have been born blind, since being born blind has massive implications on the cortical restructuring of our brains.

Of course, we can blindfold ourselves and simulate what it might be like to lose the gift of sight, but trying to simulate Terry's view of the world is pretty much an impossibility. His relation to space, objects, colors, visuospatial reasoning, and all the rest is just too foreign for a sighted individual like myself to even imagine. Much like the difficulties in knowing what it must feel like for the Canada goose to navigate the Earth using magnetoreception, we'll never truly know what it's like to be born blind. Even though we might be able to design tools, like a compass, which may help us to reach the same end, the Canada goose doesn't need a physical tool to know which way north is because they can just 'see' it in their minds and then orient themselves accordingly.

Or consider that polar bear that can smell a seal on the ice twenty miles away?[4] Even if that seal is swimming underneath a sheet of ice three feet thick![5]

Could we honestly imagine what kind of implications a sensory awareness set-up like that might have?

I think the answer is almost certainly *no*, because for us, vision is the primary sense. That withstanding, polar bears actually have good vision too, but on account of them living in an environment where everything looks the same, there was significant evolutionary pressure for their noses to become hyper-sensitive.

What might the implications of having a nose that strong be?

The world's leading evolutionary biologist, Richard Dawkins, and former contender, J.B.S. Haldane, have both made comments on the matter. Haldane, that dogs might be able to order, rank, and classify certain smells along some kind of a spectrum.[6] Dawkins, that dogs and other smell-oriented animals might even be able to smell in color. Dawkins even goes so far as to suggest that bats might be able to see in color too, arguing that colors are really just an internal labeling system used by the brain to assist the organism in navigating 'its world' whatever that 'world' happens to be.[7]

***Notes from the Editor:**

It should be specified that color doesn't actually exist out in the real world. Indeed, different wavelengths of light exist, vibrating at different frequencies, but color itself is an internal construction of the brain. A kind of labeling system brains have devised for discerning between different wavelengths of light.

—Dawkins' argument, goes something like this: "the world model that a bat needs in order to navigate through three dimensions catching insects must be pretty similar to the world model that any flying bird—a day-flying bird like a swallow—needs to perform the same kind of tasks." He then goes on to say, "The fact that the bat uses echoes in pitch darkness to input the current variables to its model, while the swallow uses light, is incidental." Dawkins even suggests that bats might use "perceived hues, such as red and blue, as labels, internal labels, for some useful aspect of echoes—perhaps the acoustic texture of surfaces, furry or smooth and so on."[8] Against this backdrop, it actually might not be too far off to suggest that polar bears might be able to smell in color too.

So what is reality then?

In a sentence: *reality is whatever your senses tell you it is.* Public intellectual and clinical psychologist, Dr. Jordan Peterson, once captured the spirit of how we perceive by likening our senses to being something like "five dimensions of triangulation."[9] Which, when you think about it, really cuts to the chase about the kind of reality we actually perceive in the first place. It isn't perfect. It isn't direct. It's our brain's best guess. And what we can sense about that reality is ultimately confined by the biology we're bringing to the table.

Sensory Substitution

The famous German biologist, Jakob von Uexküll, once encapsulated this idea about the *thin slice of reality* our brains are even able to detect with his term "umwelt" (o͝omvelt). In German, the word might mean "environment" or "surroundings," although in ethology, the word has come to mean the thin slice of reality that a particular organism inhabits. Not just its physical environment, but what that organism can sense about its environment.[10] Jakob thought two animals might share the same physical environment but still inhabit different "umwelten" (plural of umwelt) based on the fact that sensory modalities differ animal to animal. As an example of this, we might consider that our umwelt is much different than a polar bear's, owing to the fact that our primary sense is vision and not olfaction. Polar bears, on the other hand, can smell a seal that's twenty miles away; and for that reason, we should expect them to have an internal representation of the world that is quantitatively and qualitatively different.[11] In a similar vein, the cave swiftlet (a cave-dwelling species of bird) might also be expected to have a different umwelt, given that they're using compressions of air waves to construct their internal model and not photons.[12]

Or how about a sensory-awareness set-up that was largely based on membranes in your nose that detect infrared radiation, like a pit viper's?[13]

It's a really fascinating concept Jakob captured with that word "umwelt," because once you really start to take it seriously, it sets us down a path of re-examining just how different the world must look to each of us. What's more, it's even possible for two species of animal to share the same physical environment, and yet, inhabit such different perceptual worlds that they don't even notice one another.

Once again, Dr. David Eagleman, helps explain the point more clearly: "The interesting part," Eagleman says, "is that each organism presumably assumes its umwelt to be the entire objective reality 'out there.'" Because it's not intuitive, he adds, "Why would any of us stop to think that there is more beyond what we can sense?"[14] And it was precisely this line of thinking that really got him thinking about what the brain was. His conclusion, that the brain is really just a "general purpose computing device," and that it could (*keyword being *could*) turn any stream of data you fed it into something useful.

In this regard, if we're likening the brain to being something like a motherboard, and we're likening consciousness to being something like the software that runs inside it, then we can begin to imagine that the motherboard of the body is really just a general purpose processor capable of handling all sorts of peripheral devices. We know this to be true of actual motherboards, as we're constantly plugging in new devices like speakers, webcams, keyboards, and the like—but can we add senses to the body?

David Eagleman and Scott Novich seemed to think so, which is why they even designed a vest for that very purpose. The idea was called "sensory substitution" and there were already even a few studies demonstrating its potential dating as far back as the 1950s.

See *Paul Bach-y-Rita*

See *BrainPort*

See *Sonic Glasses*

Basically, David and Scott reasoned that if you fed the brain some kind of data stream that was useful, that eventually, the brain would just figure out how to incorporate it into its internal model.[15]

What might that look like?

Dr. Eagleman shows us from the TED stage. Taking off his button-up shirt, David reveals a slim white vest adorned with just a few techie modifications. Namely, thirty-two vibratory motors embedded all over that are capable of vibrating extremely complex patterns of vibrations, essentially turning any kind of data stream you could think of into a rich tactile experience. As one might expect, the first time you slip on this vest, these patterns don't make a lot of sense.

Matter of fact, they just seem like random noise.

But after training with this vest—even for just a couple of days—the brain starts to figure out how to make use of them; and at that point, these patterns start to take on a whole new meaning.

Demonstrating one way of using the vest, Dr. Eagleman introduces us to a guy named Jonathan. Jonathan's never heard a song or a single word in his entire life. And yet, after training with this vest for only a few days time, the whole world watches as researcher Scott Novich puts him to the test. Covering his mouth, he says the word "you" into the microphone.

The microphone transmits the data into a pattern of 1s and 0s; the vest transmits the data into a tactile experience; and then Jonathan writes:

you
The researcher says "where"
and Jonathan writes:
where
The researcher says "touch"
and Jonathan writes:
touch

Sensory input = Consciousness output. Change your sensory inputs and your consciousness is bound to change, right?

Standing on the TED stage, David shows us one form of sensory substitution that's possible (substituting vibrations on the torso for vibrations on the drum of the ear) effectively providing Jonathan with a new way to hear. Then he turns to the audience and poses the question, "How do you want to go out and experience your universe?"[16] And immediately, I'm thinking about hooking this vest up to a magnetometer, so I can feel the Earth's magnetic fields all around me. Granted, I wouldn't be able to control them, like Magneto, but with enough training, I will have enhanced what it is I can sense about this world, thus expanding my umwelt.

Why bring any of this up?

I bring it up because this a real-world application to everything we've been hinting at so far: (i) That consciousness is *a layered system*, and that it's layered with whatever sense organs are currently 'plugged' in, and (ii) that once these patterns actually reach the brain, whether they originate from vibrations on the cochlea or vibrations on the torso, the Subconscious OS still has to turn them into something useful. But— and here's where that asterisk comes into play—*we aren't just passive decoders of information*. Sensory input would equal consciousness output if we just happen to be surveillance robots that took in streams of data and managed information. But, as fate would have it, we're not just mindless robots; we're emotional creatures that want, feel, think, and *believe*. And what we believe about this life (our prior expectations) actually changes the reality that we find there.[17]

Chapter 9
What is Reality?

SUPPOSE YOU'RE ON a coffee date with someone you haven't seen in a really long time. You're seated outside, slipping into an altered state of consciousness we call caffeinated, when one of you starts musing about the nature of reality and umwelts. Now, of course, you wouldn't see this but there'd actually be tiny ripples dancing across the surface of your café au lait as you spoke softly to your friend. You'd notice the ripples if a T-Rex were walking by, or if someone drove by with a trunk rattling system in their car, but all those tiny ripples emanating from your vocal chords, you wouldn't. Technically, you might *see* them, but you certainly wouldn't *perceive* them, because as previously discussed, what our eyes 'see' isn't always what ends up making it into the realm of conscious experience.

Remember, there's a difference between seeing and perceiving. Hearing vs. listening. Being present vs. being aware.

From our most recent discussion about umwelts, our brains have already been primed to start thinking about reality in a completely different way because, as it turns out, any organism's "reality" is ultimately dependent upon what that organism can sense about its environment (i.e., what are its sense organs). Apart from that, we also learned how two creatures might even share the same physical environment, like sharks and octopi, but possess an entirely different inner world based on the varied sensory inputs they receive (e.g., the shark is able to detect electric fields given off by prey that the octopus cannot). So, even though they may share the same external world, their perceptual world is bound to be different considering the differences between their inputs.[1]

Just the same, boa constrictors, pythons, and pit vipers would also likely inhabit a different perceptual world, than say a mouse, given that they have specialized infrared-sensitive receptors sitting on their face that a mouse does not.[2]

Taking all this into account, I think it's rather safe to conclude that whatever reality is we really haven't got a clue, because in all honesty, we're not evolved to 'see' reality. We're evolved to survive and reproduce.

We evolved to see fitness payoffs, not reality itself, and never was this point made more clearly than by MIT researcher Abe Davis when he gave a TED talk about his motion microscope. In the talk, Abe's lecturing about how there's motion all around us—it's just too subtle for our eyes to detect. He then demonstrates this concept by showing the audience two videos side by side: One, a close-up of a person's wrist; the other, a sleeping infant. Although, if Abe never mentioned either of these were videos, nobody in the audience would've been the wiser, as there was literally no motion to speak of—or was there?

Closer examination actually reveals that if your cameras are sensitive enough, what you could see is the rising and falling of the baby's chest, or the slight pulsation of blood on the person's wrist. For humans, perceiving that small amount of motion isn't really practical. Sure, we might be able to detect photons of light vibrating at a particular range, and of a particular speed, but below that threshold, everything just appears to be static to us.

Take as an example the fact that we're blind to the movements of most plants.

Plants clearly move with intelligence. They explore and react to their environments in highly interesting ways. It's just that humans only see reality at a particular temporal resolution or "frame rate," such that, we can't really appreciate their movements until we use an instrument that alters our perception of time. Using a camera, we can start snapping up photos of a plant every hour or so, and then, once we play their movements back, suddenly, a new world starts to open up.

What might be possible if we could perceive reality at a higher frame rate?

I suppose this is the question Abe had in mind when he started singing "Mary Had a Little Lamb" to a bag of potato chips. But honestly, we may never know. What we do know is that he and his team recorded the bag of chips being sung to—at thousands of frames per second—and then, fed all those nearly imperceptible vibrations into a computer. After that, they used a specialized algorithm to analyze the footage of the bag of chips, and were actually able to reverse-create the sound that caused those vibrations.

From the TED stage, Abe plays back the audio they've created, and it's very clearly "Mary Had a Little Lamb," but as you probably might've guessed, it is by no means the kind of quality you'd expect from a CD. Still, when you consider that their "microphone" wasn't a Brian-approved Sennheiser MD421 but an empty bag of potato chips, the sound they've created is both impressive and scary. And while it's never quite stated outright, the implication is, just about anything could potentially be a microphone, provided you have the right camera and software.[3]

Nevertheless, the real question under consideration here is: *Did these vibrations happen?*

Although, as is traditionally the case with consciousness, the answer depends on who you ask. According to your CS5, the answer would be, "No, the bag of chips didn't even budge." And yet, if we were consulting the camera footage recording at 1500fps, the answer is, "Yes." So what's going on here? And who should we believe, our instruments or our eyes?

At this stage in the discussion, I think the best answer we can give is that, whatever reality is, it really exists in the eyes of the beholder, since it is dependent upon, not just what you can sense about the environment (the sense organs or tech you're bringing to the table), but also upon whatever your Subconscious OS has deemed necessary for you to see in the first place. It will be recalled that we're not perfectly designed robots, but rather, evolved creatures just trying to survive. And during this eternal struggle for existence, survival and reproduction were the only two things our ancestors had to get right in order to make it into the next generation. Implying that any beneficial adaptations we've accrued over the millennia never had to be perfect solutions, they just had to be 'good enough.'

How does one define good enough?

If it aids in the survival and reproduction of that organism, then that mutation can be said to be 'good enough.' More to the point, our eyes aren't perfect and they never had to be. They don't capture reality *as it is*. Honestly, we don't even know how to define "as it is" in the context of that sentence. What has happened over the course of evolution is that our eyes have captured 'enough' reality for us to survive and reproduce (the visible spectrum) and that's what we're allowed to see.

We don't see all there is to see.

We don't hear all there is to hear.

We just see enough, hear enough, and smell enough to survive.

In many ways, constraint, necessity, and death are the driving forces behind evolution, and because we never needed to perceive reality at thousands of frames per second, we never evolved the capabilities to do so. But perhaps even weirder than that is the idea that we probably could have evolved them if we really had to. Because whenever we look out across the animal kingdom, not only do we find numerous examples of sensory enhancement—like an eagle's supreme gift of sight, or a polar bear's supreme gift of smell—but we also find entirely new modes of perception that we can't even begin to conceptualize yet.

What's it like being able to locate a fish using the electrically sensitive pores located on your snout, or being able to track a rabbit's body temperature simply by using the heat-sensitive pits sitting near your eyes?

These questions all bring us to an important point about what reality is because when you really start to take the concepts of umwelts seriously, you begin to understand that the term "objective reality" doesn't even make any sense. Objective to whom? Or to what? Because so far, we've captured reality at trillions of frames per second, and yet, we've still never found some kind of limit where all this information caps out.[4]

So again, you ask me, "What is reality?"

And the answer seems to be: It depends on who you ask, and what kind of sense tech they're bringing to the table. Because if your "eye cameras" are strong enough, and your tech is sensitive enough, then reality appears to accommodate.

Eye Cameras and Vision as a Layered System

If you've ever played with Photoshop before, then you're probably already aware that most any good Photoshop creation is actualized by layering images and effects one on top of the other. Most any billboard, promotional photo, or piece of web art, is created using a Photoshop-type program, and is constructed layer by layer, piece by piece.

Layer 1 (the background layer) would, of course, be the photo in question, while the next layer might be the photo it fades into. Followed by a layer for brightness, a layer for text, a sepia filter, a vibrance layer, blah blah blah. Layer upon layer, all these tricks stack up with the final image being something you might find printed on the front of a book.

A highly stylized and picturesque creation.

And what I'm putting forth to you is that Consciousness 5.0 works in much the same way. Meaning, after visual information streams in through your eyes and flows back into your brain, each system gets to add its layer of encoding, such that, what you end up perceiving as happening in this "movie" is generally a pretty useful portrayal of what's going on out in the real world. Although, it should probably be stated plainly, that while we might have holes in our head, these aren't direct portholes into our brain.

Eyes may be the proverbial "windows to the soul" but they aren't actually windows. Better to say, these are little cameras that snap up photos of the world, which are then sent back into the occipital lobe for processing.[5] Then, and only then, does the brain actually get to process the raw two-dimensional images and render them into a three-dimensional model for you (the experiencer) to experience and attend to.[6]

———

Another way to think about the way we see might be to consider the way movies are shown in 3D. Next time you're at a movie being shown in 3D, twist your head back towards the projection booth and what you should notice is that there are actually two projectors running simultaneously. Both are projecting the exact same movie, at the exact same time, it's just that one of these projectors is deliberately angled off-center, so when you go to look through those special, polarized lenses, your brain interprets depth and 3D dimensionality.

Take the glasses off and not only is all the magic gone, but the film tends to look rather odd considering one of the projectors is slightly askew. Yet, it's only because one of these projectors is off-center that the whole illusion works in the first place.

In a similar vein, the only way we're able to perceive our environment three-dimensionally is because we have stereoscopic vision. Our eyes are located about three inches apart. As a result of this, each eye is actually receiving a slightly different two-dimensional feed of images—and it's the difference *between these images*—that allows our brains to make an educated guess about the depth and three-dimensional structure of objects.[7] Therefore, under the Photoshop layers analogy, your left eye would be one layer, your right eye would be another, and a network of neurons responsible for interpreting depth would be yet another. Followed by a whole slew of networks responsible for detecting motion, edges, shadows, objects, and color. Layer on top of layer, these systems of neurons all end up contributing their two cents' worth of processing, with the ultimate outcome being: this miraculous, unified construction you see right now.

> Most of vision is an internal process happening completely within your brain, and the information dribbling in through your retinas is just a small part of what you're actually perceiving. So it's about 5% of the information of your visual stream is coming in through your retinas, and the rest is all internally generated given your expectations about the world.
>
> - David Eagleman (neuroscientist)[8]

Don't worry, you read that correctly, 5% of what you're seeing is coming in through your retinas. Everything else is just a product of your brain. Your little eye cameras are snapping up 2D image after 2D image, and then sending those back through the optic nerve and into an organ that's processing, interpolating, and stitching those images into a 2½-D sketch or internal model—and about 95% of that—is coming solely from your prior expectations (i.e., what you *expect* to see, not reality itself).

This fact brings home three important points: (i) the VR simulation that is your conscious experience is only 5% retinal data and 95% internal generated imagery, (ii) the exterior world contains much more visual information than your conscious self could ever possibly attend to, and (iii) the human brain is the most complex thing in the entire universe.

The Internal Model

We already briefly touched on the idea of the internal model, but now, armed with our Photoshop metaphor, I'd like to make an addendum to that analogy by proposing that this layering process also includes a ton of predictions.

We covered how visual information comes in through your retinas and is immediately broken down into neuro-electrical impulses (the brain's equivalent of 1s and 0s), and how it is then reassembled inside the visual cortex. We talked about how there are layers of encoding, like motion, position, shadows, color, and the remainder. But it should also be clarified that when this data finally gets re-assembled, it isn't done in a linear fashion. It's not as if each brain system gets five microseconds to add their coat of paint on the image before having to send it out to their neighbor, because in reality, the visual system is much more loopy than that. Each system does get to add its layer of processing, but this processing makes loops forwards and backwards on account of the fact that your brain is always making predictions about what it sees and what it expects to see. In light of all this, I think it's rather important to acknowledge the real reason we have eyes in the first place, which is, to tether our internal model to this plane of existence.

This being the case, if you've ever interacted with someone and thought, "Geez, what planet are they living on?" Consider two things: (i) they probably said the exact same thing about you, and (ii), that your perceptions, whatever they happen to be, create *your reality*. In that sense, we don't see reality as much as we see a world filtered through the lens of our experiences.

Notes from the Editor:
19th century German physicist and physiologist Hermann Von Helmholtz was the first to advance the idea that the brain was a prediction machine, and that much of what we perceive is actually an unconscious inference or "best guess" about what's out there in the real world. Suggesting that our brains and our prior

expectations are contributing much more to the visual experience than sense data itself. For more on this subject, see "Bayesian Brain" in the Glossary.

—Here's something fun to try. Next time you're doing some people-watching, pick someone to stare at and try and see if you can count the number of times their eyes move a minute. Full disclosure, you might want to grab a paper and pen because the human eye moves about three times a second.[9] Even when you're staring at something stationary, your eyes never stay fixed for very long. One or both of your eyes are always saccading and moving about capturing 2D image after 2D image and then sending that information back to the visual cortex where it can finally be used to update your brain's best guess about what's going on 'out there.'

Don't get me wrong, it certainly feels like our eyes are these super detailed cameras capturing *continuous* streams of video. However, if we only took an extra second to imagine what a video stream like that would look like, we'd quickly intuit how nauseating that would actually be. And filmmakers use this trick all the time. One moment soldiers are storming the beach in a wide, the next, the camera switches to a first-person POV— and it's just a soldier's boots sloshing through water. Jerking the camera this way and that, the camera operator attempts to simulate just how jarring and disorienting wartime can be. Although, personally, I can only stomach about a minute of this before I actually start to get nauseous.

Luckily for you and me, the visual system does not work like this.

Even though our eyes move about 180 times a minute, our subconscious only cherry-picks the images it likes and then uses that data to modulate its own internal predictions about what's going on, effectively grounding the hallucination of waking consciousness in some semblance of reality. The end result of this process being, a perfectly stabilized, perfectly edited "movie in your mind."

Frame Rates of Reality & the Soap Opera Effect

Think back to the days of channel surfing and those fear reports they used to run like "Sitting is the New Smoking" or "Ten Things That Could Kill You in an Office." If you ever found yourself watching one of those, then you probably would've noticed this strange thing that happens whenever you try to film a computer monitor with a camera.

A big black bar flickers across the screen that no one ever talks about. If you ever wondered why the camera sees this, and you don't. The short answer is because the rate at which the computer screen is being redrawn doesn't match the frame rate of the recording.

So, let's suppose you were trying to film a computer monitor that refreshes the image on its display 60 times a second, but you were trying to film it with a camera that only shoots at 24fps. What this means is, there are going to be moments when the camera catches this refreshing in action, hence the black bar or flickering effect. In any event, the real important thing to be mindful of here is that your *conscious self* doesn't perceive this black bar because, according to your CS5, it never even happened.

Things that happen too quick or too slow are deemed non-essential by the Subconscious OS, and are subsequently screened out of existence. Just as we're blind to the movements of most plants, we're blind to that flickering black bar.

While we're on the subject of cameras, we should probably go ahead and address this rather annoying phenomenon that happens whenever you show older movies on newer TVs. For context, it should be pointed out that human beings obviously see things at a much higher "frame rate" than 24fps, but that 24fps has always, and hopefully will always, be the industry standard for motion pictures.

Why do I say hopefully?

Because with 24fps the images are flashed just fast enough for the brain to assume causality and interpret motion, but it isn't so fast that it starts to confuse the brain with reality. Take *The Hobbit* as an example—a movie that was shot with double the frame rate. And in case you don't recall, the reactions weren't exactly positive. You see, the reason we like 24fps isn't just because we're used to it, it's because at 24 frames a second we're very aware we're watching a movie. And because we know we're watching a movie and not observing real life, we seem to allow ourselves to fall more deeply into the story; suspending disbelief, as they say. On the other hand, because the footage in *The Hobbit* looked so real, instead of audiences feeling like they were watching a movie, some said it felt like they were on set with the actors watching them act. Personally, it felt like I could see the hot lights bouncing off the bald spot on one of the actor's heads, so consequently, I never fully achieved the suspension of disbelief. Oddly enough, people don't seem to mind their sports being shot at 30fps or higher, which is where the whole "Soap Opera Effect" even comes into play.

—

To cut a long story short, there's a frame rate war going on inside the processor of most newer TVs. And despite the fact that most movies are still shot at 24fps, a lot more television is being shot at 30fps. Due to this discrepancy, a lot of newer TVs have this little computer inside them that forces whatever source material their watching (DVDs, TV shows, Blu-Rays) to display at 30 frames a second, regardless of whether or not they were initially shot that way.

So here's the scene: You're pushing your cart through Target—strolling by that enormous wall of TVs—when out of the corner of your eye, you can't help but notice they're playing one of your favorite movies. So you push your cart over to admire what's left of *Saving Private Ryan*, when suddenly, you think to yourself, "Something looks off here."

Instead of having that grainy film look, everything looks all cheap and plastic. Almost like someone went back and reshot the whole thing using a Sony Handicam. Don't get me wrong, it's extremely bright, and it's displayed on a TV you might expect to find hanging in God's house. But, for whatever reason, the movie looks rather artificial, when it used to look all "gritty" and "real."

Am I misremembering? Why does this look so strange?

In a few words, you're not misremembering, this is just called "motion interpolation," and it is without a doubt the first thing I turn off whenever I'm watching something on a newer TV. The reason *Saving Private Ryan* now looks as if it was shot on Hi-Speed Video is because the movie was never shot at 30fps; it was shot at 24fps. Implying, that there's an additional 6 frames per second that the television's processor is just creating out of thin air.

Where does it get these new frames?

Essentially, by creating a hybrid of the frame that came before it and a hybrid of the frame that came after it. And if your eyes detect this artificiality, and it bugs you too, then thankfully you can probably Google how to turn this setting off.

Why do I bring this up?

I bring this up because this is what happens when your TV takes something that was shot at 24fps and pushes it to 30fps. By artificially creating 20% of the frames you see, this has been deemed "the Soap Opera Effect" on account of the fact that it makes good, well-produced movies look like they belong on daytime TV.

That said, if you can imagine what it looks like when the TV's processor is creating 20% of the frames using its best guess about hybridizing frames, then close your eyes and try to imagine what it would look like if your TV generated 95% of the frames.

Now open your eyes because this is what 95% artificial frames looks like. And no, you can't turn this setting off.

Chapter 10
Emotional Coding

I'LL NEVER FORGET where I was when I first heard the news. It must've been about 5 AM when she told me. And when she first said it, I called her a liar. Half asleep, I told her she was wrong. That I just spoke to him last night, so clearly, she must be mistaken. After that I hung up the phone and tried going back to sleep. This was hours before my first day, and I simply did not have time for this.

Lying back down in my bed, I remember pulling the covers up to my neck and closing my eyes, but I didn't cry.

Two days later is when all the calls really started coming in. Of course, this was back when people still called instead of texted. Back when we still had flip phones. I'd feel this thing vibrating in my pocket, and without even looking, I already knew who was calling and what they were going to say. Almost like a clairvoyant. And to prove my newfound power, sometimes, I even mouthed the words along with them.

Patiently waiting for my turn to be silent.

Of course, I couldn't actually hear voices, more like, everyone who called was reading off the same lousy script:

we're so sorry for your loss
we know how close you two were
if there's anything we can do, don't hesitate to...

—These were like those telemarketers, only worse. At least with telemarketers you can download an app that pre-screens your calls, but not this week. No, this week, I was forced to listen.

Some kind of social contract.

People brought me their tears, and then I regurgitated the details. Quid pro quo.

I'll never forget the moment when everything started to sound the same because I was sitting on the floor when it happened. Staring through our beer-stained carpet when it suddenly occurred to me that even my own answers were starting to sound rote. Their impersonal, canned responses were causing me to go into canned responses, and even though I was feeling the most lost I'd ever felt in my entire life, I still knew what my next line was supposed to be. I was supposed to say, "Thank you for calling," but instead, I improv'd this little bit where I threw my phone against the wall at sixty miles an hour.

Most of my life, I've had a rather calculated and logical disposition to Being, but for this one brief second, I completely let go of everything and gave into this deep well of unadulterated anger. The four-year-old in me cried out, "It isn't fair!" as he launched my phone into the wall as hard as he could. And after watching it shatter into a million tiny pieces, for the first time in a week, I was free.

Destroying something gave me power in a moment when I had none. Because for that one fleeting second—I chose what lived and died.

Granted, the hole in the wall wasn't solving any problems, but the four-year-old in me didn't care, all he wanted was justice. To destroy something perfect.

Really, he's the one who threw it. Me, I just picked up the pieces.

If This, Then That

Everywhere you go, everywhere you carry around the skull that houses your brain, you're coding whether that be consciously or unconsciously. Whether you realize it or not, your brain's constantly taking statistics, finding connections, and making associations.[1-3]

To put it bluntly, you're an unconscious learning machine.

You're a conscious learning machine too, but since we're exploring the unseen spectra of consciousness, I should probably go ahead and tell you that beneath your sphere of awareness, your brain is consistently forming heuristics, confirming expectations, and because we're emotional creatures, that means making rules.

Ava used to always say, "When you get down to it, most any computer program is just a long list of "If *this*, then *that*" commands.

If (user presses "A") execute: reward graphics and sound
If (user presses "B") execute: wompwompwomp.wav

And the picture we're trying to paint here is that the human brain isn't all that different, just complicated. In all honesty, we're making "If This, Then That" rules all the time, it's just we're not conscious of it. Generally speaking, we only experience the aftermath. To explain what I mean, just think about the things that really drive human behavior: food, sex, shelter, temperature regulation, social interaction, self-actualization, emotional well-being.[4] These can all be likened to being something like the driving forces behind our "core code." Handed down to us through billions of years of evolution, these weren't bits of our psyche we had to consciously work out for ourselves. And yet, there is a sense in which we get to code the way we feel about the events that happen to us, and it's this type of interpretive coding or emotional coding, that I'd like to focus on now.

———

One way to think about this life might be to say that we're really just an amalgamation of all the interactions we've ever had. And while most of these interactions are rather benign, leaving no discernible trace, some of these interactions carry the potential of changing us in ways we'll never forget. Rewriting the rules of our source code, modifying the way we think and behave—potentially, forever.

We can think of these painful interactions as being something like the seeds of experience that give rise to a particular belief. Link a few of these traumatic experiences up in a chain and not only do you get a particular belief-filter about the world, but you also get a "pre-thought" or "pre-coded" method for how to get through. That way, when a frightening but familiar situation pops back up, there's already a plan in place.

You're not thinking, you're on automatic.

However, in addition to the rules we create for ourselves, it's also worth noting how we're embedded in a complex web of societal rules.

We're constantly putting ourselves into contracts with other people, and whenever we do that, we simultaneously open ourselves up for the the potentiality of experiencing loss, betrayal, and deceit. Stated differently, somebody always winds up in physical or emotional danger, and whenever this happens, the brain ends up trying to protect itself in ways we can't even begin to imagine. On the grounds that 95 to 99% of our cognitive processing is unconscious, that means we're not even aware of most of the learning that shapes our behavior.

To pick an extreme example of this, I'll never forget the first time a girl cheated on me. The suspicions leading up to it. The interrogations with all the friends involved. This was one of those multi-month betrayal episodes that could really rewrite the source code of who you were going to be for the rest of your life. So when the girl did finally tell me, after months of suspicion, I physically felt pain.

Like a black hole was collapsing in the center of my chest.

Not many people get the opportunity to lose three friends over the course of one phone call, but driving home from school that day, I sure did. Their stories weren't adding up and my interrogation was wearing her down, so finally, she confesses to everything.

At first, I can't believe the depths a person would go to cover up their lie, but then I'm just mad I wasted the last eighteen months of my life with this person.

So what did I do?

I did the only thing I knew how to do; I destroyed something.

I was driving home from school at the time, so naturally, I started punching my steering wheel. Driving as fast as I was, I shouldn't even be here. But in that moment, nothing mattered, because all I really wanted to do was feel something else. So I punched. And I punched. And I punched. Even after most of the casing had come off. Even after bits of the horn were flaking off into my lap. I kept slamming my fists down until the damn thing really broke, exposing the cold, hard metal underneath. Needless to say, but about the fifth time my hands came down, the jagged metal plate won that round and my hand started leaking out blood.

Correction: *pouring*

I'll never forget glancing down, seeing my tendons and bone catching their first glimpses of sunlight, and then realizing the degree to which I'd fucked up. The knuckle of my middle finger took most of the damage, but apparently, the cold, hard metal had cut through the tops of my hand with such force that it folded the skin back in on itself.

Something like the dog-eared page of a book.

This torn flap of skin, formerly known as my knuckle, was now tucked underneath the tops of my hand in a way that meant I could see bone. I remember seeing this, instantly pressing my hand into my lap to try and stop the bleeding, and then pressing both feet on the brakes hard. My car screeching to a stop. My book bag slamming against the dash. Pens flying everywhere. Without even thinking about it, I take my good hand and roll the wheel as far as it'll go left. I pull an illegal U-turn, right in the middle of a highly trafficked road, and then floor it heading east.

My right hand throbbing, leaking more blood into the stupid khaki shorts our principal used to make us wear to school, I end up driving myself to the hospital.

Now, whenever I look at my writing hand, I'm forced to stare at this big hook of pink skin. A cute little reminder of the time I got exactly what I deserved.

I hurt a girl and she hurt me back.

But sadly, because I was seventeen and stupid, on the way over to the emergency room, I started making myself promises.

Promises like: *I am never going to end up here again.*

And to make sure I kept that promise, consciously and unconsciously, I began making up rules…

Rule #1 *Don't ever let a girl get close.*
Rule #2 *Maintain your power at any cost.*
Rule #3 *Don't punch things when you're mad.*

All of that actually happened by the way. There's no hypothetical lens we're hiding behind on this one. That Friday, I felt the kind of pain you wouldn't wish upon your worst enemy, and then I consciously made the choice to never feel that way again.

IF you ever end up loving a girl again…THEN never let her get close.

My entire life's future reduced to one line of code. Black and white. Something a child might do. Write a little line of code that would protect the part of himself that died that day. Subconsciously, I probably concocted a dozen or so rules to help ensure I'd never end up feeling that kind of loss ever again, but honestly, I only remember the three. And those rules actually worked too—that is, until the fall semester of my sophomore year came around and my best friend took a turn going ninety. Because that's when I had to make another rule:

Never let yourself have another best friend.

'Cause if you never let yourself have another best friend, what are the odds you'll have to deliver two soul-crushing eulogies?

Was this a good rule to have created?

Look, none of the rules I created the day someone put thirteen stitches in my hand, or the night I called my mom a liar were rules for *living*. These were rules for surviving, and surviving's not living.

Sure, they might've helped me survive the initial pain of the incident by giving me some kind of emotional roadmap for how to get through, but they kept me closed-off and distant for years.

The point is…*things happen to us*. We live life, we get hurt, and then we make promises to ourselves.

"Whatever else happens," we tell ourselves, "I'm never gonna end up here again," but instead of dealing with the pain directly, we make contracts with the future. Promises we couldn't possibly keep. Instead of being honest with ourselves about the kind of pain we're in, we bury and deflect. Instead of discharging about how we really feel with a friend or a trained professional, we unconsciously cast our pain upon others.

Personally, I retreated. Which is really just a fancy word for "ran away." Instead of facing my problems directly and potentially resolving this deep hurt, I drank beer and smoked a bowl.

The Adult Version of *sucking your thumb*.

"Self-medicate" ought to be one of those esoteric terms you have to explain to people, and yet, it's such a common practice, everyone already knows what you're talking about. Considering this method of coping, does it surprise you to learn that for the next six years, I didn't have a best friend? I mean, how could I? I wouldn't let myself.

Now, I want to pause for a second to point out, that this wasn't something I was consciously doing. In fact, it's only because of hindsight, time, and an altered state of consciousness that I'm even able to look back and retrospectively analyze my own behavior. So when that day did finally come—the day I realized why I was the way that I was; why I kept people at a distance; why I had trouble making guy friends—I realized it was because, deep down, I was scared. So afraid I was going to get hurt, I didn't dare let anyone in.

Subconsciously, my friend's death was still operating me. Now it's a story, but it used to be my story if that makes any sense. A piece of pain I unconsciously clung to the way a kid clings to a blanket. Left to our own devices, human beings will run away from most things; and despite what the cute little saying says, but time *does not* heal all wounds. Only truth does. Which includes being honest with yourself.

That day, standing behind the pulpit, the four-year-old in me promised himself, "Whatever else happens, I'm never gonna end up here again." And he kept it too. But sadly, that meant James was going to become closed off and bitter. That he was going to inflict his pain upon others. That somehow, in a Machiavellian sense, he was going to get his justice.

———

Growing up where I grew up, crying was somehow seen as less human. You couldn't be a boy that cried. You could be a "girl" or a "nance," but you couldn't be *a boy* that cried. Apparently, those things were mutually exclusive. The wisdom of the time was: Girls cry, men got angry.

Turns out Nature's pressure release valve—that thing you instinctively do when you're experiencing an emotional overload—was only sanctioned for use by women and children. This is one of those societal rules your dad either taught you early on, or you risked finding out the hard way. The wisdom of the time was detach and avoid. Ignore and repress.

Confront a bully; ignore a feeling.

Of course, you could always be the one who tried to challenge the status quo, but all that ever really amounted to was you finding out just

how savage a pack of nine-year-olds could be. There was a societal contract in place, and it's always been that way, who were you to change
things? Girls cry. Men got angry. That was the deal.

From the dugout, you could hear the kids preaching the gospel, "Walk
it off...*Get up!*" Sure, you might be paralyzed on the ground with the
wind knocked out of you, but you could still hear them parroting back all
the wisdom they'd absorbed. "Come on...get up!" Their pre-pubescent
voices starting to harmonize, and if you were lucky, the coach would run
over and tell it to you straight.

"Look," he says, "There's people watching."

Between his teeth, he says, "Come on...get up." The line drive to
your chest didn't hurt that bad. Come on. Get up.

So you do. You stuff it down. You take your base. What else can I say,
that's life. Girls feel, men deal.

—

The moment after everything started to sound the same is the moment I
decided to flip the script. The moment after the girl tells me, "Everything
happens for a reason," I pull the phone away from my face and just stare
at it.

Twelve inches from my face and I can still hear this girl's voice say,
"Look." Lighting up a cigarette, she says, "I know how you feel, alright."

But instead of saying, "Thank you for calling," which is what I'm
supposed to say, I leave the phone open and lean back on one leg. Twisting
at my torso, I pull my right arm back as far as it'll go. And like a rubber
band, I snap. I pitch the phone against the wall as hard as I possibly can,
and I follow through. Coach would've been proud.

Hurdling at 60 miles an hour into our shared townhome wall, I watch
as my only connection to the outside world explodes into a million pieces.
A piñata of dry wall and plastic.

Then I clench my fists and scream so loud I break my voice. The light
fixtures shake. The neighbor's dog barks. But still, I don't cry.

Finally going over to the hole in our townhome wall, I snatch up the
biggest pieces of dry wall first. Them, I chuck in the trash can just as hard.
The rubber keyboard, the plastic antenna, I shove all that shit in my good

hand. Bending over at the waist, I'm picking up all these small shards of glass with just my bare fingers, but those really tiny bits, just keep getting pushed deeper and deeper into the carpet. Almost like they're running away from me. I'm actually bent over for so long, I'm starting to see stars. Next thing I know, my face gets all hot and my nose starts to run. Insistent on getting every last piece, I go to wipe my nose with the backs of my hand, but when I do a huge breath falls out of me. An image of my friend flashes through my mind and there's so much pain I have to shut my eyes and scream. Except when I go to scream, nothing comes out.

What's worse, no air is going in either.

I'm completely frozen; unable to speak or breathe, move or sit still. Collapsing to the ground, I try and force some air down, only when it finally makes it in, something's wrong. It's all staccato. Something like a reverse laugh of breath. Followed by a sound no man's supposed to make. Finally, after letting myself feel the loss of my friend with the deepest part of my soul, I drop everything I'm holding and take my place on the floor.

There, on the ground, I cry for twenty minutes straight.

Part II

Altered States

Chapter 11
Cuckoos Part II

THIS OTHER TIME, Ava and me are watching a different documentary about cuckoos, when it all starts to make sense: arms races. Evolutionary arms races.

We're back on the red leather couch again. The couch I hated so much. This gaudy "L" shaped monstrosity Brian and I dragged into our apartment the day we moved in. Up three flights of stairs and through a regular-sized door...I still have no idea how we did it. Somehow, we managed to shove this gargantuan-sized sofa into our apartment. Although, the second it finally made it through, we both knew, there was no going back. The couch lived here now.

Which wouldn't have been too much of a problem if it didn't shed like a wild animal, but it did. Meaning, we had to sweep up after this thing four to five times a week.

At one point, I'm sure this couch cost a fortune. But apparently, when the roommate before me went to clean it, he laid on some kind of chemical the couch didn't like, and ever since that day, it's never quite been the same. On the back of the packaging, I'm sure it probably read—*Not for use on red leather couches*—but the guy used it anyway, causing it to dry and crack. This was months before I met Brian.

By the time we actually moved in together, the damn thing looked like a crocodile that had been trapped in Death Valley. The leather was cracking so bad, whenever you got up to grab something from the fridge, a little piece of couch went with you. Stuck to your calves. Stuck to the back of your legs. Your elbows. Before you knew it, there'd be flecks of couch in the bed. On the table. In your cereal. Slowly but surely, this couch was disintegrating piece by piece, almost like the couch had dermatitis or eczema.

This particular night, I'm peeling tiny pieces of couch off Ava's back, getting ready to give her this massage, when another piece of the puzzle just falls into my lap.

The way we're seated, Ava's got her back to me…my back's against the couch…my elbow's pressed between the knot in her left shoulder blade, when one of the most important tenets about evolution finally starts to sink in. On the TV, it's the same narrator going on about cuckoos and hosts, only this time, the focus is more on the African cuckoo finch and the tawny-flanked prinia.

"The cuckoo's egg must match in size, color, marking variation, and marking dispersion to avoid being chucked out," explains the narrator. It's just the images we're seeing on TV seem impossible. I mean, how on earth is a cuckoo supposed to match its egg across all these domains of analysis?

But, of course, I'm not thinking about this like an evolutionary biologist, because right now, I'm just thinking about it like a guy on a couch.

Sipping Pacifico and chasing it down with a little chips and salsa, apparently I'm completely failing at my job. On a personal level, I know I'm supposed to be giving Ava this massage, but on a subterranean level, that isn't what's happening at all. Somewhere in my brain, a hunch was beginning to form, I just wasn't smart enough to see it. And that's how knowledge works sometimes. You hear something beyond your comprehension level, and often times, the light bulb doesn't go off until months later. Like the concept of evolutionary arms races. I heard the narrator say the words. I understood what each word meant. I just didn't realize that an evolutionary arms race is the kind of thing that plays out over the course of lineages, not lifetimes. Eventually, we both got there.

And when we did, riding sidecar, was the realization that these cuckoos didn't have to do anything at all to get their eggs to match in size, color, pattern, or signature.[1] At least not consciously. Because it turns out that an evolutionary arms race isn't the sort of game played by a conscious entity. Hell, you couldn't even call this a subconscious game, because really, this is the kind of game that plays out *across the generations*. Again lineages, not lifetimes.

———

You ever see a hawk with razor-sharp talons stalk a beige bunny that almost looks indistinguishable from its background and wonder how it is they even spotted a hare with that great of camouflage?

Enter the concept of evolutionary arms races.

It was Richard Dawkins and Nick Davies who first introduced me to the idea. And to do this topic justice, there really is no better example to cite than those crazy cuckoos. In all fairness, an evolutionary arms race isn't all that different from a traditional arms race, it's just instead of escalating firepower, it's evolutionary advantageous mutations.

For instance, during the Cold War Space Race, it was the Soviets who first put a satellite into orbit. An impressive feat back in 1957, because the Russians weren't just saying, "Here's a hunk of metal that goes beep." They were proving to the entire world their advanced understanding of rocketry and space technology.

The underlying message of which being: We've harnessed the power of satellites and intercontinental ballistic missiles...*your move*.

So the following year, we put up four satellites.

They put a dog in space; we put a chimp. They put a man in space; we put a man on the moon.

Now, if we take those same concepts of progressive adaptations and apply them towards the natural world, it becomes rather easy to see how adaptations on one side of the predator-prey skirmish (or the parasite-host skirmish) literally *force* counter-adaptations on the other side.

By way of illustration, let's just say a couple hawks in a given area gain slightly keener eyesight through some random genetic mutation. This genetic mutation increases their fitness by allowing them to see farther and pick off any prey faster than any of the other hawks in the surrounding area. With an evolutionary advantageous edge like that, it's highly likely these hawks would start snatching up any and all prey that either stayed out in the open for too long or weren't very well camouflaged, finally gaining the upper hand. In accordance with this, we would expect all the slow-moving field mice—and all the mice with poor camouflage—to be the first ones to become dinner.

So what happens next?

Nature takes it course. Which is really just a polite way of saying a lot of mice most certainly died. Nevertheless, any of the mice that did end up surviving would have had the chance to pass on whatever variation that helped them survive, be it stronger legs, a predisposition for not staying still, or perhaps a darker shade of fur that makes them indistinguishable from a rock 500 yards away.

So what happens next?

Those very mutations which allowed the mice to get a foothold in the evolutionary arms race (better camouflage, better evasive tactics, etc.) ultimately end up forcing counter-adaptations to evolve on the other side. After that point, the only hawks that can survive and reproduce are the ones with eyes that can see 600 yards away.

Bit by bit, each side starts to become progressively more adapted to do battle with their opponent. And when you let these sorts of evolutionary games play out for long enough, like thousands and thousands of generations, you end up with a species of bird that looks as if it were "perfectly designed" to kill a mouse, and a species of mouse that almost looks as if it were "perfectly designed" to evade a hawk.

***Notes from the Editor:**

To avoid having all subsequent uses of the word "designed" appear in quotations, it is necessary to establish that the narrator is not trying to suggest that an intelligent designer designed these creatures. Quite the opposite, actually, given that the goal of these particular chapters is to enhance our understanding behind the mechanisms of evolution. On this understanding, all uses of the word designed going forward will not feature these quotations.

—With this in mind, it's important to clarify the speed at which these sorts of games play out, which isn't over the course of a few lifetimes, because really, an evolutionary arms race is the kind of 'ding-dong' battle that plays out across decades. In fact, a better way to imagine this battle might be to envision gene pools at odds with one another. Of course, each organism would have its own genotype, containing some kind of unique variation, but each gene pool would ultimately contain all the genetic variants currently in play. In this way, as the gene pool of the hawk starts to become more and more populated with copies of the genes that code for keener eyesight, the of the mouse starts to change as well. On account of increased predation, all the lighter colored mice, and in turn, all the genes that produced white fur, begin to become less and less popular. At the very same time, these increases in predation cause the genes coding for darker colored fur to become *more frequent*, leading to better and better camouflage, which successively makes the gene pool of the field mouse become more adapted.

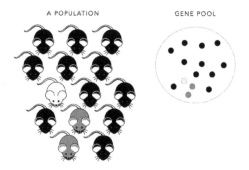

Flash forward to today and what do we see when we look out at the world? We see a highly connected, highly interdependent web of life filled with predator-prey skirmishes, parasite-host battles, organism-germ interactions, and millions of animals that seem perfectly designed for a given task, whether that's swimming, running, hopping, gliding, or in the case of cheetah, taking down a Springbok antelope.

If Terry were here, he'd probably tell you that a cheetah can go from a dead stop to running at speeds of 60 mph in about three seconds.[2] Then, when you went to fact-check his claim, he'd probably add something like, "That's faster than a Porsche 911 Carrera S," because that's what Terry's like. But then when you consider what it is they're actually chasing, we find an animal that almost looks as if it were perfectly designed to evade a cheetah.

So what's the answer here?

Co-evolution, of course. These animals *co-evolved* together. One to run, the other to run for its dinner. A point which is most accurately summed up in one of Aesop's Fables:

The rabbit runs faster than the fox, because the rabbit is running for his life, while the fox is only running for his dinner.

As Aesop informs us, there is an unequal cost of failure here, in which, the prey really has to get things right in order to survive, leading zoologist John Krebs and evolutionary biologist Richard Dawkins, to term this the 'life/dinner principle.'[3]

An interesting concept because apart from aiding our understanding about the predator-prey dynamic, it also helps us envision what this game might've looked like 40,000 to 50,000 years ago. If, let's say, we could wind the clocks back 50,000 years, one question we'd be dying to know is, "Could both of these animals still run at 55-60 mph?"

However, before we answer that, let's reconsider the parasite-host game real quick.

The Age of an Arms Race

Back in England, with the Reed Warblers vs. the Cuckoos, it appears as if we've stumbled upon an evolutionary arms race that is rather old. By all accounts, it's certainly not as old as the cuckoo finch and the tawny-flanked prinia we'll be discussing in just a moment. But back in the marshy reeds of Wicken Fen, where the cuckoos have been preying on these little reed warblers for thousands and thousands of years, there's actually some highly sophisticated tricks we haven't even begun to explore yet.

As you'll recall, the female cuckoo lays its egg in the reed warbler's nest. The cuckoo chick has a shorter incubation time than that of its foster siblings, and the cuckoo chick has to toss out all the other eggs to ensure its own survival.

But what about the reed warblers? Do they have any defenses against cuckoos? And if so, what sorts of mutations have they been forced to evolve in order to avoid being parasitized?

To answer these questions, we first need to dip briefly into the concepts of egg signatures and egg rejection (two forms of host-defense the reed warblers *had to evolve* in order to survive being parasitized).

First and foremost, let's take into account that the eggs of a reed warbler are spotted, green, and small, whereas the eggs of a cuckoo vary in size, color, and shape depending upon the species of host they happen to be parasitizing. Strange as it may seem, but brood parasitism isn't just something that happened on this planet once, because in all fairness, it's an idea that Mother Nature has stumbled upon several, several times.

Independently, I might add. Which certainly underscores the utility of such a practice.

Phrased in another way, Mother Nature has *converged* upon the idea of brood parasitism multiple times, therefore the idea of popping eggs into another bird's nests isn't unique to the common cuckoo at all, seeing as just over 100 species out of some 10,000 do it.[4] And contained within just one of those species, the common cuckoo *Cuculus canorus*, can and does parasitize a whole range of different hosts, including meadow pipits, tree pipits, wrens, skylarks, dunnocks, blackcaps, yellowhammers, redstarts, great reed warblers, regular reed warblers, pied wagtails, willow warblers, bramblings—I forget the rest, but you get the idea. Technically, this one species of bird is capable of parasitizing such a large number of hosts, however based on the nests these cuckoos found themselves in at birth, they tend to continue parasitizing on that particular species of host. So, cuckoos raised in a meadow pipit's nests tend to specialize on meadow pipits, cuckoos born in willow warbler's nests tend to parasitize willow warblers, so on and so forth. This being said, it just so happens that with each new arms race (Cuckoos vs. Reed warblers or Cuckoos vs. Bramblings, etc.) the cuckoos specializing on that particular host have been *forced to evolve* different sizes, different colors, different shapes, and different signatures on their eggs to avoid having them chucked out.

How does one species manage to parasitize all these different hosts?

Basically, cuckoos have certain races, although "races" isn't quite the right word, given that it's only the females who belong to these races.

Calling them a sub-species get us a little bit closer, but the technical term here is gens. Plural gentes. And the only reason I bring this up is because of what this fact implies.

In effect, that we have multiple cuckoo-host arms races playing out all across the animal kingdom and in various stages of development.[5]

Back on the Couch

Looking at the kinds of forgeries the cuckoo finch has had to evolve to survive in Africa, at first you might not believe your eyes.

I mean, can you even spot which one is the cuckoo's in this photo?

Fig 1.1

Image Credit: Spottiswoode/Stevens

If you guessed that it was the entire inner circle, then you are correct. While the entire outer circle is completely comprised of tawny-flanked prinia eggs (each from a different female), the six eggs on the inside are the cuckoo finch's response to that signature.[6]

The first time Ava and I saw pictures like these, we had a hard time keeping our mouths closed. Such elaborate forgeries seemed almost too good to be true. If only there were some more intermediate examples, the story would be more believable, right? Yet, according to Nick Davies, the guy from which most of these facts are derived, the reason these signatures and their corresponding forgeries are so advanced is because this particular arms race is rather old. By contrast, if we now head back to England and reconsider the evolutionary arms race going on between the cuckoos and the dunnocks, you may wonder how it is the dunnocks ever got tricked (See Fig. 1.2).

So what's the answer here?

Well, according to the world's leading authority on cuckoos, Nick Davies, we've happened upon an evolutionary arms race that is rather new; and for the most part, there just hasn't been enough time for the hosts to evolve defense tactics like egg rejection or egg signatures.[7] In addition to this, Professor Davies also insists upon the idea that the dunnocks might not have been parasitized 'enough.'

Or stated a bit differently, that there hasn't been enough *selection pressure* for the hosts to start evolving host defenses like egg signatures or egg rejection.[8]

Evidently, dunnocks will accept any egg.[9]

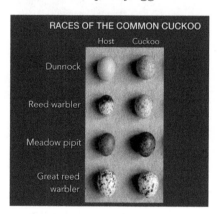

Fig 1.2

How do we know?

Because Nick's been popping model eggs into the nests of birds all over the world and recording their acceptance rate. Perfectly matching eggs, eggs of the wrong shape or size. Eggs with spotting, and eggs without. Essentially, Nick's been popping every variation of egg pattern you could think of into the nests of other birds to see whether or not they'll accept. Seriously fascinating field research Nick's been up to. Because when you really dive deep into what he's been studying, it starts one down the path of wondering just what kind of umwelt these birds are actually living in.

Put it this way, when you see a mimetic egg, like the cuckoo's, being popped into a reed warbler's nest, it's easy to understand how a bird with a brain the size of a nut might get tricked. But when you see a dunnock accept an egg that's not even the right color—you can't help but wonder just how it is these dunnocks could ever have been so foolish.

Can't they see the egg isn't blue? That it has spots?

Sitting at home noshing on chips and salsa while sipping from our lime-topped beers, Ava and I found ourselves yelling at the TV on multiple occasions.

Can't you see that cuckoo chick looks nothing like you?
Can't you see that cuckoo is so big you have to stand on its back to feed it?
That it's eight times your own weight??

I won't lie to you, Tom, things got pretty heated. Because in many ways, watching this documentary was just as suspenseful as any live-action murder mystery we'd ever seen.

The dramatic irony of knowing what the foster parents don't; the suspense of watching when the mother comes back, only to find that one of her eggs has hatched. "And often times," the narrator says, "the mother will just sit back and watch."

"The Mother will just stand by and watch as one the chicks destroys all of its would-be brothers and sisters."

Putting the Pacifico to her lips, Ava goes, "Sounds like a couple parents I know," and tips up her beer.

And that was the moment things really started to click for me. Watching this poor mother watching her own kids getting chucked out one by one, I have an epiphany.

These birds have an awareness, it's just nothing like what we experience.

Just because we can eat chips and salsa and work out which chick is the cuckoo's doesn't mean the reed warblers can too. And after spending some serious time combing through much of the research, the literature suggests—they really can't. Which is really something worth scratching your head over, isn't it? Especially when you consider that they seem to use visual discrimination to reject eggs?

But then, once the chicks have finally hatched, it's almost as if the reed warblers fail to notice they're feeding a chick that looks nothing like them. As Richard Dawkins puts it, these parents get manipulated by cuckoo trickery into feeding chicks that: (a) look nothing like them, (b) grow so big they destroy the nest they've been raised in, and (c) sound like an entire brood of reed warblers.[10] That last part is particularly interesting, because if the cuckoo didn't use aural manipulation in this way, it would surely die. The cuckoo has to manipulate its host parents using its supernormal orange gape and its impressive begging cry, or else it'll starve to death.[11]

On top of that, some clever experimentation has even demonstrated how instrumental this begging cry is. Because if you just pop a blackbird into a reed warbler's nest, which is about the same size of a cuckoo, the warblers will still feed it, they just won't feed it as much. So it's not just the visual cue of an over-sized chick that stimulates them into working harder and harder; it's the begging cry. Because other experiments have shown that if you pop a blackbird into a reed warbler's nest, and you simultaneously broadcast a cuckoo chick's begging call, *then* the reed warblers end up working hard enough to feed it.[12]

So what do the reed warblers see when they look into their nests?

Honestly, it could very well be that they see an entire brood of chicks. Of course, this is just a speculation. But when you see for yourself how this supernormal orange gape stimulates these hosts into dropping food into it—in combination with the aural manipulation of their begging call—it's really not that far off to suggest.[13]

How stimulating is their supernormal orange gape?

It's been observed that other birds flying overhead (heading back to their own nest, mind you) will sometimes see the super stimulus, fly down, and drop food into it. Almost like it were an addictive drug that hijacks their nervous system and controls behavior.

Which is precisely the way Richard Dawkins conceptualizes it in his groundbreaking book, *The Selfish Gene*.[14]

In regards to a host species being duped into raising a parasitic bird's chick, Charles Darwin called this whole debacle "a mistaken instinct."[15] Suggesting that hosts get tricked into raising other birds by following their programming: *Raise chicks you find your nest*. However, Richard Dawkins has a more salacious way of phrasing it. Rather that their nervous systems are being "controlled" through extended phenotype.[16]

Or, put in another way, genes reach outside the body in which they reside in order to manipulate the world around them (including other organisms). Which brings us back to the view that Richard Dawkins and Konrad Lorenz both espouse. Mainly, that genes can have effects (phenotypic expression) *outside* the body in which they sit, either in the form of fixed action patterns, as Lorenz pointed out; or in form of manipulation, and the construction of artifacts, as Dawkins pointed out.[17]

So a bird's nest is an organ. It's an organ in just the same sense as a heart or a kidney is an organ, but it just happens to be *outside* the body, and it happens to be made of grass and sticks, rather than being made of the cells that contain the genes.
- Richard Dawkins (evolutionary biologist)[18]

Notes from the Editor:

As a reminder, a genotype is the set of genes inside our DNA that are responsible for a particular trait. A phenotype is a set of observable traits expressed in an organism through the expression of its genotype. For example, blonde hair is the phenotypic expression of the genes that code for blonde hair; blue eyes are the phenotypic expression of the genes that code for eye color, etc., etc.

—After the extended phenotype idea, the entire world started to change. No longer were beavers just beavers. They were survival machines, built by their genes, who, in turn, built structures which were not only the result of Darwinian evolution (i.e. these dams and lodges evolved). But also just happened to have massive downwind effects on their entire surrounding ecosystem. Hence the name *extended* phenotype.

Just the same, bees were no longer just bees, and compass termites were no longer just compass termites. You couldn't talk about bees without mentioning those massive hives their genes force them to construct, nor could you talk about compass termites without citing those huge tombstone-shaped mounds their genes force them to erect.

13 ft tall mounds perfectly aligned along the north-south axis.[19]

In any case, we have to insist these termites aren't building these structures along the north-south pole axis because they 'want' to, or because one of them 'thought' it might be a good idea, they do it because they were programmed to do it.

They do it because their ancestors did it.

What's the advantage of aligning all your insect apartments north to south?

Well, apparently, whenever these mounds are oriented along the north-south axis, the broad side of their structure isn't facing the grueling midday sun. And down in Australia, where all these compass termites reside, this orientation appears to make the most sense. This way, their mound still catches the early-morning and late-afternoon sun, it's just not getting baked by the relentless midday sun.[20]

Reasoning from these facts, it can be understood that it isn't the conscious brains of these organisms that are assembling these artifacts; it's the genes themselves. Through highly organized and highly stereotyped behavior. In much the same way genes can instruct a body to "build" a liver or "construct" a kidney—including all the chemical reactions which must take place inside that organ—they can also instruct an organism to build an artifact, like a web or a dam. Of course, the layperson could be forgiven for assuming that spider webs, dams, and the nests of potter wasps were all just learned behaviors constructed in a top-down fashion, but not the biologist. And really, that's the difference to appreciate here. That one of these genetic processes is happening on the inside of their skin, where it's still comfortable for us to imagine genetic processes occurring, while the other is happening on the outside their skin, where most people just blanket-assume it's all conscious volition.

Clearly, this is not how behavior works.

Behavior isn't entirely generated from the top down, nor is it entirely generated from the bottom up. Reed warblers don't intrinsically know what their own eggs look like, they have to learn what their own eggs look like the first time they breed. And the way they accomplish this is by imprinting upon the eggs they lay that first season. In this regard, reed warblers have an instinctual drive to learn what their own eggs look like (bottom-up); but they still have to observe what those marking patterns are (top-down).

In light of all this, it seems as if Konrad Lorenz really did make a profound discovery. Because by observing that certain patterns of behavior were innate, and then figuring out that these fixed action patterns might be better conceptualized as anatomical organs, biologists began to see genes and behavior in a radically different way. Then Richard Dawkins comes along with his gene-centric view of life, and works out that these behavior patterns are actually just an example of extended phenotype.

***Notes from the Editor:*

The real beauty behind Dawkins' extended phenotype idea is that it gets one to start considering "the long reach of the gene," as the book's subtitle suggests. Before this book, genes comfortably sat inside bodies and really only affected the bodies in which they sat. However, after this publication, a new view surrounding the extended action of genes was stamped into place, and we've been shading in the details ever since.

—Regardless, we were asking a question about what these birds "see" when they look into a nest that's been parasitized, and the one sentence answer is: *we really have no idea.* Because reality is really whatever your senses tell you it is, which is really just whatever your brain models it to be, which is really just whatever your brain has evolved to see. In that respect, if reed warblers can't "see" that their child is really just a parasitic bird that looks nothing like them, and goslings can't "tell" that their mother is really just a pair of striped Wellingtons, isn't it reasonable to assume that human beings might be suffering from the exact same fate? That reality is one way but we mistakenly believe it to be another?

Let's find out.

Chapter 12
Non-Ordinary States

HERE'S THE SCENE: I give you a couple ear plugs, tell you to get naked, and then step into a bathtub that looks like it was designed for an alien. Once inside, I instruct you to lower yourself down into the tub and shut the door behind you. From the outside of the tank, the door looks like it might've been ripped right off a nuclear sub: Gunmetal black with a thick stainless steel bar running across the middle. Opening the door just a tad, I click on my penlight and let you take a look inside. At first glance, it almost looks like a shower from the future, only everything's black, and there's absolutely no lights.

No fans. No shower heads. No little spot to set your soap. Just a big shallow tub.

When the door first opens, a waft of ocean mist hits your face, and when it tickles those olfactory receptors sitting in your nose, I turn back and say, "You smell that?"

"That's the magnesium sulfate," I call back.

There's about 1200 pounds of that stuff dissolved in the water. "So when you're in there, and you're lying down, you won't even have to think about floating." Then I kneel down and stick two fingers in the water saying, "Here, feel."

The water, it's been warmed to about 95.9 °F. The same temperature as your skin. So while you're in there, you're not going to be feeling too much of your body either. After this, I hand you a towel and tell you to tie a knot around the inside door handle.

"This is for your eyes," I say, placing the towel in your hands. "You don't wanna get any of that Epsom salt in your eyes, so just use the towel to wipe your eyes if your face gets wet." And this is the part where you start to get nervous.

You're standing all the way up now, putting your hands in your pockets, when you ask, "So, what do I do in there?"

That's when I take the backs of my hands and knock on the thick, nuclear submarine door sitting behind me—the same way a salesman knocks on a product to show it's got durability, that's the way my knuckles hit this tank. But when the sound doesn't knock back in the way you might expect, I look to my ear and point a finger in the air, "You hear that?"

The entire chamber is light-resistant and sound-resistant, so you won't be able to see or hear anything; and since you'll be floating weightlessly in a solution that's essentially the same temperature as your skin—you're not gonna be *feeling* too much either.

As I explain the concept, I'm pulling back one finger at a time saying, "No sight, no smell, no sound, no taste, no touch." My voice echoing through the chamber, making everything I say sound more ominous, you take one final look inside before turning back in my direction.

"What the hell is this thing?"

Smiling big, I glance up from my clipboard, "We call it 'the float tank' and you get two hours inside."

The only question left being, "Do you do it?"

If you're like me, the answer is an emphatic, *hell yes*. But you'd be surprised at the amount of people who say no to this. And not even for any good reasons either. I mean, the door's unlocked; you can get out at any time; there's literally so much salt in there you couldn't even drown if you tried, so what's the hold-up?

At first, we thought it might be the name "sensory deprivation tank," so then we started calling it "the float tank" instead, but still, the undeniable fact is: most people just can't stand the thought of being alone with themselves. You can talk all day about how float therapy reduces anxiety, staves off depression, decreases stress levels—and they'll just say, "Nope." You can mention how Navy SEALs and athletes use them all the time to help off-line the self, speed up recovery times, improve mental performance…yada, yada, yada…and they'll still just stare back at you with their eyes closed, shaking their heads.

And, look, I get it. I get how sensory deprivation might not be everybody's thing. But between you and me, it's one of the more interesting non-ordinary states out there.[1]

Interactions with the Unconscious

As previously mentioned, our conscious selves only get a tiny sliver of data to experience each and every moment. Although, most of that is burned up just telling you what you're feeling, what you're seeing, and what you're hearing. If we could somehow take all that away—and then hand you back your mind—what would you do with it?

In other words, if your brain wasn't constantly preoccupied with sampling the environment, maintaining a sense of self, and processing loads of conflicting sense data, what might it get up to?

For a moment, just think about most any problem you've ever had, and then ask yourself the question, "Where did the solution come from?" Did it come from your conscious toiling, or did it come from somewhere else? Somewhere *outside* your sphere of awareness.

Consider those moments of clarity and those flashes of insight, and then ask yourself, "Where does it all come from?" This needn't be anything revelatory either, just think about those moments when you've forgotten the name of the lead in your favorite movie. Again, thanks to metacognition, you know you know the information, so you smack yourself on the forehead and say, "Ughh, what is her name?" Consciously, you can't think of it, but underneath your sphere of awareness, something is still searching for the answer. Then, as it usually goes, about ten to twenty seconds later, the name of the actor magically surfaces in your mind and you blurt out, "Natalie Portman!!"

We've all had these kinds of interactions with our subconscious before (these evanescent flashes of genius). But often times, these interactions don't come from our conscious deliberation, they come when we're doing something else. On a walk, on a drive, brushing our teeth...taking a shower. There's actually an entire subreddit called "Shower Thoughts" because when we're in the shower, in some ways, it's like we've put ourselves in a make-shift sensory deprivation tank. The white noise of the water, the limited visual awareness, the ritualistic washing of the body. All this distracts a part of your mind while another part is left to roam free and make connections.

Neurologically speaking, what's actually happening is that your brain waves are switching from predominantly alpha and beta, to theta, and perhaps, just a little flash of gamma.[2,3]

Those fleeting moments just after you've woken up, where you can still remember your dreams in vivid detail—that's theta. Those intervals of time where you've been driving for twenty minutes but have no recollection of any time passing...theta. Those moments when you're in a debate with someone, arguing about religion or politics: beta. Then, finally fed up with Uncle Frank, you slump down on the couch and let out a groan. Spacing out, but with eyes open: that's alpha.

Ten minutes later, in the midst of a food coma, when you're fast asleep...delta. Alpha, beta, theta, delta, and the less common gamma— those are the brainwaves we got.

Are they as clear-cut and simple as the way I've just described?

Of course not, seeing as we have waves on top of waves all the time. However, when you're primarily in delta, you're asleep. When you're primarily in beta, you're intensely focused. And when you're primarily in theta, you're likely "zenning out" (aka meditating) or you're in the float tank.[4]

***Notes from the Editor:*
The inventor of the isolation tank, John C. Lilly, actually performed experiments where he would take a psychedelic and then go float weightlessly in the tank. A true explorer of the human condition.[5]

—That magical land just after you've woken up but before you've officially started your day, when your dreams are still fresh and lateral thinking is heightened (this is where some of my best thinking gets done). Eyes closed but still awake, I always try and stay here as long as I can. This is why long drives don't bother me and showers are my friend. This is also why I love sensory deprivation tanks!

Removed from the ball and chain of processing sensory information, your mind is finally able to roam free, untethered. Without a sense to anchor you to this "now moment," sometimes, you forget where you are, slipping in and out of consciousness freely. Sometimes succumbing to those involuntary muscle twitches you get when you're first falling asleep, your body jerks awake, you feel the weight of the water wash over your skin, and then you remember, "Oh yeah! I'm still in the tank."

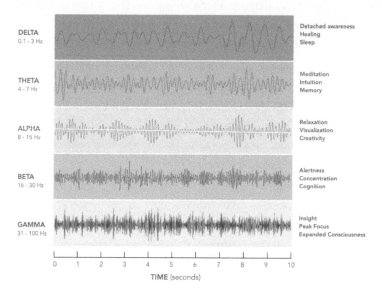

Lenses and Filters

One way to think about altered states of consciousness in relation to your baseline, wakeful conscious state (CS5) is to consider consciousness to be like a camera. So we're all born with the same metaphorical camera (an awareness that's on and recording); but the second after we're born, it's as if all our cameras have started diverging away from one another in the following ways:

From our earlier discussion, we already know that experiences get coded into the brain as memories, ultimately shaping the way we see this world. However, for the sake of this next analogy, I'd like for you to imagine that these memories are like those aftermarket filters you thread onto the body of a camera. Warming/cooling filters, linear polarizers, graduated neutral density filters...all these little add-ons change the way light streams into the camera. And to fully appreciate this next analogy, I want you to imagine that with each new memory we make, explicit or implicit, we're essentially fastening on another lens to the front of the last one. A belief-filter, if you like.

In order for us to really understand how memory affects consciousness, we need to pretend that you (the real you) is like that camera. So we all come out of the womb with the same camera, and it's our experiences—more specifically, the interpretations we make about our experiences—that end up shaping the way we see this world. In this way, we can begin to imagine that with each new experience, we're essentially fastening on a new filter to the front of that camera we're then made to look out of.

Correction: *forced to look out of.*

Because, as you'll recall, before conscious Tom ever gets to see something it's already been processed by the Subconscious OS first, implying that you never actually get to see reality without your memories or beliefs. Once a memory takes hold, a belief sets in place causing you to see the world colored by that expectation. As a consequence of this, what you see when you look at a cat depends upon your entire life leading up to that cat. Has a cat ever bitten you? Are you allergic to them? Did you ever have one as a child that ran away…et cetera, et cetera. Thus, to bring this all full circle, my old roommate Brian once told me a story about why he hates birds so much.

He told me that when he was about five, he and his mom were walking through this pet store when they spotted this beautiful, green parrot perched above some stale newspaper. Deciding to get close, Brian and his mom approach the bird, and he watches as she lightly pets him with the backs of her hand. So then Brian extends his hand out. Palm up. Completely non-threatening like. When the damn thing latches onto his finger and draws blood.

Years later, at a different pet store, it was a macaw that bit him. Then, at age twelve, a lovebird. By the time Brian hit puberty, there was a firm commitment in place:

I'm never going to get bitten again.

Nowadays, whenever Brian looks at a bird, he tells me, it isn't with kind eyes. Apparently, there's still a little bit of residual fear evoked. So whenever he hears someone say, "Aww, look at that bird," Brian looks for doors. Baseball bats. Exit routes.

Realistically, unless Brian has a bunch of positive experiences surrounding birds in the very near future, it's highly likely he'll never experience the same feelings of awe you and I might get whenever we catch an albatross doing its famous head-bobbing courtship dance. Moreover, it should be specified that whenever a birdwatcher spots a blue-footed booby *Sula nebouxii*, or pair of great tits *Parus major*, it means more to them.

Hiding behind foliage, they've probably spent hundreds of hours on the small side of those binoculars memorizing where they nest, how far down they can dive, and what their mating calls sound like. And because they've spent such a long time training their brain to notice all these different marking patterns, whenever they do end up spotting a blue-footed booby, their brain actually releases a little hit of dopamine. Whereas, for Brian, it's just cortisol (the stress hormone); coupled by a release of adrenaline in the body and noradrenaline in the brain.[6]

Now, this example might seem playful, and perhaps even a bit innocuous. But you could swap out the word "bird" for almost any other thing (strangers, dogs, bridges, elevators, car accidents, policeman) and extrapolate just how different the world must look to each of us. And not just in a visual way, but on *visceral level*, seeing as whenever a birdwatcher spots a new fledgling, they feel pleasure, where Brian just feels a wash of anxiety and pain.

Same stimuli; two different physiological responses—that's the kind of subjectivity built into the fabric of CS5.

Notes from the Editor:

According to the National Institute of Mental Health, about 6.8% of people will develop post-traumatic stress disorder at some point in their lives.[7] In light of this, it's worth highlighting that just an experience on its own, like an emotionally-charged traumatic event, for example, is enough to alter brain chemistry and behavior. In a previous chapter, the narrator explored ideas about brain trauma altering personality, however it can't be emphasized enough that the brain hardly makes a distinction between physical pain and emotional pain. In either case, the same stress hormone is released; and if the event is traumatic enough, it can lead to PTSD, which often results in a hyperactive amygdalae and a hypoactive medial prefrontal cortex.[8]

—Because Ava's a coder, whenever you show her a website, she can rattle off a whole list of reasons about why it's not a very good website, while the best I usually got is, "This one feels…clunky." So even though she and I might see the same website, it doesn't mean we notice the same things. And the main reason for that is, of course, that Ava has a completely different value structure than the one I do.

You see, the value structure she unconsciously imposes on the world has everything to do with the genetics she was given, the adventures she's had, the interests she's fostered, her age…the books she's read. Because, as you're already aware, we don't see with our eyes—we see with our internal model. This is a topic we'll have to pick back up again soon enough. But for now, it'll suffice to say: that your subconscious is basically like the ultimate firewall; that your experience of reality is much like that custom-tailored version of Facebook; and that the brain isn't some passive decoder of information, but is actively engaged in creating your internal experience.

—

In the beginning of our talk, we posited a guy and a girl who were rather opposites; a girl who was more into fashion and photography, and a guy who was more into stories and people. Then, we just briefly contrasted their experience of the same party. Now, to take that point a bit further, we should preface that the world is almost infinitely complex. It really is impossible for you to focus on everything at once, which means that your brain has to prioritize certain elements of your conscious experience at the expense of almost everything else. In order to knock down that 11 million bits into something more manageable, the brain has to construct a filtration system, something you might call attention.[9]

We 'select out' a few bits of information to focus on and attend to—but again—I hesitate saying "we" here because *you* technically don't get a say in the matter. In light of this, let me rephrase. Your Subconscious OS selects out a few bits of information for you to focus on and attend to, and this selection process (or filtration system) is dependent upon your genetics and your entire life leading up to that moment.

Because Brian's a video editor, whenever he watches a movie, he judges it on an entirely different set of criteria then someone like me. By virtue of the fact that he's a skilled video editor and I'm not, he's trained his eyes to watch, not only the foreground, but the background as well. So if a shot lingers for too long, his face starts to twitch.

If it jumps too quick, he cocks his head.

Because noticing all these tiny details is his job, he's pretty much the most annoying person to watch films with. And he'd probably tell you just as much. Because the kind of stuff his brain festers on is the kind of stuff my brain just ignores. Which is all to say that memory and attention affect the way we see this world in ways you couldn't possibly imagine.

Access Points

Going back to our camera analogy, if we can imagine that there was some kind of "normal" way of looking at the world, then we could attempt to visualize a camera with 35 years' worth of lenses tacked on front.

Lenses and belief-filters that narrow your view.

But sometimes, with the help of certain non-ordinary states of consciousness, it's like we're able to see this life with a wide-angle lens, or perhaps even at a different frame-rate. There are countless ways in which these lenses could shift and change, and for that reason, cultures all over the world have worked for millennia to pioneer and refine methods of achieving these states, whether that be through some kind of practice, like movement meditation, chanting, yoga, breathwork, shamanic drumming, fasting, body-piercing, fire-walking, dehydration, or sleep deprivation; cultures have even harnessed the power of psychoactive plants, like tobacco, cacao, magic mushrooms, ayahuasca, DMT, 5-MeO-DMT, peyote, ibogaine, amanita muscaria, Nepalese psychoactive honey, alcohol, morning glory seeds, Hawaiian baby wood rose, and cava—but why?

And why is there such a tradition of achieving these 'visionary states' or 'peak experiences' that some cultures will even go to rather extreme lengths to get there, like burning a frog's poison into their skin (something I've actually done); or consuming hallucinogenic fish to induce dream-like visions (something I'll likely never do).

Some cultures have even found that by smoking the secretions of a certain frog, like the Colorado River toad, they've even been able to tap into other realms of Being.[10-22] Realms that seem somehow "outside" of this one.

And when you investigate this question seriously, which I have, you soon find that cultures all over the world—*and throughout all of time*—have appreciated and cultivated these different access points because they have long since realized the therapeutic and healing powers that altered states provide. Because often times, it isn't someone's advice that helps us out of a bind; it's ourselves. And altered states of consciousness (aka non-ordinary states of consciousness) do just that. By temporarily removing these lenses and filters from the body of our camera, we come to see our place in the world in a new and novel way.

How is this possible?

It's possible because, even though our brain is like a computer, and even though our consciousness is like the software, our Experiencing Selves aren't given the level of access you might expect. We don't have admin privileges. Instead, it's more like we're logged in as "guests" in our own house. Perpetually forced to boot-up in some kind of Safe Mode, our Experiencing Selves aren't allowed to look under the hood and check out all those background processes. And yet, under certain environmental circumstances, technologically-induced circumstances, or drug-induced circumstances, it's as if we're given a different set of access keys. Or if you prefer, it's as if we're given a different set of lenses. Our filters fall off; our depth of field widens; and suddenly, we're able to see the problem with a new set of eyes.

Take it from me, the guy who's dying, the guy who's experienced almost every NOSC out there. I'm not going to waste your time describing the more commonplace altered states like caffeinated, drunk, high-fever, hypnosis, dehydration, fasting, sleep deprivation, extreme physical pain, awe, panic, day-dreaming, lucid dreaming, child-birth, depression, mania, schizophrenia, holotropic breathwork, social isolation, sexual euphoria, sweat lodges, withdrawal states—instead, we're gonna reach for the top-shelf of altered states: the near-death experience and the psychedelic state.

The field of expertise Ava and I focus on.

Chapter 13
Unlearning

I WISH THEY'D asked me if I wanted to listen better. I wish they'd told me it'd lower my stress. If only the guy had asked me if I wanted to fall asleep faster, I might've said yes. But as it turns out, that's not what he said. He could've told me that I'd have more energy. That I'd be able to focus better. Learn faster…study longer. But when it all went down, the guy didn't say any of that.

Instead, what he said was, that I, quote, "needed to" meditate.

Scratching his cheek, the barista goes, "You know what your problem is," then turns to his buddy and mumbles something catchy they must've heard on TV.

"Analysis paralysis," the barista says, and they both share a nod.

This was at my usual cafe. Just moments after I'd ordered my drink. When one of us decides to veer off the socially-approved script, and the next thing you know we're caught in a real conversation.

From this side of the counter, it's mostly just a pair of eyes I'm talking to. Eyes sitting above a tiny row of espresso cups. But the second I say something about suffering from anxiety, the barista shakes his head and trades glances with his friend.

"You don't have anxiety," the barista explains, pouring milk in a silver tin, "you're just addicted to thinking."

The sound of a screeching espresso machine momentarily deafens both of us, until he plunges the wand deeper into the tin and finishes delivering his prognosis.

"You're addicted to worrying—*that's* your problem. A lot of people are," he adds, trying to soften the blow. Then, over the sound of frothing bubbles and the smell of fresh coffee, he issues his edict. That I...*needed to*...meditate.

Now, to give credit where credit's due, this guy was actually right. It took him less time to size me up than it did for him to make a dirty chai latte. But his execution. Dear Lord, Tom, his execution was terrible. And I do use the word "execution" and not the word "tact" because to use the latter would be to imply that he actually had some.

He didn't.

First of all, telling someone they need to do something hardly ever works, and in this particular case, it actually ended up having the opposite effect. Instead of being intrigued by meditation, I got turned off. And even though this wasn't the first time I'd ever heard the word, after interacting with this guy it was certainly the last time I ever wanted to. Which is really rather sad when you think about it. Because that means before I even truly understood what it was, I'd already formed a belief about it.

Perhaps even more embarrassing, is the fact that this wasn't some kind of isolated incident either. At least a dozen people must've tried to strong-arm me into meditation over the years, which should probably tell you something about the analytical nature of my mind. But for some reason, this only seemed to make me want to double-down on my initial decision. To be perfectly honest, it wasn't until I saw the live scans of an fMRI for myself that it even started to make sense.

This was probably only a couple years after the barista incident. Maybe just a couple months after Ava and I got together, when our university decides to conduct an experiment with long-term meditators. So they invited a small group of monks and meditators to come sit in our machine.

The first guy we had come in, we slapped some electrodes onto his head, and then had him lie down in our fMRI so we could simultaneously track his cerebral blood flow. The electrodes strategically placed on his scalp, we used those to measure his brain's electrical activity (electroencephalography or EEG).

And, believe it or not, but that's what finally did it for me.

It wasn't being told I should do it, or that I ought to, it was getting to see the neurological effects first hand. More specifically, it was getting to watch how this man's meditation practice was suppressing his *default mode network.*[1] A network of brain regions that's so crucial to the proper functioning of our brains, we really have to spend some time focusing on it.

Flow States

You know that feeling you get when you've been painting, writing, or otherwise caught up in some activity with so much intensity that you've lost track of time?

Well, picture that as the antithesis to the default mode network.

That state of mind where you're jamming away writing lyrics, coding, or whatever activity that gets Tom "in the zone" is a state of mind we all crave. It's a state of mind we all yearn for. It doesn't happen very often, but when it does, years could pass and I'd never know. Time seems to take on this Dali-esque, melting-type qualia during one of these flow states, and usually, I'm so immersed in my work, it feels like there's nothing that could stop me. The house could be on fire, pillows of smoke billowing underneath my office door, and not looking up from my laptop, I could just hold up a finger as if to say, "One more minute," and the fire would move on, seeming to understand.

Call me crazy but I feel more powerful when I'm in this state. More productive. Less critical. Although, the one thing really worth keeping in mind here is that this is *not* the brain's default state of operations. No, this is what happens when the default mode network (DMN) gets switched off.

Notes from the Editor:
Flow states are actually a well-documented and well-researched ecstatic state of consciousness, a NOSC seemingly capable of occurring in the midst of almost any creative endeavor. From poetry to figure skating, dancing to musical composition, as far as NOSC go, flow states are actually quite common. The pioneer behind much of this research, Dr. Mihaly Csikszentmihalyi (pronounced Me-High Chik-Sent-Me-High) describes flow as being, "completely involved in an activity for its own sake. The ego falls away. Time flies. Every action, movement and thought follows inevitably from the previous one, like playing jazz. Your whole being is involved, and you're using your skills to the utmost."[2]

140

—Flow is the reason we watch sports. It's the reason we have skaters, boarders, rock climbers, and surf rats. Once all lumped together under the pejorative term "adrenaline junkies" there's actually something more interesting going on here. To get technical for just a moment, it isn't just adrenaline these people are after, it's a whole slew of neurotransmitters like dopamine, anandamide, serotonin, norepinephrine, endorphins, and oxytocin.[3] Needless to say, but it's a very rewarding state to be in. What's more, if you actually read anyone's account of what a good flow state feels like, you'll understand just why they get hooked. Aside from it physically feeling pleasurable, you're actually pushing yourself so far past your current capabilities that your Narrative Self goes off-line. Evidently, the narrator of your thoughts, the interpreter of your feelings, is a luxury (if you could call it that).

Or, to put the matter in another way, the inner critic we all have talking in our heads isn't a necessary function of the brain; it's a calculation. A calculation which can be performed under normal circumstances. But under abnormal circumstances, say, the types of conditions that generally give rise to flow states, the Narrative Self can go completely off-line leaving just the activity. Action without judgment. Knowledge without commentary. Consciousness without *self-consciousness*.

Another way to imagine these peak states of performance might be to say that you couldn't possibly be more engaged by the moment you're in. Your attentional resources are stretched to the max and you're performing so far beyond your current potential that your brain's not even able to continue calculating time or the Narrative Self.[4]

Whenever I find myself explaining this concept to people, the first thing they always respond with is, "That has a name?" Followed by an account of the last time it happened. But when I tell them about how this can happen to a whole group of people, like at a concert, or on a stage, or on a sports team, that's when people usually look me in the eye and say, "James, that's what I live for." It's for those moments when they lose track of themselves in a crowd, or at some kind of event (group flow). When their attention is so wrapped up in whatever it is they're doing that luxuries like time and narration just melt away.

"Your thoughts crawl away to the back of your mind," explains mountain biker Colin Gray, "and your body flows along the trail without effort or voice. Time changes, tension disappears...controlling your bike becomes effortless." It's in moments like these when you realize, you've just entered the "magical state of 'flow.'"[5]

———

For years, I just assumed Buddhists and meditators were wasting their time. For so long I'm embarrassed to admit, I honestly believed these people were lazy and lacked ambition. Turns out, they were actually busy cultivating the thing that's at the center of all our experiences: our awareness. Not just their ability to tune out distractions, but their ability to strengthen those executive powers and direct that spotlight. And the way this all shows up on an fMRI is, we see decreases in activation to the default mode network. It's this part of your brain...

Correction: *Parts of your brain*

That are activated whenever we're thinking about ourselves, or whenever we're engaged in any kind of autobiographical thinking (which is damn near always).[6] But this monk we had lying in our fMRI machine was such a skilled meditator...such a cultivated focus junkie...that he could basically flex a muscle in his mind and silence that network in an instant. Once I saw those little needles monitoring his brain's electrical activity go from fast sketches to slow rhythmic patterns, this guy became my best friend. His EEG was off the charts in the sense that it wasn't off the charts.

Put it like this: normal brains hooked up to an EEG always look like those classic seismic readouts with all those pens scribbling fast and high because an earthquake just hit a hundred miles out, but the Monk's EEG didn't look like that at all. Sure, he might've had some mental chatter happening when we first hooked him up, but once he started meditating, you barely saw those pens move. His ability to switch off the part of your brain that's always worrying, planning, and obsessing was extraordinary. So much so, that when we finally pulled him out of the fMRI, I was asking him every question in the book. You couldn't shut me up. I was taking notes, telling him he could stay at my place—*anything.*

That night, over dinner, trying to be funny, I went, "So if you accidentally run a guy over with your car, is that still considered premeditated murder?" And then, silence.

Thunderous silence.

At first I'm thinking, well maybe monks don't laugh, but not a moment after that, Ava stuffs a forkful of penne in her mouth and rolls her eyes so hard she loses a contact.

Catching the tail end of that exchange, the Monk breaks into a fit laughter so contagious I can't help but join in. Laughing at myself. Laughing at Ava trying to see. The Monk's high-pitched laugh. For at least three whole minutes, not one of us could look at the other in the eyes without starting it all up again. At one point, it got so bad, I could barely breathe. I could barely suck enough air into my lungs to expel back out again, but there was never a point during those three minutes when I was thinking about the scar on my face or what I was going to do that weekend. Ava, she wasn't worried if she'd overcooked the asparagus, or if that time she cheated on her AP History exam was ever going to come back and haunt her, because for those three minutes, the past didn't matter and the future didn't exist. For just three minutes, we were all so consumed by the moment we were in that our ability to project ourselves into the future or think about the past, simply wasn't available.

I seriously can't tell you the amount of times I've chewed on that particular memory, and right now—stuck in this place—I'd trade almost anything to get back there.

The Deep Now

Picture it, you're having a conversation with someone, you say some sort of trigger word, and the next thing you know your friend starts to get quiet. You notice them looking down and to the left, but from the way their eyes are set, you can tell they're not looking at the floor. In fact, they're not really looking at anything. Rather, they're lost in a memory or thought. Without a doubt, their brain might be sitting there in front of you, but their mind is miles away.

So what do we always do?

We wave our hands in front of their face and shout "Hey" snapping them back into the deep now. But then, because you're the curious type, you have to lean forward and ask, "Where'd you just go?" pointing at your head.

Simply put, your friend was just outside themselves. Outside their kinesphere. Something only possible because we have a system of memories and a network of brain regions we call the default mode network.

Why's it called the DMN?

Basically it's because this is what your brain *defaults* to whenever it's not heavily engrossed in some task.

To illustrate what I mean, let's consider what happens when we put participants in an fMRI machine. When we give them a task, what we notice is less activity (less blood flow) going to the default mode network.[7] But whenever we have participants just lying there, awaiting their next instructions, their minds begin to wander. They start traveling in time. Not literally. But mentally, they start to untether and think about something else (what they're going to fix for dinner, or what they should've said in yesterday's meeting). Lying down in our fMRI machine, they start mulling over their performance at work, wishing they would've spoken up more, followed by a worrying thought like, "What if I get fired?"

Now, you could ask, what's the one thing all these thoughts share in common? And the answer is, they're all involved in self-referential thinking, or self vs. other modalities of thought. And whenever this happens, we see increases in blood-flow to the default mode network, leading many to believe, that this network might be the neural substrates of the ego.[8] Your sense of self. The "I" you feel when you say, I am Tom.

***Notes from the Editor:

One of the nodes of the DMN, the posterior cingulate cortex, actually receives 40% more blood flow than any other region in the brain.[9] Combine that with the fact that your brain is already siphoning off 20% of your total energy intake, and it's no wonder that a major player in this field, Dr. Robin Carhart-Harris, would liken this network to being "like a capital city in a country."[10] What London is to the U.K. A worthy analogy given this network's dense connectivity and hub-like structure.

—Look Tom, this story…this plea for help I'm writing, it's a roadmap not an anthology. I'm not trying to bog you down with unnecessary details, like the specific nodes of the DMN, because if I did that, you'd just get lost in the details. Anyone would.

Sometimes, when you're too focused on the details, you can't see the forest for the trees, and for our purposes, this level of magnification isn't really all that important.

What is important is what this network does—what it controls and why, when it's switched off, consciousness expands. It's been called "the control center of the brain" by Dr. David Nutt. "The orchestrator of the self" by its discoverer Dr. Marcus Raichle, and an "incredibly important integration hub" by Dr. Robin Carhart-Harris. Although we're still investigating and paring down the exact functions of this network, the one thing we can say for certain is, it's the most likely candidate for "the ego" we've ever seen.[11-15]

***Notes from the Editor:*
In colloquial speech, when someone thinks too highly of themselves, we might say they have a big ego, whereas in psychology, the term is actually more indicative of the part of your psyche that manages the demands of the external world and is self-conscious. To avoid confusion, the narrator has seemingly chosen to use the term Narrative Self most often.

(The idea of 'the Kinesphere' as defined by Rudolf Laban)

Chapter 14
Pinning it Down

AT THIS POINT, you might be wondering where consciousness happens. At this point, you might be asking yourself, "So, where am 'I' anyway?" I meaning the you that you feel. But sadly, the answer to that isn't as straightforward as the question. For that, I blame folk intuition.

Don't get me wrong, it certainly feels like there's this movie of your life playing up somewhere in your skull, but how it looks is; there's nobody home. Don't get me wrong, it certainly feels like there's just gotta be some kind of "master control center" where the miniature version of you watches/controls the movie of his life, but how it looks is; there's no real you to begin with.

So what does all this mean for the study of consciousness?

Specifically, it's that Consciousness 5.0 gives us the feeling (the sensation) that there's this movie up in our heads, and that it's 'playing to' someone. Except, whenever we go looking, we can never seem to find him.

First you make an incision down the midsagittal plane; you separate the left hemisphere from the right; and just when you think you're about to happen upon some little homunculus sitting up there piloting around this meat skeleton…you don't.

Truth is, there is no king neuron or "synthesis cortex." No one specific spot you could point at with your fingers and say, "This right here, this is where it all went down."

To make a claim like that, you'd have to just point to the body as a whole, because as it turns out, consciousness isn't easily localized to one particular area of the brain, and this is evidenced by the fact that we've cut open hundreds of thousands of brains so far, and yet, we've still never found that little guy.

So what's going on here? Are we not looking hard enough, or what's the deal?

Perhaps, the answer is, that consciousness is a bit more sophisticated than how it's presented on TV.

That special place where all the contents of consciousness merge together and are 'projected to' is what the philosopher Daniel Dennett derisively calls the "Cartesian Theater." He then jokes about even if there was a Cartesian Theater, "and there was a little man in there, then we'd want to look in *his* brain," and then *his* brain ad infinitum, making the case that: (a) it just can't be Cartesian Theaters all the way down, and (b) this theory has no legs to begin with. At some point, Dennett argues, we have to break away from this idea of there being a "homunculus at the center of the controls" and distribute all the work we're imagining him to do, in space and in time.[1] Which, by the way, is exactly what we're up to. Layer by layer, piece by piece, we're out to demystify this subject together.

That said, now that we've displaced the homunculus idea, we can start moving away from this conception that consciousness must 'play out' somewhere or 'to something' and trade it in for a theory that's a bit more sophisticated. Like embodied cognition, for example. For too long, we as a society have been entranced by this idea that the brain and the body (or mind stuff and material stuff) are fundamentally different things. When in reality, the body produced the brain; and embodied cognition takes this fact rather seriously. Instead of treating the body as if it were somehow peripheral to our understanding of the mind, embodied cognition takes into account the evolutionary backstory of bodies and then builds upon that story by reinterpreting all higher-order cognitive processes, like math and intelligence, as something which evolved out of that. In this way, our ability to reason and be rational isn't some abstract thing neurons can do once you cram enough of them together. Rather, our ability to reason evolved *out of* our ability to move and to think about moving.

To decide between two courses of action.

Pre-planning motor acts, predicting the movements of others, determining the consequences of our own actions upon the environment: the second you start doing any of that you need a generative model in your head that makes predictions. And one of the ways we define intelligence is by asking the question, "Can this thing make predictions?"

> If you think about it, there's very little that you can do, apart
> from secretion, without *movement*. Without your body. Speaking,
> looking at different parts of the world, perambulating, moving
> around, redeploying limbs—nearly everything depends simply
> upon moving your body—so, the only way the brain can talk to its
> environment is through its body.
>
> - Karl Friston (neuroscientist)

It's actually quite a very profound idea that Dr. Daniel Wolpert and Dr. Karl Friston first introduced me to. Because once you take into account the fact that most species on this planet don't have brains and subsequently can't produce movements like us, you soon realize that fine-motor control and the ability to produce complex and adaptable movements truly is what sets us apart.[2,3]

Localization & Lateralization

Next time you're prying off the shell of a computer, try and see how long it takes you to spot the motherboard. Then, once you've found that, look for the little chip that has the words "Intel Inside" etched on the back, and in one fell swoop you've just pinned down where all the main processing occurs. On the contrary, when it comes to the brain, it's important to recognize there is no central processing unit. No "Intel Inside" equivalent.

Oddly enough, each of the 86 billion neurons in your head is, in its own right, a little CPU. But perhaps even weirder than that is the idea that not a single one of them actually knows who you are. An entity named Tom may *emerge* out of their interactions, but you won't find the "you" that you're feeling located inside any one of them.

In our section about brain trauma, we just briefly glossed over the idea that different parts of the brain (lobes, cortices, and networks) each add to the richness and complexity of the conscious experience; we also mentioned how each one of these parts could be off-lined due to lesions or stroke. And now, I'd really like to round out that argument by re-examining just what happened to Dr. Jill that day. It will be recalled from our earlier discussion that a good chunk of her left hemisphere got destroyed but that she, herself, didn't die. A big part of herself might have gotten wiped out that morning, but her consciousness didn't just wink out of existence. Nor did Phineas Gage's for that matter. Although, both did become seriously altered. The blood clot in Dr. Jill's brain might have taken out most of her language centers, but her essence never disappeared.

Rather, her ability to articulate and define herself.

At the current moment, I'd rather not derail our conversation with too much talk about bi-hemispheric specialization, or the fact that many species throughout the animal kingdom have lateralized brains with an asymmetry of function. But just know that we've performed entire hemispherectomies on patients before and people don't just disappear. The brain is a highly plastic organ, and if performed early on in life, people can still go on to live rather normal, healthy lives, even with only half a brain.[4,5] Which isn't to say that each hemisphere is like a back-up copy of the other one, and you only really need one to survive.

Basically, all we're trying to draw your attention to is that consciousness doesn't seem to be localized to one hemisphere or the other. Because if either are removed, consciousness still persists. Demonstrating, quite remarkably, that either hemisphere is capable of sustaining consciousness on its own, lest the other gets damaged.[6,7]

In addition to this, it should also be mentioned how each hemisphere is highly specialized. So there's an asymmetry of function, not only in 'what' they do, but 'how' they do it. That withstanding, losing the left or the right hemisphere will subsequently result in a different kind of loss, especially if it happens later on in life. For now though, the last point I'd like to make about the two hemispheres is that, back in the seventies, there were a lot of articles produced about how the left hemisphere does math and language while the right hemisphere does emotion and creativity—and almost all of this has since been debunked. In actuality, both hemispheres are involved in almost everything that we do (reason, language processing, and all the rest). It's just the way they contribute that's different, and for ease of conversation, let's wait just a little bit longer before we dive back into this highly nuanced discussion. Make no mistake, this is a conversation that needs to be had, it's just we're not quite there yet.

Why bring it up now?

I bring it up because there are those who may wish to claim that consciousness must exist in the left hemisphere or it must exist in the right, when the reality is—it's just not that simple. Case in point, we've even severed the thick band of neurons that connects the two hemispheres, the corpus callosum, and even then consciousness doesn't just vanish. Although, things do become quite strange.

Nowadays, corpus callosotomies are generally thought of as a kind of last resort procedure to stop chronic epileptic seizures, but all the research involved in those that have had these "split-brain" operations suggest, that in many ways, there are now two consciousnesses living inside one body![8]

Employing clever techniques like flashing words to one side of the visual field and not the other, Michael Gazzaniga and his team were able to smuggle information into one of the hemispheres without the other one even knowing. And because the two hemispheres were no longer communicating, you could ask each one a targeted question and get a completely different response.[9]

Really mind-bending stuff.

For the time being, we're going to have to put a pin in this conversation. But before we do, let me just reiterate, once more, that we have a left hemisphere and a right hemisphere; and that these two hemispheres actually see and process the world in different ways. To borrow Dr. Iain McGilchrist's description, they each provide a different 'take on the world,' and they certainly work better together, rather than apart.[10] But supposing you met someone who had one of these split-brain operations, unless they specifically told you about it, you may honestly never know.

Putting all this together, consciousness doesn't appear to be localized to one particular hemisphere or lobe. Hence, if all your brain systems are intact and functioning properly, you get Consciousness 5.0 (our self-defined baseline, waking conscious state). On the flip side, if one of these systems gets off-lined, like your occipital lobe or your auditory cortex, then everything would theoretically be the same, it's just you would no longer be able to see or hear, respectively.

Nevertheless, if I had to point to the "you" that you're feeling right now, I'd point to this picture. The little yellow bits, those are the nodes of the default mode network. The red, green, and blue bits—that's their connectivity. This is what it looks like when what you're feeling merges with what you're thinking and remembering. When what you're sensing integrates with what you're predicting and expecting. Somehow, this network of brain interactions is you.

Tom; the conscious agent of your brain.

Observing the Mind

The first time I sat down with the Monk, he told me I had to "control my mind." For twenty minutes straight, I was just made to sit there, Native American style, in the living room of my crummy Koreatown apartment while the merry-go-round of shame played itself out. As one might expect, I tried not to think about anything, but somehow, the simple act of trying not to think only seemed to make everything worse.

Going into the experience, I had felt rather confident about myself. But once I shut my eyes, it was like an uninterrupted stream of thoughts.

A firehose of shame, anger, and sadness.

The same way your phone might explode with notifications after a month-long hiatus, that's the way my mind felt for the first fifteen minutes straight. After that, the Monk puts his hand on my shoulder and goes, "So how many thoughts did you have?"

Looking around the room, I get up to stretch my legs and tell him, I'm not sure. "Maybe a billion," I say. Then, rubbing temples beneath tired fingers, I add, "It's kinda hard to know where one thought ends and the next one begins."

And at this the Monk just smiles. Letting out the kind of laugh like he's heard all this before. "Next time," he says, "we're going to get you down to nine hundred and ninety-nine million." And while I physically hear him telling me this, all I can really do is shake my head and stare at the floor, because what he's saying is impossible.

"I am literally a meta-analyst," I tell him, "It's kind of what I do."

But then the Monk hands me a weapon.

Next time a thought comes into my head, he says, I should try and send it away with the thought, "Oh well, peace."

And then, we go again.

This time, armed with a mantra of sorts, I finally had something to wield against my untrained mind. Whatever thoughts popped into my head, I did as the Monk said and tried to send them away.

This is stupid.
This is pointless.
I could be studying right now…*Oh well, peace.*

Then about one to two seconds later:

I wish I never agreed to this. I'm bored. I have real work to do. This guy doesn't know what he's talking about…*Oh well, peace.*

Then about five seconds later:

I could be doing laundry.
I could be washing my car.
I could be organizing my…*Oh well, peace.*

Now, I'll be the first to admit, there were numerous stretches of time where I'd get taken in by a thought and completely forget what I was doing, but overtime, I did get better. Eventually, I began to notice *when* my mind was wandering, and whenever that happened, I did as the Monk told me to: I acknowledged the thought; I accepted it as non-judgmentally as I could, and then I recommitted to the moment I was in.

"What you want," the Monk would say, "is to focus on an object like your breath." He'd hold his stomach and push out his hand so I could see, then he'd bring it back in and look back up at me holding up one of his long bony fingers.

"You want to *watch* your breath, not control it."

Observation without interference. Awareness without control.

Then, at the end of our next session, the Monk leans forward and asks, "Now, how many thoughts did you have?"

Notes from the Editor:

The word "meditation" is almost as non-specific as the word "exercise." It's one of those catch-all, umbrella terms under which there are various subdisciplines: mindfulness meditation, mantra meditation, focused or concentration meditation, Transcendental Meditation, only to name a few. The type of meditation the narrator seems to be advocating most closely resembles a hybridization of mindfulness meditation and mantra meditation. At any rate, reduced activation in the nodes of the default mode network have been associated with a number of different meditation practices.[11]

—And that was the Monk's way. He never rushed anything. Never told me I had to do anything. He met me where I was and helped me learn an invaluable lesson. Incidentally, the same exact lesson Dr. Jill learned when she lost half her brain.

Namely, that you are not your thoughts.

Sure, your thoughts are a part of you, and who knows, maybe they're the loudest part of you right now. But it only takes a little bit of time "zenning out" in this way, for you to realize that thoughts really are like breath in the sense that there's an automaticity to them. They go on...and they go on...and they go on, while you—the real you—can actually just sit back and watch. Without any interference or guidance, the thoughts will carry on completely of their own accord. Granted, there can be a voluntary aspect to thinking. We can direct our thoughts to a certain extent just as we can force ourselves into taking giant, lungfuls of air. But, at the end of the day, if you watch your thoughts long enough, you end up noticing at least two things: (i) how involuntary thinking can be, and (ii) that we really do have the option of sending these thoughts away. And really, that's the benefit of meditation.

Contrary to whatever the word sounds like, you're not trying to "levitate off the floor," or become such an elitist that you start passing off snap judgments to everyone in your cafe. All you're really trying to do is lower the density of thoughts that pass between your ears, commit to the moment you're in, and grow your prefrontal cortex.[12]

But sadly, the barista didn't say any of that.

Chapter 15
Disentangling the Self

THIS ONE NIGHT, I'm just trying to make headway on my ████████ ████████, when Brian calls me into his room to show me something. Of course, he's sitting at that big black desk of his. The kind of desk you could have a good nap on. All leant back in his fancy ergonomic chair, staring up at his multi-screen monitor set-up, when suddenly, Brian swivels around and says, "Wanna see something," while arching his eyebrows up and down.

Me, I'm standing in the doorway with my back against the door frame. In the middle of crossing my arms, when Brian spins back around and puts up two fingers in the air, motioning for me to come closer. Hearing my footsteps, knowing I was hooked, Brian turns vaguely back in my direction and goes, "So when's the last time you played Command and Conquer?"

Mumbling, I try to ask which Command and Conquer, but I'm so tired I can barely get the words out.

"Red Alert 2," Brian says, swiping open a second screen.

But what Brian's saying is impossible.

Mostly because neither one of us owns a prehistoric computer, circa 2001, but even knowing this, I bite. I walk my torso over to his desk, and the second I'm close enough to see what he's up to, Brian opens up a program that reads "VirtualBox" at the top.

Now, this might seem a tad bit obvious, seeing as Brian's a video editor and all, but he really was a whiz when it came to computers. Days, when Ava and I would be off trying to finish up our degree, Brian would be at home becoming a master of his domain. His computer set-up was something like modern day shrine to Apple. This guy could edit on Adobe, Avid, Final Cut, After Effects, Photoshop—you name it. Naturally, his preference was to edit at home on his beastly Mac Pro, however there were at least a few times when he needed to run something on a Windows or Linux-based platform, and whenever that happened, Brian used VirtualBox. A program that allowed him to create a *virtual machine* inside a physical machine.

In other words, VirtualBox is an app that allows you to create a self-contained virtual computer inside your already running physical computer. His physical computer was, of course, the Mac Pro, running whatever OS they just happened to be putting out that week. But inside of that, he could be running a virtual machine (or VM) adorned with an entirely different operating system like Windows or Linux. This little detail seems rather important to point out, because apart from being able to run two different operating systems at once, this also meant that one of the operating systems was actually nested inside the other. So while they might be using the same resources, the Guest OS (Windows, in this particular example) would not be sharing information with the Host OS (aka the Mac).

This is actually such an important point, I'm going to stress it again. *They don't really share information.* Which is actually a feature, not a bug. By creating a virtual machine inside a physical one, this means you now have the ability to test new programs inside of a safe container. Or, if you're Brian and me, then it means you can run one of your favorite games of all time, right there, on a modern-day Mac without having to dig up some dinosaur-sized computer and some impossible-to-locate software.

As one might expect, it takes a decent amount of computing power to run a VM inside a physical one, but the main point behind all of this VM talk is that virtual machines aren't literal machines; they're calculations. Calculations which can be performed under normal conditions, but under abnormal conditions, like when the Host OS is hogging up all the resources, a virtual machine isn't going to run very well. If it runs at all, it'll be glitchy at best. And to make a quick parallel here, this is exactly what we see with flow states.

During your bona fide flow state, when attentional resources are stretched to the max, and you're performing at your absolute peak, trying to generate virtual machines like "the Narrative Self" or "the Remembering Self" isn't really possible, which is also why people often report the loss of their inner critic during them.

Going a bit further here, under your psychedelic state, when a subject has ingested enough psilocybin or lysergic acid diethylamide for their DMN to become *dis*integrated, their ability to articulate and define themselves becomes rather problematic. Impossible even.[1] Being that the "orchestrator of the self" is no longer conducting the way information flows, the reports about this state are wildly chaotic. And before we drop into this subject full force, we really need to address the multiplicity of selves that exist within each of us.

——

So far, we've already covered the idea that you are not your thoughts, and how there's an involuntary aspect to thinking—I even went so far as to point to a picture of the default mode network and say something like: *This is the 'you' that you feel.* And on one level of analysis, that was completely true.

I mean, technically speaking, it was a very accurate statement to make (that is the "you" that you feel). However, on a deeper level of analysis...it was a little misleading. Truth be told, there are layers of self just as there are layers of truth, and while it may be tempting to believe our identities could solely exist inside this network, we actually know from all our studies concerning flow states, meditation, and psychedelics, this just isn't the case. More to the point, it actually can't be the case because in all three conditions, it's this network which gets suppressed.[2-4] Or in the case of high-dose psychedelics, obliterated.

Not permanently, but temporarily, this network is no longer able to fire and conduct in its usual way. In light of these facts, let me revise my previous statement. Even though it may feel like "you" (Tom) are the product of your default mode network (i.e., you may feel like you are your Ego/Narrative Self), when all that neurocircuitry gets wiped out, something still persists, and that something we are now going to call "the Awareness."

For a moment, just consider the idea that the sense of self you have now isn't the same as the one you had when you were a baby, and this goes way beyond the fact that babies can't think in language. Hell, babies don't even have object permanence until about four months old.[5] Toddlers, for instance, aren't even aware that other people can hold views different from their own until about five or six years old.[6] Reasoning from these facts, we have to intellectually grapple with the idea that your ability to articulate and define yourself is something that unpacks and develops over the course of your life.[7]

Just as we learned to see, we learned to see ourselves. And after this section, we will have learned to see our *selves* just a little bit better.

The Developing Brain

Think of it this way, the wetware you're given at birth isn't the same wetware you're running right now as it's undergone so many different software, firmware, and structural wetware updates. So much is the case that it's physically impossible for you to think about yourself the same way as you could when you were a baby. Even if we put aside all the structural changes that have occurred (neural pruning, decades of memories, that firmware update that realigned all of your goals and perceptions around your twelfth birthday)—even ignoring all that—there was still something that changed everything about your mental life and it happened before you even knew how to stand on two legs.

You learned a language.

According to psychologists Katherine Nelson and Robyn Fivush, it may have been this acquisition of a language that effectively "reformatted" the wetware of our minds, giving rise to our inability to recall most memories before the age of three.[8] If they're right, that means infantile amnesia might be the result of how language reformats the memory system. But even if they're wrong, the undeniable underlying fact is: The way we're even able to think about ourselves has gotten progressively more complex.

Sure, we might come out of the womb rather simple, relatively speaking. But over the course of our lives, we become the types of creatures that can think about thinking (metacognition); think about what other people are thinking (theory of mind); and all this arises through (a) our brain's natural growth cycle, and (b) our interactions in childhood. Playing with other kids, rough-and-tumble play, imaginative stories: these are all essential components in sharpening our ability to think about what other people are thinking—in addition to the way we think about ourselves—because, as you'll recall, we're not quite transparent with ourselves.[9] We don't have direct access to what we believe, or what our true intentions are; we have to make inferences about these things based on our actions. And this is the thing that Dr. Professor Roberts guy was trying to explain to me the day I spewed half-digested eggs all over his bathroom. A concept called *theory of mind*. We generate models of what we believe others to be thinking; and the way we do that is by closely monitoring ourselves.

By closely monitoring the way we think and behave, we can assume others might behave in a similar fashion too. So, in its most simple incarnation, I might see you reach for a cup that's empty and infer you might be thirsty. I may think to offer you another water, but then, after observing the way your feet are angled towards the door, ask if you wanted one for the road instead. You see, it's little body language cues like that which always give away our true intentions (how we hold our body, and where we point our eyes). Not to mention all those little micro-expressions that leak across our face when we're trying to mask how we really feel.

"Oh my god Tom, that's such great news!" [Forces smile].

But if you never saw the micro-frown that leaked across her face for as little as 40 milliseconds, then according to your theory of mind, she seemed to be happy.[10] Bearing all these facts in mind, we can now turn our focus to something that's a bit speculative, but at the same time, highly informative.

A model for atomizing the self, pieced together from legends like Carl Jung, Sigmund Freud, and Nobel laureate Daniel Kahneman. Although, as we're about to find out, even these giants didn't quite see eye to eye. Don't get me wrong, they all agree that we should divide up the psyche—it's just nobody agrees about what to call it—or how many there are for that matter. For example, the neuroscientist Anil Seth chops it five ways, while legend of the field Carl Jung has only chopped it four ways. Sigmund Freud famously chopped it three ways, while psychologist Daniel Kahneman, author of *Thinking Fast and Slow*, has only made two divisions: one in thought, and the other in self. First, Kahneman makes a sharp delineation between the two types of thinking our brain gets engaged in (System 1 and System 2), then he divides our experience of being a self into two separable entities, he calls "the experiencing self" and "the remembering self."[11] To cut a long story short, they all tried to make sense of the human psyche—and they all made some really good points—but since I'm still dying, let's fast forward to at least ten years after you read them; jumping straight into the moment you realized they all missed something.

You were on the elliptical at the time.

You'd already spent time with the Monk, you'd already spent years thinking about the nature of consciousness and what it means to be a self. You even went ahead and took it upon yourself to undergo as many non-ordinary states of consciousness as one person could possibly endure, when something finally clicked. An image formed. A way to think about unique, separable selves without going completely insane.

Of all the arguments you read, you were only seduced by a few. Of the few, you tossed out the weakest parts. Of what remained, you imported into a Venn diagram.

Supposing you did all that, and you really spent a few years thinking about the nature of being a self, this is what you might come up with:

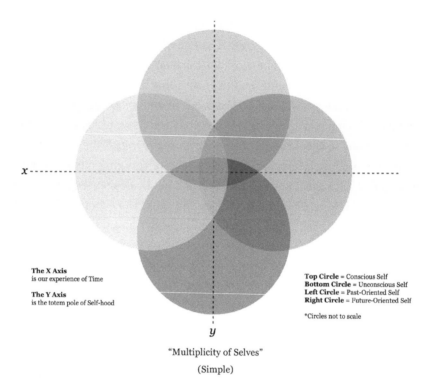

"Multiplicity of Selves"
(Simple)

—To make some sense of this, let's consider the example of someone who tells you they really want to lose some extra pounds. Of course, they might say all the right words, buy the best clothes, and join the best gyms, but as we're all quite painfully aware, weight loss requires more than just paying lip service and a gym fee. It requires a firm commitment by at least a couple, if not all, the aforementioned selves.

Taking this into account, we can see that the unified self or Tom, is best divided up into four main selves: a self that's predominantly focused on the future, a self that's predominantly focused on the past, a conscious self, and an unconscious self.

In this regard, if all of your selves are not in alignment, you're probably not going to be walking in a very straight line, metaphorically speaking. Just as well, you may notice us describing these selves almost as if they were their own people, because in many ways, that's the appropriate way to conceptualize them (as distinct sub-personalities with their own ideals/agendas). Or, if you prefer, as complex networks of neurons whose interactions combine to form something approximating virtual machines, who in turn have their own "core drives" that are distinct enough to warrant their own schema.

Nevertheless, you may have noticed that the previous diagram was a little too simple for someone trying to crack the mother of all mysteries, so I suppose we'll have to chop the self up even further.

An octotomy.

That way, when we finally do get the subject of altered states we can discuss their effects with surgical precision.

"Multiplicity of Selves"
(Detailed)

The Persona: This is the public image you try and project to the outside world, your 'social mask' as a student of Jung might say. You may have multiple personas that you can pull out in various situations, such as a public persona, a work persona, a home persona, a social-media persona, the persona you have around your parents, and so on. However, unless you specifically chose to broadcast your private thoughts or inner experience, the Persona would not include this information.

The Narrative Self: Also known as the narrator, the interpreter, the inner critic, the executive self, and the Ego Self. According to Freud, it is the ego that develops and matures once a child becomes more aware of his or her own individuality. Freud once analogized the relationship between the ego and the unconscious as being something like a driver and a horse-drawn buggy.[12] Personifying the unconscious in the form of a horse, and the ego as being the driver, Freud envisioned the ego as sitting up on its throne—riding crop in one hand, reins in the other—just waiting for another chance to assert its dominance and control. The wonderful underlying idea being, that the horse has a mind of its own. Typically, we might conjure up an image featuring blinders on the horse, but as we shall continuously uncover, the blinders are really on the driver. Much like an eight-sided die, the ego is just one of our faces (the face we associate with the most). Although, it should be reemphasized that the Ego Self, hereafter referred to as *the Narrative Self*, is really just a low-resolution snapshot that is sheltered, airbrushed, and sanitized.

The Remembering Self: Originally introduced by psychologist Daniel Kahneman, the Remembering Self can either be your greatest ally or your worst enemy. It can either imprison you with regret, or rapidly assist you with complex decisions. Being a natural storyteller, the Remembering Self cares about memories, patterns, meta-narratives, and experiences. Using its intimate knowledge about the past to help drive decisions about the future, the Remembering Self ultimately tries to steer the Experiencing Self away from pain and towards pleasure.[13] Likes to reminisce, ruminate, mythologize, and disconnect. Likes what is familiar and known.

The Experiencing Self: Living in a tiny window that only spans seconds on either side, the Experiencing Self is rather hard to describe considering we're always stuck here.[14] Having said that, one way to think about what the Experiencing Self is like, is to recall the way it feels when we get knocked out of it. Either by being pulled into one of the Remembering Self's memories or into one of the Prospective Self's daydreams, the one thing we can say about this self is that craves novelty and information; it hates boredom. Sitting in the proverbial "theater of consciousness" the Experiencing Self only has the present moment to work with, and by virtue of that fact, imbues everything it sees with an entirely different value structure, than say, your Remembering Self's desire for a good story. What may excite the Remembering Self (an anticipated memory about skydiving) might actually terrify the Experiencing Self. Apart from that, when you want front-row tickets to see your favorite band, what you're actually doing is anticipating a better experience, and a better *memory of the experience*, which is all you really get to keep anyway. What's more, to actually get this memory, your Experiencing Self is probably going to have to wait in some big, long line to even get these tickets. Does the Experiencing Self want to wait in this line? Probably not. But because your Remembering Self is anticipating a memory that's infinitely better than the nosebleeds, it's actually able to drive behavior in the present. Lastly, it should be acknowledged that the bulk of what the Experiencing Self experiences never gets passed on to the Remembering Self. Just think about what you had for breakfast three days ago, or what the cashier's eye color was at the store last night. Most of the tiny details about our moment-to-moment experience (smells, sounds, sensations) leave no discernible trace. Which isn't to say they don't affect our implicit memories, but evidently, it's only the really emotionally salient moments of our lives that tend to get remembered.[15] Just as the majority of our dreams fade hours after waking, the bulk of what we experience never gets coded into explicit, recallable memories.[16]

The Prospective Self: If you don't give into every hedonistic impulse you have, you likely value the idea of yourself in the future. Regardless of the fact that our literal self in the future is something like an ever-receding mirage that we never actually get to, we certainly treat it as real. We plan for it. We save for it. We make sacrifices for it. But the Prospective Self isn't real in the same way that the Experiencing Self is real. Like the Narrative Self, the Prospective Self is a calculation too, only this VM is more concerned with what's to come, not what is. Remember, memories are only evolutionarily advantageous if they serve to guide *future actions* and *future behavior*. Therefore, memories aren't really about the past as much as they're about the future.[17] For this reason, we can imagine that our brains and our bodies, which are really just instruments for orchestrating movements through space, must always have a division that's oriented towards that end. Whether that be monitoring wishes or forming goals, revising expectations or tracking reciprocity, the Prospective Self is highly concerned with how our actions affect the future.

Deep down, I think we all know when we're not living up to our full potential, and perhaps when we're giving in too much to our impulsive side; and the Monk had an interesting way of phrasing this. He once told me that there were two paths in life: the path of pain and the path of listening. "Depression," he said, "was your body's way of telling you 'something is wrong.' Like a check-engine light." He said, when you've chosen the wrong path in life, those who actually respect you enough to tell you the truth will let you know.

"And if they don't," he used to say, "then don't worry, because your body will." Manifesting as either depression or anxiety, your body will often let you know when one of your selves is out of alignment; and I always admired him for saying that.

The Awareness: Sitting on the edge between the Experiencing Self and the Bodily Self, this is the part of you that persists even after everything else fails. It is the last remaining piece of your psyche to get wiped out. Due to the fact that we haven't introduced this term until now, I've been referring to this self as "True Tom" or "the real you." There is a 'false self' (the thing we think we are); but once we dismantle the DMN through meditation, flow states, or psychedelics, we notice that there's actually a deeper layer to our sense of self than previously realized. To dust off an old image, the Awareness is what sits at the base of that totem pole and is forced to 'look out' through all the memories, expectations, and belief-filters piled on top. It's the type of consciousness you might expect a baby to have (pure, non-judgmental, and unbiased). Raw intellect untarnished by memory. In a way, the Awareness is the state of mind that meditators are always trying to get back to, as it's a state of mind that's completely devoid of wishes, wants, hopes, desires, categories, and lists.

The Unconscious Self: Subconscious OS territory. The place where most of our cognition happens. Makes associations, draws conclusions, and forms snap-judgements. This is the part of yourself you consult when you've forgotten the lead in your favorite movie, but more importantly, this is the part of yourself for which you have no conscious awareness. It may house repressed feelings, unconscious desires, unfulfilled wishes, instinctual behavior, and false beliefs. Apart from this, the Unconscious Self would also contain the Jungian concept of "the Shadow." For Jung, the Shadow was the darkest part of our psyche (that place where the 'light of consciousness' does not touch).[18] Correspondingly, if the ego is the biased, photoshopped view we have of ourselves, then the Shadow would include all those unsavory characteristics we've decided to have scrubbed from view. Unlikable traits and behaviors, patterns we'd rather ignore. Instead of confronting these things directly, Jung believed we projected our shadows onto other people. In other words, the characteristics we cannot stand in others are actually the parts of ourselves we care not to see.

The Bodily Self: Formally introduced to me by neuroscientist Anil Seth, the idea of the Bodily Self encompasses not just that we are embodied entities operating *particular* bodies in space, but that these bodies contain an unimaginable complexity and richness of intelligence that we're just beginning to understand.[19] The day-to-day operations that our bodies are involved in range from growing our hair, making our heart beat, digestion, respiration, cellular restoration, hormone regulation, immunological responses, and so many other complex, life-sustaining processes, of which, we have no real access or control.[20] No one can *think* their way out of a headache anymore then they can *force* themselves into falling asleep. So, even though we may think of ourselves as being like the driver of our vehicle, the vehicle itself seems to have its own agenda. Of course, we can override some of these needs, like eating, breathing, or sleeping, for a short amount of time, but if we really want to be honest with our situation, we're all slaves to the Bodily Self's wants and desires.

Experience vs. Memory

Daniel Kahneman has a nice way of explaining the value differences between the Experiencing Self and the Remembering Self by first getting us to realize the difference between "being happy in your life" vs. "being happy about your life" (i.e., the difference between experience and memory). He shares a tiny story about his favorite vacation—and how much he's thought about that memory over the years—then he invites us to do a rather funny thought experiment. He says, "Imagine that for your next vacation, you know that at the end of the vacation all your pictures will be destroyed and you'll get an amnesic drug, so that you won't remember anything." Then, he follows up that proposition by asking the question, "Now, would you choose the same vacation?" "And if you would choose a different vacation," he says, "there is a conflict between your two selves, and you need to think about how to adjudicate that conflict."[21]

Aside from this, it's essential to add that the Remembering Self is prone to many different biases concerning the way it remembers, recalls, and interprets events.[22] For instance, we've all either gotten back with an ex, or watched someone who has, only to realize, rather quickly, just what kind of dirty trick the Remembering Self has played. Instead of choosing to remember this person in their entirety, the Remembering Self has been rather selective about the details it recalls. Apparently, it only remembered the happy dates, and the laughter. This angel of a person. While conveniently forgetting all the details that drove you to cut ties in the first place. Being surrounded by memories, meta-narratives, and patterns, the Remembering Self seems to like the familiar, even if the familiar is terrifying. Personally, I'm not one for basing my entire life around a single quote, but for the Remembering Self, "Better the devil you know than the devil you don't," seems to be its modus operandi.

Practical Examples

When you've been driving for thirty minutes straight with no recollection about any of the drive passing, you've either been taken in by the Remembering Self's memories or the Prospective Self's daydreams. While the Unconscious and Bodily Self have been busy manning the wheel, your Experiencing Self has been so caught up in some form of contemplation, it's failed to log away any memories about the last half hour.

—When you're in a sensory deprivation tank, in many ways, what you're experiencing is a loss of the Bodily Self. Not a complete loss, but significant enough, which generally has the effect of enhancing your mental life. Given the fact that the VMs for the Bodily Self, and the Persona no longer need to be generated, the mental resources normally allocated to these selves is freed up, allowing for some very interesting NOSC to occur.

—When you're at the cinema and you're watching a movie that really grips you, you may momentarily forget you're in an auditorium with other people, implying that the Narrative Self, the Persona, the Remembering Self, Prospective Self, and Bodily Self might have all gone off-line (even if only temporarily). Nonetheless, when filmmakers have really done their job right, you may notice your Experiencing Self getting lost for extended periods of time. Lost in the emotions of the characters, caught up in the story, even temporarily suspending disbelief about the implausibility of the narrative. Add all this together, and you can see why stories are so fun. Aside from them being powerful devices for delivering information, they can also stimulate our own cathartic emotional release (Bodily Self), even if we know we're only watching a fictional character's story.

—After you've been put under a general anesthetic and the doctor tells you to count backwards from ten, before you ever hit zero, you tend to wake up, post-op, without any sense of time passing. Unlike sleep, where there's at least some semblance of the passage of time, under a general anesthetic, the transformation from pre-op to post-op is practically instantaneous.[23] As far as you're concerned, the "you" that you're feeling disappears; therefore, under this state we can imagine that the Narrative Self, the Persona, the Awareness, the Remembering Self, Experiencing Self, and the Prospective Self, have all gone off-line, leaving just your Bodily Self and Unconscious Self with no awareness whatsoever.[24]

Unless, you just happen to be one of those people who doesn't go under. Something they call *anesthesia awareness*.

Kind of like sleep paralysis, where your mind wakes up before your body. Except in this particular situation, you're on the operating table unable to signal for help. And, given that the paralytic that's hindering your muscles is still active, your mind wakes up, only you can't seem to motion to the guy holding the bone saw to stop. What's worse, the pain and sleeping meds have both seemingly worn off, otherwise you wouldn't have been able to achieve some sort of awareness in the first place. Fortunately for us, anesthesia awareness is actually quite rare, with only one or two cases occurring in every thousand.[25] But even still, the fact that it can happen tells us something about the nature of our minds. Namely, that parts of yourself can wake up (or be conscious) while other parts are not.

See *Comas*

See *Persistent Vegetative State*

See *Locked-in Syndrome*

If you think Clive Wearing's situation was bad, then don't even bother Googling that last one. Because locked-in syndrome makes anesthesia awareness look like a summer in Belize. Sleep paralysis, like a weekend getaway.

In any case, the crucial idea behind all this atomization isn't to polarize you and say, "Look how divided you are." Rather, this is a call to action for us to start carving out those activities which help us to unionize the self and work towards wholeness and balance.

The first step of which is, realizing you're subdivided.

So if you asked me again to point to who you really are, at this stage, I'd point to that blended edge between the Experiencing Self and the Bodily Self. What Ava and I have taken to calling *the Awareness*, because, at one's core, that's who you really are. An awareness sitting underneath the multiplicity of selves, forced to 'look out' through all the overlays, memories, and belief-filters stacked on top.

Notes from the Editor:

One detail that seems necessary to enter on here is that these selves seem to have limitations based on the memory system they have access to (e.g., the Experiencing Self only has working memory to operate with). Consequently, the type of attention paid by each self is bound to be different.[26]

—Just as a bit of recap, if we examine the detailed Multiplicity of Selves diagram once again, you'll probably notice that the Unconscious Self, the Bodily Self, the Awareness, the Experiencing Self, the Narrative Self, and the Persona all sit along the same vertical axis. Reason being, these selves build upon one another like a hierarchy of complexity, which is actually mirrored, to some extent, in the structure of the brain. Along the X axis we see how the psyche deals with time (a division of the unconscious dealing with events that already happened; a division of the unconscious dealing with events that have yet to happen); but all along the vertical axis, we get to see how sensory information builds in complexity.

To give you a sense of this, pretend your phone is vibrating in your pants pocket across the room. In this scenario, the phone vibrating is just a faint sound that your Unconscious Self becomes aware of first. If the sensation is significant enough to warrant conscious attention, then that sensation swiftly graduates from being just a sub-perceptual sensation to that of a bona fide conscious perception. In turn, that means it becomes something for your Experiencing Self to experience, your Narrative Self to comment on, and if the perception is strong enough, something your Persona may have to mediate, as your emotional reaction is likely to be different depending upon who's watching, what time it is, and what the relationship is like.

From this perspective, moving upwards along the vertical axis we have:

Awareness —> Experience —> Reflection —> Bodily projection

Final Thoughts

In regards to the Multiplicity of Selves diagram, the last thing I'd really like to harp on is that these selves intermingle and intertwine in ways that couldn't possibly be articulated in just one section. For example, the Remembering Self must be in constant dialogue with the Prospective Self, or else you couldn't possibly make plans for the future.

How would you know who you are, if you didn't already know who you were?

How would you know where you're going, if you didn't already know where you've been?

It's from patients with profound impairments to the memory system that we see just how much our identities, even our imaginations, are intertwined with the memory system. Because with patients like Clive Wearing or K.C. (the patient on which the movie *Memento* was based), the future just looks like "a big blankness."[27] On account of their hippocampi being so profoundly destroyed, whenever these patients were asked about the future, they literally couldn't picture it.[28]

What this tell us is, that imagination requires the ability to bust apart certain recalled events and then to recombine them in such a way as to create a new scene.[29] And because Clive and K.C.'s ability to recall specific details were so severely impaired, their ability to imagine and envision the future was affected as well.

Chapter 16
Mental Disorders

I HEARD A kid once say, "When you're asleep, you're pretty much awake...but you don't know it," and to that, I'd pretty much agree, except I'd want to add that when you're awake, it's like you're asleep, but you don't know it either. And what I mean by that is the fact that most people walk around carrying this a priori assumption that *their experience* is the end-all-be-all of experience, that *their memory* is infallible and absolute, and perhaps worst of all, that *their view* about life is the only correct one.

This, I submit, is the collective dream we're all embedded in.

So far, we've been making the case that when it comes to consciousness, (a) there really is no such thing as an objective experience, (b) that the brain isn't some passive decoder of information but that it's actively engaged in creating our inner experience, and (c) that it not only cherry picks two-dimensional frames and just runs with them, but that the majority of what you see (95%) is actually just internally generated imagery.

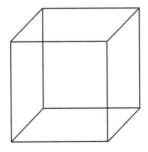

What's more, this inner movie we all have running inside our heads certainly gives us the sensation that it's some kind of authority about reality, when the sad truth is, all we've ever seen is just an interpretation. One interpretation amidst a myriad of interpretations that are possible.

Notes from the Editor:
Stare at this Necker Cube for more than just a few seconds and you'll start to understand why they call it a multi-stable stimulus. Given that the cube is drawn rather ambiguously, with no clues to the cube's true orientation, the brain doesn't just choose one orientation and stick to its guns. Rather, it wavers between the two that are possible every few seconds.

—And when you simultaneously consider that the only difference between the rich visual experience of a dream and the one you're having right now is only about 5% retinal data, it becomes increasingly clear why Carl Jung thought we were dreaming all the time.[1] Because, in many ways, we kind of are.

Not bad for someone born a year before they invented the telephone.

Admittedly, we could only speculate about the malleability of the internal experience back then. But now, armed with as many researchers and tools for introspection as we have, psychologists, and neuroscientists are finally starting to uncover just what Jung meant.

To hear a modernized interpretation consider neuroscientist Anil Seth's musing about hallucinations. "If hallucination is a kind of uncontrolled perception," says professor Seth, "then perception right here and right now is also a kind of hallucination, but a controlled hallucination in which the brain's predictions are being reined in by sensory information from the world." He then goes on to say, "In fact, we're all hallucinating all the time, including right now, it's just that when we agree about our hallucinations, we call that reality."[2]

Dysfunction and the Default Mode Network
The problem of consciousness is really a problem of choice. It's a problem of either staying present and engaged, or checking out and thinking about something else. With apps like Screen Time, we're just beginning to see how often people disengage with reality and check out their phones. But

there's actually a mental equivalent to that (thinking about the past or the future) that many would argue is hell of a lot worse.

Just for context, there's a study from Harvard, published in the journal *Science*, which found that people spend about 47% of their waking day thinking about something other than what they're doing.

The name of the article, "A Wandering Mind is an Unhappy Mind."[3]

Now, we should probably go ahead and clarify that a certain amount of reflection is normal and natural. We need autobiographical thinking and self-referential thoughts to help us navigate through our immensely complex social environment, so that's not really the problem. The problem is, most people have a screensaver of self-referential thoughts that toggles on every few seconds, and they don't know how to turn it off. Meaning, not only are they quite literally stuck in contemplation all day, but they're actually caught up in thinking about themselves which, as you can imagine, is an exhausting place to be. In fact, this is a type of cognition that enhances anxiety and depression, not suppresses it.

From a technical point of view, both of these conditions seem to share the same underlying cause (a hyperactive default mode network).[4,5] And seeing how depression is a condition which affects *hundreds of millions* of people worldwide, we can't really afford to keep sweeping this issue under the carpet any longer. Irrespective of whether or not this subject is fun to talk about, not talking about it would do us such a disservice, that we really have no other option. Because what we're about to talk about is what's at the root of most, if not all, psychpathologies.[6,7]

———

In the same way that our section about brain trauma might have taught us a little something about the inner-workings of our own minds—specifically, the modularity, adaptability, and compartmentalized nature of brains—this next section will advance some of the latest trends in neuroscience and psychiatric well-being while, at the same time, enhancing our understanding about intrinsic connectivity networks like the emotional salience network (SN), the central executive network (CEN), and our good, old friend the default mode network (DMN). We already mentioned how the brain is better thought of in terms of large-scale networks rather than being wholly modular; we advised against the idea of taking modularity to its extreme (phrenology). And now, to zoom out just a little bit further, we can say: *If* you have over-activation in one of the aforementioned networks, *then* you will likely get dysfunction. It's as simple as that.

In the last few sections, we've been mainly focusing on the default mode network, as it really is like the MVP of the brain. Although, the one thing you never want in your star player is for them to be a ball hog. Nobody wants that in sports and nobody wants that in the brain.

***Notes from the Editor:*

Hyper-connectivity in the DMN is now being studied as a potential biomarker for major depressive disorder.[8]

—Imagine if someone told you "I'm a cutter" or "I'm a self-mutilator." Pulling their sleeves down, they scream at you to leave them alone.

"That's just how I am," they say.

What would your intuition be? That some people are just luckier than others, or that there might be some underlying and preventable factors associated with a given psychopathology?

Consider the following psychiatric disorders: OCD, panic attacks, major depressive disorder, hypochondria, agoraphobia, generalized anxiety disorder, post-traumatic stress disorder.

Now, you could ask the question, "What do all these 'disorders' have in common?" And the answer seems to be: *worrying*.

Worrying is at the epicenter of all these mental illnesses, and as you're probably already aware, but worrying is another form of contemplation that enhances anxiety, not suppresses it.[9]

To quote an old Swedish proverb, "Worry often gives a small thing a big shadow." It's the thing that creates stress in our lives—practically out of thin air—and then people let all these hypotheticals, false beliefs, and *what ifs*, dominate their mental experience to such an extent that it actually generates a physical response.[10] In addition to this, people who suffer from post-traumatic stress will also start exhibiting physiological changes in their neurobiology, like a hyperactive amygdala and a hypoactive prefrontal cortex.[11] Physiological changes stemming from something they either *saw* or *did*.

The obsessive compulsive *believes* he has to wash his hands five times after touching a door, and because of this belief his hands are rubbed raw. The agoraphobic *worries* so much about what could happen if they leave the house, they never do. Someone with panic attacks *believes* so strongly that they're about to die, that they physically get pain in their chest and start to hyperventilate. Someone with recurrent major depressive disorder *believes* that their life isn't worth living with such intensity, that they end up taking their own life.

Worry. Fear. Belief. Repeat.

So how do you stop worrying?

Before we answer that, let's pry off the hood here and really examine this one with a fine-tooth-comb. Because, believe it or not, there's actually something more fundamental than worry going on, and that's our ability to direct focus. To choose where we put our time and energy. In other words, we're back to attention again.

Now are you beginning to see why I invited a monk to come stay at place? I mean, if you knew that hyperactive large-scale networks like the emotional salience network or the default mode network generally cause dysfunction, wouldn't you do everything in your power to stop it?

Even in disorders like ADD and ADHD, you could ask, "What's the underlying issue here," and again, we're back to attention.

Of course, you'll read that the real culprit behind depression is something they call rumination (replaying events over and over in your mind's eye).[12,13] But then if you really start going into what "worry" even is, you soon realize that worry is nothing other than a form of *forward-thinking* rumination. Only, it isn't past events people are replaying over and over, it's hypothetical situations that haven't even happened yet.

Situations that statistically couldn't happen.

—

The Problem with Time

Back in 2017. During one of our fireside chats. I can still remember the way in which my mentor broke down the entire "mental health crisis" inside of ninety minutes. Ava and I were trying to grapple with the dangers of rumination and cognitive rigidity, when all of a sudden, my mentor starts citing Dustin Hoffman's performance in the movie *Rain Man*.

In the film, Dustin Hoffman's character is an autistic savant, so in some respects, he's absolutely brilliant. He can count cards, memorize impossibly large volumes of baseball stats, but he has almost zero cognitive flexibility. Things have to be a certain way, or else all hell breaks loose.

"That's cognitive rigidity," he explained.

And while it might be easy to point out in exaggerated cases like ASD, it really should be a household name, seeing as we all have some degree of this. "This combined with worry and rumination," he argued, "were the three pillars of psychopathology." Championing the idea that it was our ability to time-travel (in our minds) that ended up causing the majority of people's anxiety and depression. No other animal can chop up an exhalation of breath into discrete utterances of sound, and no other animal can perform the same operation with time.

"Not only are we aware of the future," he used to say, "but we can actually suffer because of it." Anticipatory regret. Anticipatory anxiety. Existential anxiety. Instead of being fully engaged by the now (the way most animals are), we can detach and unlink, existing solely in our heads. Physically, we might be at a party, but mentally, we could be miles away, wishing we were someplace else.

Instead of being engaged with what's real, he said people would rather put themselves on the cross about things that might've been.

Counterfactuals, hypotheticals...histories that don't exist.

"People take all this onto their shoulders," he'd say loading up invisible logs onto his back, "and then they go and think about it."

Not just once either, he'd call back, because the depressed mind is a loopy mind. The anxious mind is a loopy mind. Meaning, these people will literally just sit there and replay the sad, unfortunate events of their life, over and over. Playing back sound bites of failure like they were on the DJ's Top 20.

"This is a perfect example…"

***Notes from the Editor:**
According to the World Health Organization, depression is a condition that affects some 280 million people worldwide.[14]

"…Of what *not* to do."

"I remember being at a party once," writes a formerly depressed James, "My roommate's party. Back at our place. When I finally realized just how alone I really was. I'd been denying it for weeks, but there I was standing in a room full of people, thinking to myself, 'I've never felt more alone than I do right now.' And I meant it."

Alone in a room full of people. If that's not depression, I don't know what is.

So how do we fight it?

Depends on who you ask. If you asked the Monk, he'd tell you that depression is just your body's way of telling you something is wrong.

But if you asked most any doctor, they'll tell you depression is just a chemical imbalance.

To combat depression, the Monk would advise you to start paying closer attention to your thoughts and actions—to look for trends.

But if you asked most any psychiatrist, they'll still tell you something vague like, "Depression is brought about by 'a major life stress event.'"

If you asked the Monk, he'd encourage you to start simplifying your life, and to cut something out, like a behavior pattern that's no longer serving you. Whereas, if you asked most any psychiatrist they'd still try to enroll you in some kind of neurochemical deficiency narrative. Then, instead of going after the real cause, doctors scratch signatures on little bits of paper. Little bits of paper you can trade for pills. Prozac. Zoloft. Lexapro. Celexa. Tricyclic Antidepressants. Monoamine Oxidase Inhibitors. Selective Serotonin Reuptake Inhibitors. Little tablets and capsules that don't go after the real issue, but instead, just cover-up the symptoms. Like a child who lazily tosses a sheet of newspaper over a mess instead of cleaning it up. Does it surprise you to learn that these drugs only work while you're on them? That they don't even work for a third of the people who take them?[15]

I hope this makes you feel slightly uneasy, Tom, because, in many ways, we're still operating in the supposed "dark ages" of the mental health epidemic, when the issue is rather simple. Seriously, when it's all said and done—and you really understand how pathological beliefs affect your identity—it'll blow your mind how this isn't the type of thing they taught you in secondary school. That said, let's see if we can't get to the bottom of this issue together by finally turning focus over to our area of expertise. Methods of treatment that are proving to be successful because, once again, they actually go after the root of the problem: *our beliefs*.

Notes from the Editor:

Please note that the word "belief" is used rather broadly to include not just the expectations we have about this world (this chair is hard, gravity pulls things together, the universe is vast), but also the beliefs we carry about ourselves. Technically, metaphysical beliefs and spiritual beliefs would also be included in this definition, however the focus seems to be more on the unconscious convictions we've made about ourselves.

—Beliefs like, I'm not the kind of person who 'gets' to take a break, or I'm not the kind of person who 'deserves' to feel loved. Even a belief as simple as, Bad things always happen to me, or I'm a bad person, are still pathological belief-filters that don't map onto reality.

And yet, so many people believe them to be true.

And because *they* believe them to be true, this conviction alters the way in which they interact with the world, see the world, it even affects what they remember about the world.[16]

So how do you get people to realize they've made a pathological belief that's holding them back?

By repeatedly exposing them to the truth. Which might take weeks, perhaps even years, depending on how willing they are to switch their thinking (i.e., how cognitively flexible they are). However, in almost the very same breath, we should at least acknowledge that our natural inclination isn't to be constantly tinkering with our internal representations of the world. Once a belief takes hold, a model of the world sets in place, and it's very rare that we're ever gifted the opportunity to go back in and make changes to the way we feel (even in session). Which is also why compounds like MDMA, psilocybin, and LSD are proving to be so successful.

Because when used under specific settings—and in conjunction with talk-therapy—these compounds have the effect of opening us back up. Cracking open that rigid belief box our subconscious has built a fire-wall around and allowing us to re-examine all of those past hurts with a little more openness and a little less certainty.

Notes from the Editor:
Of the Big Five personality traits, all three of these compounds have been shown to have long-lasting effects on personality by increasing the personality trait "openness."[17-19]

—Just like the ambiguously drawn Necker Cube, which has two ways of being interpreted, we have to be willing to accept that even our own life story might carry the same sorts of ambiguities. Then, when you add into the mix that we're always viewing life through a prism of interpretations, using imperfect sensory organs—*and a brain that's riddled with cognitive biases*—the question isn't if you made the wrong interpretation, the question is where, and what were the down-wind effects?

Chapter 17
MDMA, Psilocybin and LSD

HERE'S SOMETHING YOU probably didn't know. At one time, MDMA, psilocybin, and LSD were all once legal and being studied for medicinal purposes, although without going into some big, long history lesson, this is the CliffsNotes version about why they were made illegal, and why, after decades of medicinal bans, they're finally starting to be treated with a little more reverence and little less criticism.

First, we'll start with MDMA.

With MDMA, it should be known that from about the mid 1970s until the year 1985, it was being used as a psychotropic tool for psychiatrists during therapy sessions, and was enjoying great success.[1] But then word got out about MDMA's empathic effects and it started making its way onto the streets.[2] Sold under the moniker "ecstasy," sometimes this drug actually had MDMA in it, and sometimes it did not. In any case, the crucial point to focus on is, at one point, this compound was legally sold in nightclubs all around the world. The most infamous of these clubs was, of course, the Starck Club in Dallas. But because we as a society didn't understand MDMA's dehydrating effects, or how much you should take, our collective ignorance meant a few unlucky people ended up dying.[3] These deaths were subsequently highly publicized in the news, forcing the DEA to get involved, and since this was right in the middle of the War on Drugs, MDMA wound up getting placed on an emergency scheduling. In one fell swoop, it was swiftly taken out of the clubs—and rightfully so—but it was also taken right out of the therapists' hands, unrightfully so. The testimonies about MDMA's medicinal value were flat out ignored.[4-6] And so, consequently, in the year 1985, MDMA was emergency scheduled as a drug with no medicinal value and a high potential for abuse, *Schedule I.*

Flash forward to the year 2017 when the FDA goes above the DEA by granting MAPS (the Multidisciplinary Association for Psychedelic Studies) breakthrough therapy designation for MDMA as a potential treatment for PTSD. Not only that, but this breakthrough therapy designation was only granted because of all the successful MDMA-assisted psychotherapy research that's been carried out since April of 2004.[7] And, just to give you some sense about MDMA's success in this area, you should know that 12 months after the treatment, 67% of participants no longer met the criteria for post-traumatic stress.[8]

LSD (lysergic acid diethylamide):

First synthesized by a Swiss chemist working at Sandoz pharmaceuticals back in 1943, this was another one of those accidental discoveries that ended up changing the world. Distributed under the moniker "Delysid" this compound was first studied in the medical community, legally, from about 1943 to 1970, definitively making LSD the most studied psychedelic of all time.[9] The studies it was involved in ranged from the treatment of alcoholism (something Alcoholics Anonymous founder Bill Wilson actually benefited from and supported) to studies with creativity, depression, and psychosis.[10-12] Interestingly enough, Bill Wilson actually wanted to use LSD to help people in AA overcome their addictions. Then again, we should probably keep in mind that this was 1956 and LSD didn't quite have the same political baggage that it carries today.[13] Remember, these were still the pre-propaganda days.[14]

Pre-"LSD stays in your spinal cord."

Pre-"LSD causes permanent insanity."

Pre-"LSD causes couples to give birth to some kind of an octopus."

Dr. Stanislav Grof, one of the co-founders of transpersonal psychiatry, really hit the nail on the head when he described LSD as "a catalyst or amplifier of mental processes." He said, "If properly used it could become something like the microscope or telescope of psychiatry." But then word got out about its 'profound effects' and this compound started making its way onto the streets.[15]

People started using it recreationally, without truly understanding its potency, or its ability to call forth an experience that is considered by many to be one of the most significant experiences of their entire life.[16]

The godfather of LSD research, whom I am much indebted to for his personal assistance and advice, puts it like this:

> Think of the things that change a person's life: You fall in love. You get married. You have kids. A parent or sibling dies. You get a divorce. You take LSD. And for some people, they never see the world again exactly the same way. Now why can a molecule do that?
>
> - Dr. David Nichols (chemist, pharmacologist)[17]

Sadly, we never really got to find out the answer to that question, because delysid got tangled up in a political movement (the counter culture of the 1960s). And as some sort of response to that, Nixon gave us the Controlled Substances Act of 1970, placing a hard stop on every medicinal study in existence. Not just locally either. Our political prowess ended up reaching through the hands of the United Nations, effectively shutting all of these studies down, worldwide.[18]

So when 2009 came around and the Beckley Foundation began its first brain imaging studies using LSD, everyone in my field threw a parade. It was like the first glimmer of hope in the long battle of unnecessary worldwide governmental regulations.

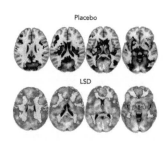

Credit: Beckley/Imperial Research Program

Believe it or not, but images like this one change the world.[19] Because photos like this are direct evidence for what we've all been saying for years: certain parts of the conscious experience are enhanced during the LSD experience; and certain parts are switched off.[20] And after studying this compound under a variety of different settings, and for a variety of different purposes, we're just now starting to tease apart: (a) what this substance is doing, and (b) how it's proving to be so effective at treating a whole barrage of different disorders like alcoholism, addiction, depression, anxiety, and PTSD.[21-24]

Psilocybin:

The use of psilocybin as a psychotropic tool for self-discovery surely predates any sort of written history, as its image has been traced back to pre-historic paintings found in the Sahara, which have been dated between 4500 and 6000 BCE.[25] In America, however, we know when the first wave of popularity hit, as it came right on the heels of mycologist R. Gordon Wasson's famous trip down to Oaxaca. It was there, that Gordon met a woman named María Sabina, and had the honor of participating in a sacred healing ceremony with the Indigenous Mazatec.

Wasson, completely blown away by what he'd experienced, ended up writing about his journey in an article called "Seeking the Magic Mushroom," which ran in the 1957 issue of *Life* magazine. The article was, of course, seen by millions. Although, before this article came out, it's worth noting that, at the time, the vast majority of people still had no idea about plant medicines. The concept that a little tiny plant could somehow cause a radical shift in consciousness was only known to the Indigenous peoples that cultivated them, and perhaps a few scholars who'd read about them.[26] So, aside from coining the name "magic mushrooms," Wasson inadvertently ended up sparking the entire world's curiosity, not to mention laying the foundations for the psychedelic movement of the 1960s.

Then, in 1958, Wasson and another mycologist named Roger Heim started up a correspondence with the discoverer of LSD, Albert Hofmann, and sent him a hundred gram sample of *psilocybe mexicana*. Hofmann diligently found and isolated the active ingredients (psilocybin and psilocin); and here's where things really start to take an unexpected turn.

Fifteen years prior to this, when Hofmann first discovered LSD, he was working with a completely different kind of fungus that grows on rye, barley, and wheat called *ergot*. Concentrations of ergot had already been used for centuries to assist midwives during birthing complications, which is precisely why Albert Hofmann was even studying it in the first place. But where Hofmann thought he might have found a compound that could help stimulate the respiratory and circulatory systems, he actually ended up discovering the most potent psychedelic in the world.[27]

How potent?

Imagine plucking out one of your eyelashes and then placing it on the tip of your finger. That eyelash weighs about 75 millionths of a gram, the same as your typical dose of LSD.

SERATONIN   LSD   PSILOCYBIN

From left to right, what we see highlighted in bold are the similarities between the molecules. On the left, we have the naturally occurring neurotransmitter serotonin, followed by a derivative of a fungus that grows on rye in the middle, followed by a similar compound that's found in over 150 species of mushrooms growing all over the world.[28]

The apparent method of action, the serotonergic system.

More specifically, a subtype called the serotonin 5-HT2A receptor, but just knowing that most psychedelics achieve their effects through the serotonergic system is good enough.[29]

———

So there you have it, on opposite sides of the world, and almost at the exact same time even, biologists and mycologists were discovering different variations on "Nature's microscope" (aka Nature's key to the unconscious). And yet, when Albert Hofmann first accidentally ingested LSD through his fingertips back in 1943, he was under the impression that he'd unearthed something quite rare.[30] That is, until about sixteen years later, when Wasson sent Hofmann another specimen. Like the *psilocybe mexicana* sample, this was another one of those plant medicines the Aztecs used as a visionary sacrament, and it's something that Indigenous tribes of Mexico still use today.

Something they call *ololiuqui* or morning glory seeds.[31]

Up until that point, Hofmann had considered the similarities between psilocybin and LSD to be quite remarkable. But when he isolated the chemical constituents inside these little seeds, what Hofmann saw he could barely believe.

The active ingredient allowing these tribes to "commune with the gods" was none other than lysergic acid amide—a precursor to LSD—something Albert Hofmann already had much experience synthesizing. Matter of fact, you could even argue that LSD is just a more refined version of lysergic acid amide, and here it was just sitting in some seeds.[32]

Produced by Nature; cultivated by people.

But then word got out about these "magical mushrooms" that could take you places and people started taking them recreationally. Instead of treating them with reverence and respect—the way they're used in a healing ceremony—they started becoming associated with certain political undesirables (aka hippies); then in some sort of failed attempt to "save America," the U.S. Government outlawed their use with the Controlled Substances Act of 1970. Flash forward until now when psilocybin is finally being studied for treating end-of-life anxiety in those with terminal cancer, treatments for cluster headaches, depression, anxiety, smoking cessation, addiction to alcohol, OCD, and post-traumatic stress disorder.[33-37]

In regards to the smoking cessation studies, Johns Hopkins University recently published a study reporting their participants achieving a 67% abstinence rate at the 12-month follow-up. To put that in perspective, the most effective smoking cessation drug we currently have (varenicline) only has a 31% success rate (12 months after treatment).[38]

So what's going on here?

Why are psychedelics suddenly being treated with respect and significance after decades of medicinal bans?

In short, it has something to do with the fact that these compounds tend to de-pattern and de-couple behavior. That would be the one-sentence answer. The more refined answer will take a bit of time to unpack. But basically, LSD and psilocybin tend to decrease communications between the regions of the brain that make up the default mode network (the alleged neural correlates of the ego).[39] And when this happens, what you experience is a complete dissolution of the Ego/Narrative Self, coupled by an experience that 80 to 90% of people describe as being one of the most profound experiences of their entire life.[40]

But what is this experience?

For the moment, I'd rather leave it vague and ominous by just calling it "a profound experience," but rest assured, whatever this experience is, it's powerful enough to challenge beliefs, break addictions, climb out of depression, stop excessive worrying, and catalyze lasting change.

Make no mistake, eventually, we're going to have to tear this experience apart so we can explain the mechanism by which these psychedelics work. Although, before we get into any of that, we have to at least acknowledge two points: (i) the elusive nature of this experience, and (ii) that whatever surface-level nomenclature people end up using to describe it, the undeniable fact is, it's difficult to articulate.

"Impossible" by most people's accounts, but then again, Ava and I aren't most people. Most people don't spend their entire lives dedicated to answering one question, whereas for Ava and me, that was our calling. While everyone else was busy hammering down the locus of action (the 5-HT2A receptor) we were busy trying to figure out just what in the hell this experience was.[41] And more importantly, how best to describe it.

"Taking LSD was a profound experience, one of the most important things in my life. LSD shows you that there's another side to the coin, and you can't remember it when it wears off, but you know it."

- Steve Jobs

As Jobs alludes, whatever this experience is…it wears off. Your access to it fades somewhat like a dream after waking. And yet, somehow, a vague memory persists that seems to restructure one's internal hierarchy of values. Ineffability truly is one of the hallmark themes of a transcendent experience, as there really is some kind of grand realization one often has—only like Jobs said, "it wears off."[42]

And here's where I ask you to consider the thought:

But what if it didn't?

What if, for argument's sake, the memory didn't wear off.

What if, hypothetically speaking, two individuals made it their entire life's mission to keep going back again and again, until they had finally figured out a way to describe the indescribable.

What then?

Well, I imagine they'd try to encapsulate whatever they'd learned into some kind of lecture or a book, which is precisely what we tried to do, except that anthology never fully materialized. And then, well, I got trapped in a room without doors or windows.

Stuck in a place where I literally have no choice left but to figure out a way of describing the indescribable.

This being the case, I should probably go ahead and tell you that almost every word, term, and story uttered thus far has been in pursuit of describing this one experience. Because if someone could describe it, just supposing someone could, then basically, they'd be describing the thing that's powerful enough to break addictions, lift people out of depression, and incite serious change. Against that backdrop, wouldn't that be something worth dying for?

Correction: *worth living for?*

Mind Manifesting

Psychedelics. The first part of the word comes from the Greek word "psyche" which means mind or soul, while the second part of the word comes from the Greek word "delos" meaning reveal or manifest. Smash those words together and you get a few compounds that do exactly what their name implies: *they reveal things.*

Like all drugs, their lock-and-key effects reveal to us ways in which our conscious experience can stretch. Stimulants like Ritalin, Adderall, and caffeine speed up the messages between the body and the brain, letting us feel the unsustainable mode of "hyper-drive."[43] While psychedelics, on the other hand, seem to be capable of exposing the unconscious beliefs that drive our actions.[44]

To further the point, just think back to almost any technological problem you've ever had. Any problem will do, since it's always the same story. First you had to dial up some tech-support hotline, then you had to wait on hold for thirty minutes. After that, they made you re-explain your problem. Then they made you re-explain your problem again. Then, after they made you re-explain your problem a third time—and you actually landed on someone who could help you—what did the person on the other end say?

"Please unplug the device, Mr. Tom."

Wait ten seconds.

"Now, plug it back in, and tell me what you see."

And more often than not, this actually works. Which is a probably a good thing in light of the fact that, most times, a simple power-cycling maneuver is all we have in life. That said, if a simple reset is all that's needed to clear out the bugs of most any technological problem, why would we expect the OS of our minds to be any different?

In all honesty, this really does get at the heart of how psychedelics work. I mean, I can explain it in more difficult terms too. But at the center of why these substances have therapeutic value, is the notion that they give you the space, distance, or otherwise cognitive reset you need to see your problems with a fresh set of eyes.

***Notes from the Editor:*

For the sake of clarity, it should be emphasized that when psychedelics are used in a therapeutic setting, doctors aren't just dolling out high-purity psychedelics and then sending people on their way. Quite the contrary, actually, considering these are highly controlled settings with pre-screened patients that are constantly being looked after. That withstanding, LSD and psilocybin are tools for disintegrating established brain networks and breaking down hardened patterns of behavior, which generally have the effect of opening patients back up to new possibilities and new ways of re-evaluating past trauma.[45]

—Honestly, one of the best pieces of advice I ever got came while I was working on a research paper with a friend. Whenever we got hung up on a paragraph, my partner didn't get flustered. He didn't panic. Instead, he just recognized that we got stuck, threw up his hands and said, "It's fine, we'll come back to it later." But while we had seemingly moved on, a part of our brain did not. Consciously, we might have been focusing on something else, but underneath our sphere of awareness, the Subconscious OS was still hard at work. Then, as it usually went, when we finally did come back, the problem wasn't even a problem anymore. Because, somehow, while we'd been away a solution had presented itself.

So what's the big idea?

The big idea is this: psychedelics can (*keyword being *can*) provide the much needed perspective-switch that time usually provides. And if you consider time in the Einsteinian sense, where space and time are conceptually linked, spending time away from something also implies that you're putting some distance from it too.

In special relativity, because space and time are inextricably linked, time is treated like a fourth dimension (a fabric that we're embedded in and move through). Therefore, when you're actually spending time away from something (taking a walk, going for a drive, getting up to stretch) what you're really doing is putting some distance between you and the problem. That way, when you finally do come back, you're able to approach the issue in a new and novel way.

Proceeding from this line of thought, one can at least imagine how psychedelics could provide the same sort of perspective-switch that can happen with a lot of time. Thus, without actually having to take a month off, it's as if you're given the same quality of mental distance that a month-long hiatus might afford.

Currently, the scientific literature is being bombarded with reports of people breaking decade-long addictive habits in just a week using psychedelics like: iboga, psilocybin, LSD, peyote, or ayahuasca.[46-49] And, seeing as these people are often able to break these hardened patterns of behavior with just one experience, it really begs the question: What specifically is it about this experience that changes people?

Of course, there were members of our team that were more concerned with the molecular structure of these compounds, but for Ava and me, that just wasn't enough. We wanted to know *how* these experiences were capable of facilitating such large transformations because, according to our line of thinking, the hero wasn't in these compounds, it was in the experience itself.

An experience which seemingly lays dormant inside all of us—that is, until the right catalyst comes along and draws it out.[50]

And while psilocybin and LSD were certainly proving to be reliable catalysts for how to get there, the fact that there was a "there" to be gotten to is really what made our heads twitch.

Or, phrased in a slightly different way, if one of these profound experiences could seemingly happen to anyone, and at anytime (with or without the aid of psychedelics) we wanted to know—why that? For us, the fact that you could take a plant medicine or a fungus derivative to get there was really rather beside the point because, in the final analysis, there was something about this "indescribable" experience that was changing these people. And really, that's the question that drove us to do what we did.

Chapter 18
Piecing it all together

ACCORDING TO THE Beckley Foundation and the *Proceedings of the National Academy of Sciences*, lysergic acid diethylamide "decreases communication between the brain regions that make up the default mode network." The way they describe the DMN, it's "Like the conductor in an orchestra policing the amount of sensory information that enters our sphere of awareness." Later on in the publication, they talk about how the default mode network "disintegrates under LSD, allowing for a magnificent increase in communication between brain networks that are normally highly segregated."[1]

In essence, once the integrity of the DMN gets compromised—and it's no longer able to perform its usual regulatory functions—the normal flow of consciousness goes out the window, allowing our brains to start reorganizing and reassembling themselves in ways that were previously not possible. Long story short, some kind of "cognitive reset" ensues, and during the chaos of the reassembly process, the you that I'm talking to now is treated to an experience so terrifyingly magnificent that it goes by many names. Which means, if you're like me, you'll call it "ego death"

...but if you're a psychologist you'll call it "ego dissolution."

If you're a Buddhist you'll call it "nirvana"

...but if you're a Japanese Buddhist you'll call it "satori."

If you're a Hindu you'll call it "moksha"

...but if you're a hippie you'll call it "being one with the universe."

Burners call it "the feeling of unity."

New Age folk call it "waking up."

Christian Theologians call it "the beatific vision."

Ava calls it the "transcendent" or "mystical-type" experience.

Whatever name strikes your fancy, just know, that at the end of the day—it's all the same experience. It's that moment when you lose your sense of self. That moment when the storyteller of your mind goes to write its final chapter, and yet somehow, after the moment when everything else seems to have ended, 'something' still breathes, 'something' still sees, and 'something' still feels.

What is that something?

"It was like when you defrag the hard drive on your computer," writes Patient 11 from the psilocybin studies at Imperial College London.

> I experienced blocks going into place, things being rearranged in my mind. I visualised, as it was all put in order, a beautiful experience with these gold blocks going into black drawers that would illuminate and I thought: My brain is being defragged! How brilliant is that!

Then at the six-month follow-up, praising the effect psilocybin had on his treatment-resistant depression, Patient 11 remarks, "My mind works differently. I ruminate much less, and my thoughts feel ordered, contextualised."[2]

Although, if we really want to do that question justice, we're going to have to dive a bit deeper. On a first pass interpretation, this 'something' is still you...it's just a different side of you. In an effort to distinguish the real deep-down you from the boisterous, blabbering Narrative Self sitting on top, we've been calling this thing "True Tom" or *the Awareness*. But our fundamental claim here is that, surface-level nomenclature aside, whatever words people use to describe this experience (nirvana, a reset, a tour of your own psyche) it's all more or less the same on the inside. Meaning, even though people may choose to use different words when trying to describe this experience, the bones of the experience turn out to be highly similar.[3] Having said that, if someone's been brought up in a religious household, they're likely to clothe this experience in religious terminology: "a God archetype told me to stop smoking," whereas, someone who's been trained in the fields of neuroscience, psychology, and evolutionary biology, might just say, "Once the neural substrates of your ego starts to disintegrate, and you lose the distinct feeling of having a body, that's when you're given the best opportunity to see whatever's at the bottom of that totem pole." As one might expect, explaining what's at the bottom of our psyche is probably going to take us some time, but on the flip side, once we nail down what's at the ground of Being, explaining what's on top will be a cinch.

Notes from the Editor:
In the original manuscript there wasn't a "cheat-sheet" attached for ease of use, but in view of the fact that there are quite a lot of data points to keep present in one's mind concerning the effects LSD has on the performance of the default mode network, it seemed like a good idea to post one here.

DMN

- The DMN is "a collection of hub centers that work together to control and repress consciousness" (Beckley Foundation).[4] "Like the conductor in an orchestra policing the amount of sensory information that enters our sphere of awareness" (Beckley Foundation).[5] Described as the neural correlates of the 'ego' by Dr. Robin Carhart-Harris (Carhart-Harris, 2010).[6]
- Involved in remembering one's past, imagining one's future, and imagining the thoughts and feelings of others (theory of mind) (Spreng, 2010).[7]
- Called the "orchestrator of the self" by its discoverer, Dr. Marcus Raichle (Raichle, 2010).[8]
- "We know that the DMN is engaged during self-reflection, so that's a very staple finding. We also know during complex mental imagery such as spatial navigation, or imagination…fantasy in one's mind's eye…you'll also see increased activity in the default mode network. Mental time-travel. So that's being outside of the moment and day-dreaming about future events or past auto-biographical events. So whenever you come out of the moment, and you day-dream in this way, you see increased activity and connectivity in the DMN" (Carhart-Harris, 2017).[9]
- Like "a capital city in a country"; "It's an incredibly important transit hub" or "integration center" (Carhart-Harris, 2017).[10]

- "The DMN has been found to be most highly active when individuals are left to think to themselves undisturbed or during tasks involving self-related processing, and less active during tasks requiring cognitive effort" (Garrison, 2015).[11]
- Those with Major Depressive Disorder show increased default mode network connectivity compared with that of healthy controls (Berman, 2011).[12]
- Increased DMN connectivity found in individuals with a high familial risk for depression (Posner, 2016).[13]
- Meditation leads to reduced default mode network activity beyond an active task (Garrison, 2015).[14]
- "DMN disintegrates under LSD, allowing for a magnificent increase in communication between brain networks that are normally highly seg-regated" (Beckley Foundation).[15]

LSD

- Decreases communication between the brain regions that make up the default mode network (Carhart-Harris, 2016).[16]
- Decreases DMN integrity (Carhart-Harris, 2016).[17]
- Decreases in DMN integrity (or DMN "disintegration") correlates with ratings of ego-dissolution (Carhart-Harris, 2016).[18]
- Increases the power and energy of brain states (Atasoy, 2017).[19]
- Increases in whole-brain functional integration (Atasoy, 2017).[20]
- Expands the repertoire of active brain states by significantly increasing the activity of high frequency brain states (Atasoy, 2017).[21]
- Increases visual cortex blood flow (Carhart-Harris, 2016).[22]
- Increases primary visual cortex functional connectivity (Carhart-Harris, 2016).[23]
- Increases openness and optimism "Ratings of the personality trait 'open-ness' (linked to imagination, aesthetic appreciation, non-conformity, and creativity) were higher two weeks after the LSD experience, but not after placebo" (Beckley Foundation).[24]
- Excites the cortex via the 5-HT2A serotonin receptors (Nichols, 2004).[25]

—Relatively speaking, that is. At any rate, it really can't be overstated how big of a problem one's identity is, or how crucial the subject of identity is to our conversation. So, perhaps another mini-recap is in order.

When you first go into the study of consciousness, we find it to be of paramount importance to begin by studying those who've either suffered from lesions or stroke. In this way, you come to see all the ways in which damage to an organ located inside your skull is capable of altering this personal baseline experience of consciousness we call CS5.

Then, after reviewing all the ways in which the conscious experience can stretch or shrink (i.e., Dr. Jill temporarily lost her ability to define herself from anything else; Clive Wearing permanently lost his ability to have a functioning Remembering Self, etc.) we come to understand a bit more about the grand symphony that's being played.

Soon after that, we were made aware about all the great psychologists and psychoanalysts who tried chopping up the self in the past; and we even posited a simplistic model that blended the majority of their ideas. First, we began by dividing up the psyche into four main selves: the conscious self, the unconscious self, the past-oriented self, and the future-oriented self, effectively giving us the cardinal directions of the mind (See *Multiplicity of Selves - Simple*).

Following this, because we desperately needed a model that was more high-res, we started chopping up the self even further. Dividing the psyche into eight selves, who all seem to have an outlook/value structure that corresponds with the type(s) of information they have access to (e.g., the Experiencing Self only really has a tiny window of memory/experience; the Remembering Self has much more knowledge and access to memories). And now, thanks to all the studies being done with altered states (flow states, psychedelics, sensory deprivation tanks, and all the rest) we're able to understand how that sense of self is really just a calculation or Virtual Machine, contingent upon which *layers* of the brain are active and functioning properly.

In layman's terms, this means we've been able to zero in on which networks are even capable of generating this sense of self—and when we do that—all the literature points toward our good old friend the default mode network. Because whenever this network gets suppressed, it's the Ego/Narrative Self that gets dissolved, not the person. And in the absence of a functioning Narrative Self, no longer is there a separate ego doing the attending, there's just the attention.

Or as the philosopher J. Krishnamurti once put it, "If you really attend there is no centre from which you are attending."[26]

Which is exactly what we see with flow states.

The inner-critic disappears leaving just the attention. Action following action. Moment following moment without any intrusion of thought.

Presently, there can be no doubt as to the importance of this network and its role in helping us to be effective agents in the world, however it should also be understood that wherever we find hyper-connectivity in this network, we also find dysfunction.[27-29] And yet, whenever we find this network being suppressed, we tend to find bliss.[30-32]

All things considered, it appears rather self-evident that VMs like the Narrative Self and the Persona aren't really necessary functions of the brain, as when the going gets tough, notions of time, the Persona, and the Narrative Self are the first to be jettisoned. What's more, when you experience enough of these flow states, what you really start to pick up on is that the Narrative Self, quite literally, *is the past*. It's constructed from the past. It lives in the past. And it's a function of the past.

On top of that, it just happens to be the case that Mother Nature produces tons of serotonin-like analogues that are capable of demonstrating the exact same thing, only in a slightly different way. And it isn't just humans who've figured this out because a fair bit of the animal kingdom seems to do it too.

Reindeer and insects consistently seek out the amanita muscaria.[33,34] Chimpanzees and birds repeatedly get drunk on naturally occurring palm wine.[35,36] Herds of elephants have been found getting drunk on fermented *anything*.[37,38]

Porcupines, mandrills, boars, and gorillas have all been known to dig the root of the iboga plant and consume it until intoxicated.[39,40]

Cats eat catnip; lemurs consume toxic millipedes; dolphins have even been found playfully passing around a pufferfish that emits a deadly nerve toxin.[41] The dolphins are careful not to kill the pufferfish, but by gently holding it in their mouths for a while, the result seems to induce a trancelike state.[42] Given that there are already several books on the matter, I won't bother entering upon any more animal examples here; however, I will just restate the general point, which is: *Organisms like altering their state of mind.*[43]

For us, compounds like LSD (which I have already suggested is really like the most refined version of Nature's microscope) is able to affect our brains by acting like a partial serotonin agonist, altering not just the speed of the messages, but the manner in which those neurons communicate. Then, if we re-consider what reality actually is, we swiftly remember how reality is just an internal construction put on by your brain. Dependent not only upon your biology, but your neurochemistry as well. Because if you ingest a substance that has some sort of effect on that neurochemistry—say a substance that's been proven to increase the power and energy of brain states, increase blood flow to the visual cortex, and increase whole-brain functional integration—then essentially, what you've done, is ingested a chemical that's made the motherboard of your body go into hyper-drive.

"Sort of like overclocking a computer," is the way Dr. David Nichols describes LSD's effects on cortical cells.[44] And if we now import that knowledge back into our Photoshop Layers analogy, we may start to gain a deeper understanding about what reality actually is, and how ingesting a certain chemical can affect the performance of that wetware.

Chapter 19
Experiences.

BY OUR ESTIMATION, the transcendent experience has at least four essential components by which it carves out the possibility for immense change, and after exposing the unified self to these four mental rites of passage, the positive health effects tend to be both far-reaching and long-lasting. Aside from it being subjectively rated as one of the five most important experiences of a person's life, it may also be of note that only the most profound healing occurs *after* a patient undergoes one of these transcendent experiences. That actually the level of entropy the brain gets into is directly correlated with the amount of "openness" and changes in personality one has.[1]

Whether it's for treating addictions, anxiety, or overcoming depression, the literature is quite suggestive that the transcendent or "mystical-type" experience is the key component for the maximization of healing. Not to discount the effectiveness of talk-therapy, or the molecules themselves, but when you combine the two—*and the subject has one of these impossible-to-describe experiences*—that's when you know you've hit upon the winning strategy for change.[2-7]

The first time we learned this, the majority of our team was a little taken aback.

Being pharmacologically-minded, most of our team was rather surprised to learn that there might be something more than chemistry going on here; and that actually, the level of healing these patients received seemed to be underpinned by an experience of "transcendence."

But what does that even mean to someone who's a skeptical agnostic?

In a nutshell, it's this: Irrespective of the words people use to describe this experience (awakening, enlightenment, a spiritual experience, a mystical experience, a religious experience, nirvana, satori, moksha, ego death, ego dissolution, the feeling of unity, the beatific vision, being one with Nature, a visionary state or 'peak experience') what happens is, the unified self is exposed to four mental rites of passage we like to call: Dissolvability, Elasticity, Malleability, and Possibility, or DEMP for shorts.

In a slightly elongated form it's: (i) the Dissolvability of Self, (ii) the Elasticity of Attention, (iii) the Malleability of the Internal experience, and lastly, (iv) the Possibility for Change. As fate would have it, the first aspect of DEMP is likely going to be the most difficult to explain, by virtue of the fact that it deals with the nature of one's identity, however, it also seems to be the most important, seeing as it's our identity which informs the way we pay attention in the first place.

Why is the Dissolvability of Self so important?

Mainly because it's only after the current regime of established brain networks are dissolved that new pathways begin to open up.

With death, comes rebirth, and with ego death, comes ego rebirth.

In a previous section, we already touched upon the idea that our psyche might be best conceptualized in terms of multiple competing sub-personalities, and now, to take that point a bit further, sometimes, what we really need most, is for one of our selves to just shut off so we can actually think for a moment.

Dissolvability of Self

Back when I first started spending time with the Monk and he began training me on how to be with my own mind, he said something to me that really solidified the essence behind what we were doing. He and I were in the middle of a session. One of our earlier ones. When some kind of massive truck heavy enough to rattle the walls of my apartment drives by. Followed by another…and another…and another. So many eighteen-wheelers or dump trucks drive past that I have to open my eyes, only to find my teacher's eyes staring back at me, just as confused.

Looking to his ear, the Monk says, "You can always use the sound of the present moment to check you back into reality."

Half smiling, he adds, "To ground you."

Now normally, the Monk would've instructed me to keep meditating amidst all the noise (to simultaneously hear all the sounds while resisting the urge to name them) but, seeing as, this was an entire parade of cement mixers, the Monk and I decided to take a much needed break. This particular day, I remember not being able to focus all that well as my health was seriously on the decline and nobody seemed to know why. What used to be only a handful of food allergies had suddenly multiplied into so many different food sensitivities that it felt as if my body was rejecting almost everything I ate.

Not knowing what else to do, I finally confided in the Monk.

I told him about how whenever I got home from work or school, I was usually so wiped that I'd eat several meals back to back.

I'm in the middle of explaining all this when the Monk start picking lint off his clothes; my cue he's heard all he needs to hear. A couple seconds after that, he asks me if I know what the phrase *Mizu no Kokoro* means, and I have to shake my head: *No*.

"It means 'mind like still water,'" he says, moving his mouth to one side. But since I still don't get it, the Monk decides to put the matter in simpler terms.

Grabbing his beard, he asks if I really want to know why I'm stress-eating, almost like he's testing me. Instinctively, I just say yes, but when he makes his eyes go small and asks me for a second time, I feel a part of my lip slide into my mouth while my eyes paint the floor. In my head, there's some kind of battle being waged between the part of myself that genuinely wants to know, and the part me that wants to keep on pretending everything's fine, but eventually, the stronger part of my psyche wins out.

Out loud, I try and mouth the words, "I want to know," only it comes out so soft, the Monk has to read my lips.

Combing his beard with his hands again, he asks me to picture a lake smooth as glass. Then he says that I should imagine my mind to be like that lake.

"When a lake is calm—with no ripples at all—it's perfectly capable of reflecting back everything around it, isn't it?" he asks.

Then, touching his temple with his forefinger, he goes, "*Mizu no Kokoro* means to make your mind like the lake." In order for you to see reality, as it truly is, he says, you have to be willing to wait long enough for the ripples to stop. It must be calm. Leaning in to touch my forehead, he adds, "*Your* mind must be calm."

The way he looks at me when he says this, it's like his eyes are scanning back and forth between mine to make sure this part really sinks in. Because this idea (that the mind can get caught up in certain patterns of repetitive thought) forms the bedrock of how mediation even works in the first place. You stop the top-down flow of ideas long enough to let the Subconscious OS slip a few data packets through, and by doing so, you end up developing an open and honest relationship with the unconscious.

"Healing," the Monk explains, "is all about having revelations,"

"It's about truth," he says, sitting back really tall. "Without revelations…without truth, there can be no healing."

How are you supposed to have any revelations if you're buzzing around all the time? Right now, you're buzzing around distracting yourself with so much work and so much school that you're pretending not to know. You're asking me why you have this compulsive eating behavior, when really, there is only one person who can answer that question.

Identity

In regards to the Multiplicity of Selves diagram we showed you earlier, there are really two points worth harping on again. The first is that these selves are immensely interconnected. The second is that they all lean on each other in ways that are by no means self-evident. For the avoidance of confusion, the diagram was drawn with thick black lines delineating the division of labor inherent to each one of these selves. Although, same as any real map, those thick black lines are more permeable than they seem. This being the case, perhaps the best way to describe how these selves interact is by way of analogy.

Next time you're watching a gang of friends attempting to take a group selfie, pay close attention to the person who takes an extra second to turn to their "good side." Watch how they use lighting, angles, and other people to masterfully play up the features they like, and then ask yourself, "Am I doing something similar?"

I mean, if the ego is really just a low-resolution snapshot our brain chooses to experience of itself—and we have the power of maintaining that snapshot—then, why wouldn't this experience of ourselves be subject to some kind of pathology? In other words, why wouldn't the Ego/Narrative Self tend to think more highly of itself in certain groups of people (think narcissistic personality types), while completely failing to perceive any of its own greatness in others?

Just flip on any reality TV show and you'll see the two extremes I'm talking about. Two channels up and it's some guy with the kind of undeserved confidence you wish you could just steal. Two channels down, and it's some girl with body issues so bad, she's about to go under the knife.

If either of these extremes teaches us anything it's that: (a) the brain is perfectly capable of misrepresenting itself, and (b) these pathologies of self-image can easily drift in either direction: positive or negative.

———

In cosmology, you'll always hear astronomers and astrophysicists use words like the 'observable universe' or the 'observable zone' for the plain and simple reason that not every part of our universe is capable of being seen.

True, we might live in an ever-expanding universe, but the speed of light is still fixed. And because of this one fact, that means there will always be parts of the universe which are hidden from us, simply because, there hasn't been enough time for light to reach us yet. Comparable to the way a horizon resets at sea, where you can only see a few miles out in any direction, there is a horizon to our conscious experience, such that, if something exists beyond that horizon, it isn't conscious to us. Which isn't to say that it'll never be conscious to us, but at the given moment, it's currently sitting outside the observable zone of our mental universe.

"We live on an island surrounded by a sea of ignorance. As our
island of knowledge grows, so does the shore of our ignorance."

- John Archibald Wheeler
(physicist)[8]

Blackouts and Light Pollution

If you've ever experienced a blackout while living in a city, then you
probably already know the experience I'm referring to. At first, you're
extremely annoyed you're reduced to candles and batteries, when it
suddenly occurs to you that you have a rooftop and a blanket. So you
go up the stairs, you push through a door marked "Do Not Open," and
there on the roof, you catch site of a thousand stars you didn't even know
were there. Thanks to the city-wide blackout, the entire night's sky has
changed. And as long as the blackout persists, you get to see your skyline
in a new and novel way.

That's the fundamental idea here.

That there's just way more to this life...*and yourself*...than your
Ego/Narrative Self ever 'allows' you to see. Leading us to ask the following question: If the neural substrates of the ego is always hogging
up all the resources of the brain—which it is—and it's always drowning
out, suppressing, or inhibiting all the thoughts it doesn't like—which it
does—then how are we ever supposed to perceive reality as it is?

"[Psychedelics] increase the permeability between the
unconscious and the conscious mind. They allow us to have access
to our unconscious, in a way that we don't normally."

- Mark Haden
(executive director of MAPS Canada)[9]

Having said that, if we are able to get our Ego/Narrative Self out of
the way for just a moment, effectively eliminating all the unnecessary
"ego pollution" it provides, then perhaps, there's a grander view of ourselves we can connect to.

Why is this important?

Because if meditation and flow states are capable of causing miniature blackouts in the continuity of self-hood, then psychedelics, by comparison, would be capable of causing full-blown city-wide blackouts. Outages so intense, one might never see themselves the same way ever again.

Self-Image & the Belief Box

Under your true psychedelic state, when the control centers of the brain are no longer performing their normal, regulatory, orchestrative functions, different access levels of knowledge are not only possible, they're probable. That would be the broad view about what happens when the ego gets dissolved. And for the small view, once again, we must return to the concept of internal models because, as it turns out, the brain doesn't just model the outside world—it models everything.

From internal maps of the body to the image we carry about our-selves, from what people say to our face to the things we've heard them say behind our backs, all of this combines to form the self-image you currently possess. And this wouldn't be a problem if there existed some kind of litmus test for the accuracy of our narratives, but as it stands, no such test exists. Even worse, not only are you the data collector, but you're also the analyst, the judge, and the subject. Saying there's a conflict of interest would be putting the matter too lightly, so instead, I'll just ask you to consider the following.

Consider that your entire life people have been trying to tell you who you are, and who you're not, when the reality is, they probably don't know who you are any better than you do. Sure they might be able to spot certain patterns of behavior—of which you're willfully blind—but they're actually in no better place to judge 'what you're capable of' than you are.

Truth is, you have no idea what you're capable of.

And yet, somehow, we as a society, seem to let other people tell us what we can and can't do all the time. Depending upon the person and what they mean(t) to you, you may have made a false impression about what you're capable of; and whether you realize it or not, you may have been carrying around that biased, pessimistic interpretation and feeding it ever since. For instance, you could've been bullied or teased so badly in middle school that you unconsciously internalized some kind of negative body image. Hypothetically, one person's off-the-cuff comment said at just the right moment in your developing brain might've been enough for you to swear off wearing shorts in public ever again, effectively seeding the body dysmorphia that still haunts you today.

Of course these are all just hypotheticals, but the point is, these impressions we glean off one another aren't law—they're just impressions. If someone looks up at a cloud and tells you they see a one-eyed chimp wearing a monocle, you might close one eye and try to see it too, but you wouldn't start basing your entire life's future around this one person's comment, would you?

And yet, when it comes to ourselves, people do this sort of thing every day.

By any measure, it isn't just what *we* think about ourselves that ends up making it inside the belief box, because we also end up taking on other people's perceptions as well. As if they weren't hallucinating their reality just as much we are. Apart from that, if we end up feeding these impressions, then they graduate from being "just impressions" to something approximating full-blown beliefs. Beliefs like: I'm not the kind of person who 'gets' to wear shorts.

These beliefs start to form the basis for our self-image, our identity, and in turn, the future we create. To make matters even worse, these impressions might've never even been true! Interestingly enough, that part doesn't actually matter. What matters is, if you believe them to be true. If you believe yourself to have gross knees and stumpy legs because some asshole said it to you in the parking lot twenty-something years ago, then you could potentially go on living with that impression until some sort of confrontation with the truth literally forces an update.

———

You know, a few sections back, I shared with you about how close-minded I used to be regarding meditation, and I only bring this now up to shed some light on how easy it is for us to form a hardened, stubborn belief based on zero experiential knowledge and only a handful of bad impressions.

What's more, I never really got to mention how this belief kept me closed off to anything remotely sounding "new age" for years.

Seeing the Monk silence his default mode network in an instant—that was my personal confrontation with the truth. Before that day, I believed meditation was a complete waste of time. But after that day, I finally understood how "zenning out" could be such a powerful tool in sharpening one's attention and dissipating excess mental energy.

In essence, my hardened beliefs about meditation had finally started to soften up.

Bearing all this mind, it should be double underscored that we have beliefs about ourselves and the world we live in; and that these beliefs are underpinned by certain networks of neurons that like to fire together. In neuroscience, we have a saying, "Neurons that wire together, fire together," and aside from the general benefit of having a technological reset, the dissolvability of established brain networks has the effect of re-opening discussions about how we negotiate with reality. Essentially, allowing us to "re-wire" the image we have about ourselves.

How does this end up helping in something like addiction?

To answer that, it first has to be acknowledged that our identities can often get so wound up in certain self-destructive patterns of behavior that we often begin to conflate the patterns with our core-self. Smoking a cigarette after meals, drinking a bottle of wine after work, Adderall with breakfast—whatever your poison is—if we make the patterns too habitual, we start to identify with the behavior itself. Believing that because we've had a huge dessert every night since before we can remember, that somehow eating sweets is a part of our personality. Instead of it being something we casually enjoy, consuming them becomes something we *have to do*, the ritual itself being almost as addictive as the substance.

Unfortunately, addiction is a problem I'm all too familiar with, and it's been my personal observation that if you talk to anyone who's addicted to something whether that's gambling, shoplifting, or a prescription drug, and then ask them to give it up, you can literally watch the wheel of excuses start to turn.

"I can't live without _____."

"I have to have _____ or I can't _____."

Or my favorite addict line, "I can stop whenever I want to," followed by a brief pause, "I just don't want to yet."

What you want to say is, "Who's the 'I' in this scenario?" Or better yet, "Which one of your selves has gotten addicted?"

If it's something like nicotine or prescription pills, it may very well be that your Bodily Self has gotten physically dependent. On the other hand, when the addiction involves something like social media or porn, it's hard to say where the addiction lies.

In any case, the underlying issue beneath all this addiction nonsense seems to be that the subject has started to associate 'a thing they do' with their core-self. And after living with few of these nasty compulsive behaviors for long enough, I can report, rather unambiguously, that it can be very hard to imagine a life without _____.

I mean who would I be if I didn't _____.

In other words, we become the behavior. The alcoholic. The klepto. The sex addict.

Instead of seeing our behaviors as something we could potentially drop, we give away our power and subsume our identity as an addict.

—

At groups, in the basements of churches and community centers on Wednesday nights, you'll find no shortage of people swapping stories about how an overdose just saved their life. The irony of them needing a literal shot of adrenaline to "wake them up" being lost on no one. Hell, any good AA sponsor knows a good trip to the hospital, or a nice run-in with the cops, might be the slap in the face someone needs to initiate serious change. Granted, no one ever wants to wish these kinds of traumatic, near-death experiences on someone they love, but against the alternative, sometimes, a good brush with our mortality is enough to remind us there's something still worth fighting for. And this is just one of the methods by which LSD achieves its healing—an "ego death" experience.[10,11]

With larger doses of 200 micrograms and up (abbreviated to 200 μg+) delysid has the effect of dissolving your sense of self piece by piece. Beginning with the top of the totem pole first (the Persona) and then working its way down, delysid has the effect of unzipping your sense of self all the way down to its most basal operating layer (the Awareness), and in the process "the deep-down you" finally gets a chance to perceive itself without all the filters.

Who are you when the Narrative Self's rose-tinted glasses are smashed into bits? Who are you when the network of neurons that tells you, "You're an American," melts away? Or how about the one that says, "You're from Ohio."

To illustrate what I mean, let's momentarily re-examine the most defining moment of your life's existence. This could be the worst thing you ever said, or it could be the greatest accomplishment you've ever been able to muster—whichever comes to mind first—I want you to tell me who you'd be if that memory no longer defined you.

Not who you'd be if that memory was deleted; who you'd be if that memory was properly processed and filed away.

If that memory was no longer sitting on the desktop of your conscious experience and causing you to feel so unworthy of love, who would you be? Who could you be if you were no longer allowing yourself to be controlled by events of the past?

I think the only reasonable answer to that question is: *We don't know.* However, we can say that there will come a time when whatever you're holding onto (limiting beliefs, unprocessed emotions) will all be taken away, because in those final moments, your entire sense of self disappears. Your memories, your fears, your ability to define…all this melts away in a process my friend calls "the shedding of the selves." The reason Ava and I prefer to use the term *transcendent experience* most often, is because there is a piece of ourselves that seems to "rise above" the death of our egos; and it's only after we've shed these outer layers of our psyche, that we allow ourselves to be treated to some of the most profound revelations.

The depressed person sheds their pessimism bias.

The anxious person confronts their deep-seated issues with control.

The addicted person learns something they knew all along but chose to forget. Namely, that who they *really* are is much bigger and much grander than some parasitic behavior pattern that's grafted itself onto their psyche.

Spectrum of Alignment

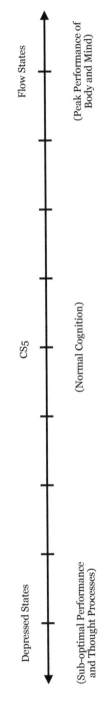

Depressed States

CS5

Flow States

(Sub-optimal Performance and Thought Processes)

(Normal Cognition)

(Peak Performance of Body and Mind)

Depressed States:

- Difficulty thinking or making decisions
- Characterized by feelings of hopelessness
- Leading cause of disability
- Tends to occur when we're scattered, unfocused, or overly focused on ourselves
- Subjects tend to have a hyperactive DMN (preponderancy to disengage with reality and the present moment)
- Not a desirable state of consciousness as subjects often report suicidal ideations, an extreme pessimism bias, aches and pains, lack of enjoyment for things once loved, etc.
- "Everything feels wrong…nothing feels right"

Flow States:

- Total concentration; choices feel effortless
- Characterized by feelings of unstoppability
- Leading cause of olympic medals
- Tends to occur when we're physically challenged and creatively stimulated
- Subjects in flow report such increased engagement with reality that their brain is no longer able to generate the inner critic
- A very desirable state of consciousness underpinned by the release of dopamine, serotonin, norepinephrine, anandamide, and endorphins (and sometimes oxytocin)
- "Everything feels right…nothing feels wrong"

Chapter 20
DEMP Cont'd

THE SECOND ASPECT of this experience that leads to the process of healing is what is known as the "Elasticity of Attention." And it's very much on par with the kind of lesson one might learn from a good flow state. Because when you're performing at your absolute peak, and you're in a flow state, you're shown just how much attention and brain power you actually have. By collapsing all of your selves into this one moment in time (and not having their attention scattered about), your execution of the task feels effortless. Temporal distortions and the loss of the Narrative Self are interesting side effects—no one's doubting that—but we would actually argue that it's this peak level of performance that hooks people.

Pleasant accompanying neurotransmitters aside, once you're exposed to the upper-limits of what's possible, you have something to strive for going forward—getting back there. Watching televised sports or the Olympics, we all secretly hope to catch one of these "in the zone" moments; and all over the globe from athletes to Navy SEALs, from billionaire executives to bankers on Wall Street, everyone is seemingly trying to hack consciousness so they can drop into these states more often.[1]

See *Microdosing*

See *Sensory Deprivation Tanks*

See *Stealing Fire by Steven Kotler & Jamie Wheal*

The only difference is, with the transcendent experience, the lesson is much more pronounced. Because once your brain stops wasting energy on thoughts that don't matter and futures that don't exist, the sheer amount of attention you're exposed to is terrifying. Neuroscientist Arnie Dietrich once captured the moment perfectly by describing it as a "mental singularity." Because once you've fully dropped into it, your sense of time is so profoundly altered, that you simply cannot extract yourself from the present moment.[2] At the very same time, increased burst firing of an area Dr. David Nichols like to call "the novelty detector" (the locus coeruleus) reintroduces you to that child-like wonder you used to have about the world, letting you actually process more sensory information than usual.[3-5]

In layman's terms, by altering your sense of the way time flows, you rediscover a new appreciation for the present moment.

Malleability

The third aspect of DEMP is a rite of passage we've termed the "Malleability of the Internal Experience," as it's only after you've dramatically altered your internal state of mind that you begin to understand just how much "reality" is internally generated to begin with.

As previously mentioned, so much of Consciousness 5.0 is just an interpretation anyway, what with 95% internally generated imagery, so by exaggerating the internal experience with a little pharmacology, we're shown just how stretchy the internal experience actually is.

Just take the Necker Cube as an example.

It only takes staring at this cube for about twenty seconds for you to realize that vision is an active process taking place inside your head. Then, if we take that concept and start extrapolating upwards, we're confronted with the idea that a multi-stable stimulus can be drawn with as few as twelve strokes of a pen.

So, the question is, if twelve strokes of a pen is all it takes to create a three-dimensional, multi-stable stimulus on a two-dimensional sheet of paper, what are the odds that your life story with all its millions and millions of variables has a different interpretation?

That somewhere along the lines—not everywhere, but at least somewhere—you made the wrong interpretation?

Don't get me wrong, CS5 certainly gives off the appearance that this "movie in your mind" is happening this one way *and only this one way.* But we also know from all our studies concerning split-brain patients and those with damage to one hemisphere and not the other, that the two hemispheres of the brain perceive the world in very different ways. Not only that, they're both capable of sustaining consciousness on their own, dealing with reality on their own, and if the corpus callosum is severed, having an independent consciousness on their own.

As one might expect, we don't perceive these tiny differences as happening, or even existing, at the level of experience. Although, Dr. Iain McGilchrist (the world's leading authority when it comes to bi-hemispheric specialization) has really done an outstanding job of laying out these differences in his book *The Master and his Emissary*. The central theme of which is: Why an organ, which is seemingly designed to make connections, would become so asymmetrical and so divided in the first place?

McGilchrist even lectures about how, "One of the main, if not the main functions of the corpus callosum is, in fact, to inhibit. To inhibit the other hemisphere." Which begs the question, just how and why the brain does this?

His answer, that inhibition "enables the two to do distinct things." "Of course, they have to work together," says the world's leading authority

> ...but usually good team work doesn't mean everyone trying to do the same role. So differentiation is very important for two elements to work together, and inhibition is one way of doing that. So effectively, the two takes on the world, if you like, that the hemispheres have are not easily compatible. And we're not aware of that because at a level below consciousness there's a meta-control center that is bringing them together. So in ordinary experience we don't feel we're in two different worlds, but effectively we are. And they have different qualities...and different goals. Different values. Different *takes* on what is important in the world.[6]

In sum, at the level of Tom's experience, things are always perceived as happening this one way. However, during the transcendent experience, it's as if we get to see what's below conscious perception. Sometimes even below the meta-control center McGilchrist was speaking of. Accordingly, when your brain finally does enter into this hyperconnected state, where more of the brain is talking to itself than normal, you begin to perceive more from each of these 'different worlds.' Leading you to the inevitable conclusion about how malleable the internal experience really is (i.e., how powerful your thoughts/beliefs actually are, and the degree to which these thoughts can distort reality).

> Your beliefs become your thoughts,
> your thoughts become your words,
> your words become your actions,
> your actions become your habits,
> your habits become your values,
> your values become your destiny.
>
> - Frank Outlaw (entrepreneur)[7]

Fundamentally, if you can change your thoughts, you can change everything.

Possibility

The fourth and final aspect of DEMP is what we may term Possibility or "the Possibility Space" because, in many ways, this rite of passage feels like a pilgrimage to a physical place. In actuality, subjects never leave the room, but anecdotally, it feels like there's this metaphysical place you've somehow entered into, hence the name. For some, it manifests as a kind of dream-space/mega-computer troubleshooting with itself.

A kind of descent into the source code, if you will.

While those without a background in coding, may choose to describe this place as being "outside time," or in the presence of a God archetype. Meaning, if you're a Christian you see God, but if you're a Jew you see old testament God. Hindus see Brahman; Buddhists see the Ultimate Ground of Being...I'm sure you remember the spiel. My point is, maybe those who don't subscribe to any particular religion just perceive their mind's OS getting debugged. Whatever the case may be, this is a problem with words, not data

Think about it like this, if you were to ask ten people in a row what love is, I'm almost certain you'd get ten different responses. And while the answers may differ in their surface-level descriptions, I'd bet there'd still be enough recurring motifs for you to extract out a few general principles about what love actually is:

> *Love isn't about possession or control; love is about giving the gift of self and not expecting anything in return. It's about connection and acceptance, not judgment and division.*

Perhaps those might be some of the common themes amongst the ten responses. And in a similar fashion, when it comes to the transcendent experience, people only tend to describe it terms of what they already know, and really, what's so wrong with that?

Consciousness is fundamental to us, therefore we can only describe NOSC in terms of the experiences we've already gathered. Personally, I'm not a big fan of inserting the word "god" into every knowledge gap science has yet to fill (See *God-of-the-gaps fallacy*) which is why Ava and I tend to use religiously neutral terms like *the Possibility Space* and the Subconscious OS.

***Notes from the Editor:**
"Ratings of the personality trait 'openness' (linked to imagination, aesthetic appreciation, non-conformity, and creativity) were higher 2 weeks after the LSD experience, and those subjects showing the greatest brain entropy on LSD showed the greatest increase in openness 2 weeks later." A direct quote from the Beckley Foundation's study on LSD.[8]

—You know, back when we first started dating, Ava told me she figured out pretty early on in life that she couldn't control what happened to her, but she could control how she felt about it, and this was hands down one of her most admirable traits. Whenever something bad happened to her or to me, she always managed to spin it in a positive light. An ability that never ceased to amaze me.

Where I would see a door slamming shut and take it personally, she'd see a new possibility and get excited.

Her ability to reframe a problem was beyond incredible. Life-saving, actually. But sadly, it took me going through several crippling bouts of depression to realize, I, too, could be a little more cognitively flexible. And after getting to visit this Possibility Space dozens and dozens of times, I can confidently report that once the floor beneath your consciousness drops out—and there's no longer any pieces of your identity left to cling to—the effect it has is something like ultimate possibility.

What does it feel like?

Imagine raw intellect unmarred by memories of failure. Imagine the speed of normal cognition, but then imagine removing all the regulators that constrain those thought processes.

What actually happens in the brain?

On a pharmacological level, it's this: LSD starts expanding the active repertoire of brain states that are possible, and it's this rapid influx of brain states (combined with increases in brain-derived neurotrophic factor) which helps give rise to neurogenesis, synaptogenesis, and cognitive flexibility (our ability to adapt and conform to new situations).[9-11] By depatterning or decoupling the brain for just a moment, your brain is allowed to reorganize and reassemble itself in ways that were previously not possible.

"Probably not since infancy," remarks Dr. David Nutt, "has the brain been *able* to make as many connections as it does under psychedelics." So instead of narrowing your focus, it opens you up. Instead of reinforcing the constraints that have been "driving you down a path to think in a very rigid, normal way," psychedelics break them down.[12]

By temporarily removing all those pathological belief-filters you have threaded onto the body of your camera, the psyche is exposed to new possibilities and new ways of thinking. Although, from a subjective perspective, the way it feels is a bit more chaotic.[13] Almost as if you've somehow descended into the birthplace of thought and are now getting to watch the brain troubleshoot with itself. Just as the executive director of MAPS Canada said, psychedelics "increase the permeability between the unconscious and the conscious mind," so when you're down there, and you're exposed to what the Subconscious OS is capable of, the effect it generally has on people is humbling.

Because the complexity you're finally exposed to is beyond comprehension. It's beyond words. And once you see finally see it, you become palpably aware of how CS5 is only possible because of it.

Conclusions about the Four Rites of Passage

In just a few words, I'd say that the therapeutic effects behind the transcendent experience seem to arise as a result of having our certainties dissolved. What we think we know often hinders us the most because instead of keeping us open to possibility, "I know" keeps us closed down to a certainty (e.g. "I already tried that," or "That'll never work.").[14]

Essentially, our mind can get flooded with certain pathological belief-filters like: *I'm not worthy of love*, or *I always need* _____ *to function*. And inside this experience, these hardened stubborn beliefs are either: (a) reopened for interpretation, or (b) finally allowed to reach their inevitable conclusion. Which is why coupling this experience with talk-therapy is so important. Set and setting (your mindset and your environment) are crucial to the success of these compounds—as is dosage—and before we continue, we really must draw our attention to this point once again.

Psychedelics are by no means a panacea. They're not some magical cure-all that could fix all the world's problems if everybody just took a little and held hands. However, if they're used in a very particular way where set, setting, dose, molecule, patient, and therapist are all accounted for, then they can be extremely powerful tools for introspection and cognitive development.

—

So there you have it, the dissolvability of established brain networks gives rise to the elasticity of attention, which shows you the malleability of the internal experience, which opens you up to the possibility for change: D-E-M-P. Calling it the mystical experience sounds too woo, calling it enlightenment sounds too elitist, calling it a spiritual experience sounds too narrow. The only name that really seems to fit the bill is the transcendent experience because, when we're finally in it, we don't cling to the past as much as we transcend the death of our ego and allow ourselves to become something more.

And now that we've covered the scientific basis for why this experience is so important, I'd like to offer you a first-person rationalistic account of what this experience actually looks like.

Taken from the memory banks of my existence, this isn't the kind of story that's going to make me look good. It's raw, it's unfiltered, and most important of all, it happened a very long time ago.

The pre-Ava days.

The "before times" as she would say.

If my intention was to sell you on psychedelics, I certainly wouldn't tell you *this* story, but, seeing as my motive is to describe "the shape of this user-illusion" while, at the same time, unifying all of the concepts we've learnt thus far, I can think of no better story to tell than my first true experience with the four rites of passage. Something my attorney would later coach me into describing as happening "well-outside the statute of limitations" for both sale and possession.

Not many people know the moment when their entire life switched directions, but for me, I've got it pinned down to the minute, second, and hour. So buckle up and lock yourselves in, because where we're about to head to next, is a place I never even thought was possible: the ground of Being itself.

Chapter 21
The 'Chair' Incident

THE NIGHT IN question, the first time we experienced nirvana, moksha, or whatever you wanna call it, happened on an evening we shall now call "Koreatown night." That night, this girl named Cali and I decided we were going to experiment with delysid at a music venue in uptown Oakland. We knew what we were doing, we knew we could handle the stimuli, the only thing we miscalculated was the dosage.

How bad?

Let's flash to about four hours later when Cali decides to throw a chair through my second story window. Hearing glass shattering and falling onto the concrete, my roommate Brian bursts out of his room, sees the broken window, sees we're missing a dining room chair, and then says quote, "What the fuck are you guys doing?"

Responding in her own way Cali grabs the glasses off his face, throws them against the wall and says, "Just accept it!"

Cut to me pacing back and forth saying, "I know. I know. I know. I know…" Over and over, like a broken record.

How the hell did we get here?

What events could've transpired for two levelheaded adults to have lost themselves so deeply? To answer that, we must start at the beginning.

———

Regulations at the ███████████ Institute are very clear: *No researcher is allowed to self-administer or participate in the studies in any way*. But, off the record, researchers only tend to get into this field after they've experienced some of the healing aspects firsthand. Which, by the way, is what you really want in a researcher or psychotherapist in the first place. What you don't want is someone trying to talk you through an experience they've never been through themselves, and well, let's just say there's only so many trials you can read about before you start to get curious. Once you hear a few stories about subjects losing their sense of self, you start to wonder—*What's it like?*

What's it like when the molecule completely dissolves your DMN?

What's it like when the conductor of your mind stops regulating the amount of sensory information you take in? We knew it was prone to happen with dosages of 200μg and up, but being that it never really happened to us before, Cali and I never seriously considered it was going to happen to us. Truth be told, this wasn't our first rodeo. We'd already had plenty of interesting psychedelic experiences before (outside of therapy) just not so much with delysid. On some level, I suppose we knew what was possible. We knew things would get progressively more confusing as you approached 200 mics, we just didn't know how confusing.

Not having any stock of the compound ourselves, meant we had to source it from somewhere else. Which wasn't hard. If anyone ever asks what I study, or where I go to school, the conversation is bound to come up. And when it does, half the time people are trying to give me some.

Although, looking back, I was unusually reckless that day.

Buying tabs from a friend of a friend, not testing them to make sure they weren't some potentially lethal research chemical like 25I-NBOMe (street name "N-Bomb"). But in my defense, the guy came on good authority. The friend I trust even told me about how this guy sometimes has his sheets lab tested for purity.

So, with that in mind, I drive to the part of the town where people lock their doors at red lights, and tell the guy, "Give me ten."

Living in Koreatown, I probably could've walked there, but this was California, and it was cold. So I pull into some dimly lit parking lot, where a guy I don't know motions for me to get in a car I don't own.

Notes from the Editor:

Typical dosages of LSD range from about 75μg to 200μg, with the average tab or drop containing around 75-100 micrograms. Additionally, while there are numerous reports of people accidentally insufflating dosages that exceed 10,000μg (100 doses), it should be made clear that there are only 2 documented deaths directly attributable to LSD toxicity. In both cases, the decedents had apparently ingested well over 100 milligrams of LSD, not micrograms. Therefore, if the average tab contains 100μg of LSD, one would have to consume well over 1,000 tabs to achieve LSD toxicity.[1]

—Sitting in that black Ford Fiesta, I never really got the sense Derek was bullshitting me too much about the strength. I mean, you gotta account for the usual dealer-hype when you're buying this sorta thing. And the degradation factor. But the tabs still looked pretty fresh. They were vacuum sealed, had a tinge of grey on the back, and on the front, a pretty pink lotus. Holding the ten-strip underneath the dome-light of his car, I remember thinking, they look so harmless. Kind of cute even. But these dealers you have to meet, they always have some big safety speech they want to run you through. And this one, well, let's just say he's about halfway through his second cigarette when he finally says, "Oh, and don't drink any water while these are in your mouth."

Pushing a smoke ring out of his throat, he adds, "The chlorine in the water will destroy the 'cid in a matter of seconds—got that?" and then flicks a lit cigarette out the window.

Okay, I say. Anything else? Hoping this was the end.

And this is the moment when Derek says something that would come back to haunt me about eight hours later. His eyes looking up and to the left, he moves his mouth to one side and says, "Look, I'm gonna be honest here." Then, tapping a fresh pack of cigarettes on the dash, he goes, "I'm pretty sure my guy overlays his tabs," and pulls a solitary cigarette out of the pack with his mouth. The yellow flame from his zippo makes his face glow gold, and while he's busy exhaling a giant cloud of smoke, he lets out one of those elongated, "Yeeaaahhhs" like there's some big story to tell. Then, either thinking about what to say next or maybe just lost in a memory, he scratches the back of his head and takes another big puff to emphasize his point.

"He tends to do that," Derek says, breathing smoke.

After that he just goes on and on about how these pretty pink lotus tabs were supposedly lab-tested. "They're the real deal," he kept saying, "laid at 70 mics a piece," but for some reason, I completely distrust this story. Maybe it's because there's only about fifteen labs on the entire planet that actually make this stuff, and I'm sorry, but Derek just didn't strike me as the kind of guy that was one or two hops away from the source. But, then again...I've misjudged people before.

—

Perhaps it was cockiness. Perhaps it was sheer stupidity, but based on only two previous experiments with delysid (both with dosages under 200µg) I somehow get it in my head that I could probably handle at least three of this new guy's tabs. So the second we walk into the music venue I tear open the vacuum-sealed tabs with my teeth and snap back three right out the gate. Cali, a good fifty pounds lighter than me, only takes two.

Flash forward to about an hour later when we both take another because...well, "We weren't really *feeling it*."

Flash forward to about forty minutes after that when we both split another one.

Chewing up that last tab, I remember thinking, "Fuck it—even the first sailors had to risk falling off the edge of the world to know it was round, right?"

Then, about an hour after that, because things we're starting to get strange, Cali and I thought it'd be a good idea if we hailed a taxi back home.

Yes, a taxi. I'm older than you think.

Notes from the Editor:
As a general rule of thumb, it typically takes anywhere from 1-2 hours before someone who's ingested lysergic acid diethylamide really starts feeling its effects. With about 90 minutes being the average come up, it should be known that set, setting, dose, and manner of delivery, are all important factors to take into consideration when dosing.

Koreatown Cont'd

As previously discussed, the way LSD disrupts the normal flow of sensory information is by temporarily disintegrating established brain networks like the default mode network. One of the side effects of which being, that new sensory information streaming in is no longer being connected to past information. So consequently, everything is seen anew, fresh, or as I like to put it: raw and unfiltered.

As you might expect, there can be numerous reasons as to why someone should always have a sober sitter present, but among them is the consideration that, after a certain dosage, you may no longer have access to your memory banks. Which, I might add, can be a little unnerving. Suffice it to say, that in the ten-minute taxicab ride home, the molecule was so prevalent in our system that we essentially forgot everything.

Not everything though, right?

Yes, *everything*. We couldn't even figure out how to work the credit card machine to pay the driver, so the cabbie takes us to an ATM across the street.

Flash forward to us standing in front of the ATM.

Flash forward to us knowing that we're supposed to get money out of this strange machine embedded in a wall, but not even knowing how to start such a process.

What is money anyway?

The idea of electronic banking was completely foreign to me. Almost as if I'd never even seen one before. So Cali and I mash buttons and exchange confused looks. Unsurprisingly, not a lot happens. So in a moment of pure vulnerability I just turn to the cabbie all fear and loathy-eyed saying, "Listen man, you're gonna have to help us because I don't know what the hell this thing is." A sentence which makes the cabbie's brain freeze.

The way he cocks his head, it's like he couldn't possibly understand what I mean. That is, until he finally looks into my eyes and registers how big my pupils are. Because after he cries out the words "Holy shit" into my face, I finally get the sense we understand each other.

"Man, I wish I was wherever you are right now," he says, pushing past me.

Then, fearing he'd overstepped the cabbie-client contract (if there is such a thing) the cabbie turns to talking about the weather as he thumbs in the first four digits that most readily pop into my head. Letting out a shiver, he goes, "Damn, it's cold!" An exchange that might've normally gone unnoticed, but given that his deflection was so artificial...so obvious...it triggers something bleak in me, and for the first time all night, I finally start taking stock in what Derek said.

What if this guy really did overlay his tabs?

Fear and Time Dilation

The experience Cali and I were about to have, one of the reasons they call it 'the unitive experience' is because, at a certain point, you may lose track of an identifiable and separable self. Traditionally, conscious states are presented to us like a story with a beginning, middle, and end. However, during this heightened state of connectivity/entropy, things aren't so neat and organized. Events can appear shuffled or out of order. Then, when you combine that with the fact that the sheer volume of information you're exposed to is so overwhelmingly beyond anything you've ever seen, you're left with an experience that is both ineffable and unforgettable.

From her New York Times bestseller *My Stroke of Insight*, Dr. Jill gave us an account of what happened the morning her ego-driven neuro-circuitry and language centers got off-lined by a stroke. She even went so far as to detail what some of the revelations of having an experience like that were (something we'll come to in a future section). Then, by outlining some of the functional differences between the way the right and left hemisphere operate, Dr. Jill was able to get the reader to understand that their sense of self is much more complex than the unidimensional narrative the Ego Self might have you believe.

In other words, we mistake the virtual machine of the Narrative Self as being all we really are, when in reality, we are the physical machine that gives rise to the virtual machine. On top of that, if you actually go read her account, or any of the other thousands of accounts regarding transcendent experiences, what you start to pick up on is that there are a few recurrent themes that keep popping their head up time and time again.

Theme # 1: **Ego dissolution,** the feeling of being a separate self melts away.

Theme # 2: **Time dilation,** a moment expands like an accordion, as is often reported by people involved in a car crash or a near-death experience.

Theme # 3: **An expanded sense of self,** the feeling that one is connected to something greater than themselves (loss of self/ego boundaries).

Now, whether or not you've had one of these experiences really makes no difference because I'm willing to bet that you've already experienced at least two of the three aforementioned themes. We've all felt those moments where we've lost track of time, or lost track of ourselves—whether that be during a dream or a flow state—and whenever these moments do occur, the "I" of experience melts away leaving us with just the experience. On top of that, I'm also willing to bet you've experienced the elasticity of time as well. Ten minutes on a StairMaster should feel as long as a ten-minute massage, but I think we all know from personal experience, that just isn't the case. Because, when it comes to time, the activity you're engaged in really matters.

And why?

Because, it turns out, that the subjective experience of time is just another calculation performed by the brain. And in much the same way that your sense of self can become altered due to the hyper-involvement of your entire being, your sense of time can become distorted as well. Matter of fact, we know that during a highly fearful event such as a near-death experience, the amygdala actually lays down a second track of memory providing you with a more detailed account about what happened.

Before 2007, it was generally believed that a life-threatening event, like car crash, might cause the brain to kick into high gear or something and process the moment faster, but this idea never quite sat well with neuroscientist David Eagleman. He even tells a story about the time he fell off a neighbor's roof and experienced that "everything slowed-down" effect firsthand.[2] Ever since then, it's safe to admit that David's been a little obsessed with time, which is precisely why he and his team ended up devising an experiment to finally test this question.

How were they able to reliably generate life-threatening experiences in the lab?

By dropping participants off a tower 150 ft tall, of course. Backwards. *And* without any safety ropes. These volunteers (*that's not a typo) then tumbled towards the earth, in complete free fall, until they hit a modified circus net below. I don't think I need to play up how utterly terrifying the Texas Tower experiment must've been, but rest assured, hitting that net traveling at 70 mph was more than enough to bring about that "everything slowed-down" effect in all that participated.

Although, the way David and his team tested for this was really something.

Seeing as they wanted to find out if noradrenaline or some other chemical was, in fact, speeding up cognition, giving rise to those familiar distortions in time everyone always describes, they rigged these volunteers to wear a perceptual chronometer during their fall.

The idea being fairly simple: If the brain were kicking into high gear, then participants should be able to read the numbers displayed on this fancy piece of tech belted to their wrist. And yet, if you comb through all of their results, you'll find no one performed better than chance. This hi-tech gear all the participants had attached to their wrists was flashing digits at them just below the threshold of conscious perception (custom-tailored to flash digits at about 6 ms below the threshold for each individual participant), such that, if the high-gear hypothesis were true, then researchers would've expected to see at least some volunteers who could've accurately reported what those numbers were.

But, as you already know, that's not what happened.

What happened was, they all *perceived* their fall as lasting longer than it did.

In actuality, they only fell for about 2.49 seconds. But when they were each handed a stopwatch after their fall and then asked to replay the event in their mind's eye, participants reported their falls as lasting at least 36% longer.[3] So, to summarize, the brain doesn't actually kick into high gear, rather, the attentional/memory system does. Leading subjects to subjectively report their experience as lasting longer than it actually did.

Why is this the case?

In plain English, it's because when we're put in a highly fearful situation there's a richer encoding of memory that's taking place. The amygdala starts laying down a second track of memory, increasing the layers of memory and the density of detail. Then, whenever the subject 'plays it back' in their mind—because of all this unusually rich detail—the event appears to have lasted longer.[4]

Why does the brain do this?

There can be a multitude of reasons but I shall give only two. First, we have to take into account that the calculation of time isn't localized to one area of the brain, and that it's a complex and distributed phenomena much like the memory system.[5] Second, we have to remind ourselves about why we even have brains to begin with, and that's for producing complex and adaptable movements through space.[6] We didn't evolve brains so that one day we might be able to read books; we evolved them so we could survive. And what survival ultimately entails is being able to manipulate the world through some kind of movement. Thus, we're either always moving towards things we want, or moving away from things that could harm us. Against this background, what the brain is really there for is to help guide future movements and future behavior.

How does the brain accomplish this?

By writing down the details about what happened in the hopes that we can simulate possible futures. The big idea being: if you can remember in rich, vivid detail about the time you almost died, then maybe you'll be less likely to end up there again. At least that's the idea.

Chapter 22
Koreatown Revisited

THE SECOND TIME I saw an ATM for the first time just happened to be the same night my former partner hurled a steel chair through my second story window. Why?

To save us from the "the Chinese" of course.

Moments like these, I can't not hear my dealer's famous last words. "I think my guy overlays his tabs."

The amount of times that phrase bounced around my brain that night, I couldn't even tell you. It had become my mantra, the thing I repeated over and over to try and remind myself I was on something. Birds chirping, broken glass and ecstasy scattered all over our living room floor and I'm thinking, "I think my guy overlays his tabs," laughing in the corner like a madman. Scrambling to put his broken glasses back together, my poor roommate Brian couldn't have been less impressed.

And sure, I could've been upset too.

I could've been mad about the broken window or the crushed pharmaceutical grade MDMA that had been tossed around and was now getting mixed in with small pieces of couch. And sure, maybe I should've been asking Cali some follow-up questions about why she threw the chair, or how it was supposed to save us from whatever was happening in her mind, but I guess I figured, whatever it was, the girl had her reasons.

Besides, if you don't recall, I kind of had my own crisis going on.

Pacing back and forth chanting, "I'll remember. I'll remember. I'll remember," I'm fairly certain I was trying to go back in time. But before we get into any of that, let's flashback to an hour earlier.

3:00 AM: That ATM was hands down the most beautiful ATM I'd ever seen. Even though I had no idea how to use the thing, I remember feeling as if I'd never seen such beauty in the red bricks next to it, before or since. It truly felt as if I could see every crevice and detail in the concrete—that is, until some beeping droid named Chase kept messing up my vibe.

Staring at those red bricks, I knew perfectly well what I was supposed to be doing. I was supposed to be paying attention to the driver walking us through some kind of withdrawal. But instead of doing that, I just became overly fixated on that wall. Of course, I couldn't be sure whose side Chase was on, but once it eats the plastic from my pocket—and the cabbie types in the first four numbers that pop into my head—this no-nonsense robot starts spitting out some kind of crisp green paper, which I instinctively grab.

Then, without even registering what kind of bill it is, I just hand it to the driver and point at my apartment. At this stage, money means absolutely nothing to me, all I care about is getting us home.

So what was going on in my mind just then?

Well on 400μg (aka 400 millionths of a gram) I was seeing raw information and way more of it than usual. Those floodgates of sensory awareness were completely blown open causing such an overflow of information that my brain was having trouble trying to keep up. Almost like it was stuttering. On top of that, this flood of sensory information wasn't being connected to what I already knew about ATMs.

Hell, it wasn't even being connected to the concept of fiat currency.

The visual information was bypassing my memory banks, bypassing all the experiences I've ever had concerning ATMs, so for the first time in years, I was finally seeing an ATM in a new and novel way.

No longer were those sensory awareness ratios, my usual 20 bits for smelling, 90 bits for seeing, 15 bits for touch, 15 bits for taste, 60 bits for hearing, because in this moment it was 140 for sight, 5 for smell, 8 for taste, 10 for touch, and 37 for hearing. The next second it was 50 for touch, 10 for hearing, 25 for smell, 3 for taste, and 112 for sight.

As you'd expect, it's quite normal for these ratios to shift and change from moment to moment, but radical chaotic shifts like this are usually inhibited.[1]

Flash forward to about twenty minutes later.

We're back in my apartment staring at these freshly picked strawberries against the backdrop of a grey sink basin, when neither one of us can quite get over the experience of red we we're seeing. Almost like each strawberry was dripping paint.

Vibrant doesn't even come close to describing the experience.

Nevertheless, if we were to compare vision to pixels again, this wasn't like seeing a strawberry in 1080p, this was like seeing it in 4K. It was the most intricate strawberry I'd ever seen, perhaps even the most intricate strawberry I'd seen since somewhere around my fifth birthday.

More than anything, when we looked at this strange piece of fruit, we saw that it was *still alive*. That it had life and vitality! No longer was this just a strawberry, rather, it was a living thing that had been picked off a bush only days ago. The shade of red, a color we'd never seen before. If Cali and I normally got to perceive reality at something approximating 75 frames per second, this was like seeing it at 400fps.

How is this possible?

Again, we can employ another one of our computational metaphors to help us explain. If we're imagining the brain as being like a computer, and let's say for argument's sake, it were normally allowed to use something like 15% of its processing power to perceive a strawberry, this was like my brain was using 70% of its processing power. So the idea of seeing this strawberry in 4K at 400fps actually holds up—if you can forgive my back-of-the-envelope calculation, of course.[2]

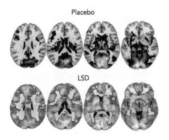

Credit: Beckley/Imperial Research Program

Technically, we already showed you this photo before, but I'm *re*-presenting this photo a second time, because now, I really want you to understand what you're looking at. Here we have a brain under two conditions. One, where the brain is rather dimly lit up (the placebo control condition of someone with eyes closed). The other, a scan of their brain after they've ingested 75 micrograms of LSD (with eyes closed again). I need hardly point this out, but the brain that's ingested about an eyelash's weight of LSD is clearly much more lit up. Indicative of the fact that much more of the brain is now interacting with itself than usual. But to be a little more specific, what you're actually seeing is more brain real estate contributing to the processing of visual information than is normally allowed.[3] Similar to a high-res camera that's able to capture more photons and then turn those particles of light into pixels, the brain is essentially able to use more of its processing power and real estate, to extract more information from the images it receives. With more power and energy allotted to dissect wavelengths of light than is normally otherwise allowed, it's quite common for people to report seeing shades of colors they've never seen before.[4,5] After all, color doesn't exist out in the real world anyway. Color is just an internal label used by the brain to distinguish between different wavelengths of light.

Why does delysid have this effect?

The answer, at least in part, is because LSD is a partial agonist. Meaning, that when it does slip into those serotonin 5-HT2A receptor sites, it has an excitatory effect on the neurons and speeds the message up, effectively changing the way we think and perceive.[6]

Patterns of Thought

Just consider the way we think and talk. People are so quick to jump into pre-programmed responses, either regurgitating something they've already said, or parroting back something they've already heard. It's actually quite rare when somebody verbalizes sentences that are wholly original. To say something truly off-the-cuff takes time and effort, so for most of our day-to-day operations we tend to lean on chunks of language that are already "pre-thought" so to speak.

See *Idioms*

See *Sayings*

See *Previously Established Literary Devices Used for Driving Home Points*

We all have catch-phrases and verbal go-to's. Word combos we like, clichés we abuse, and more often than not, we're not even aware of most of them. For example, it wasn't until my sister told me that I say the words "you know" a lot, that I even began to take notice. A couple days after that, I could barely even stand it.

Turns out I said it all the time, I just had no idea. No awareness. It was an unconscious pattern I was completely oblivious to. And so, when I finally did try and stop myself, it felt like I was stifling yawns all day. The phrase had become something like an unconscious reflex.

My period at the end of sentence.

Hopefully, this little digression about the way we use language provides us a little insight into the way we *think* and *see*, because in reality, thinking and seeing have undergone evolution just as much as the eyes themselves. In evolutionary terms, we might say there's a cost-benefit analysis to be weighed when considering an adaptive trait; and as a corollary, thinking and seeing are no exception. There is a cost to seeing just as there is a cost to thinking, and this cost is paid in the form of energy (metabolic costs), brain real-estate (at least 20 to 30% of the cortex is devoted to visual processing), and, of course, the time it takes to process all this information (we live around 500 ms in the past).[7,8]

Without question, human beings *could* reconstruct reality at a much higher "frame-rate" than CS5 normally allows, but that level of detail is really just unnecessary, and in most cases just a nuisance. So instead, what the brain does, is engage in activities that aid in the conservation of time and energy. Which includes but is not limited to: running familiar neural loops, creating internal models, writing "free throw" programs, and always, always, always connecting new information to past information. That way, whenever you see a chair, your mind recognizes *Chair*...conjures up all the pre-programmed information it already knows about chairs (they have four legs, they're hard, they can be tossed out windows) thus preventing you from truly seeing all of its uniqueness.

Instead of treating each chair as a unique and novel encounter, which might be beautiful but comes at a high cost, your brain recognizes it, says, "Well, I know what that is," and then sends up a tiny fraction of visual data about this particular chair while phoning in all the rest.

Why does it do this?

In short, because this saves time and energy, and in past environments, efficiency tended to ensure survival. Currently, we live in a time when access to food is the highest it's ever been. But back before the dawn of agriculture, when selection pressures weren't largely mitigated by technological innovation, those who failed to make these fast judgements were swiftly taken out of the gene pool.

Granted, the archaic or classical view of consciousness would have you believe that the brain is just a passive filter of information that resides in the skull and is fed streams of data, but as we've already uncovered, that isn't even half the story.

First, it will be remembered that the brain isn't passive, it's active. It's consistently using its body and its sense organs to actively sample the environment. Although, it is also worth noting that with this massive sweep of data comes a lot of noise, which is where the benefits of internal models and Bayesian inference comes into play (prediction and expectation). Second, if you actually did try to recreate consciousness using a bottom-up method (building up the world from scratch each time) it would never work. Which is exactly what happened in the field of robotics.

Engineers tried to build a robot that pieced together its world using an outside-in approach, but the robot was just too slow.[9] So, Mother Nature, being the crafty mistress that she is, figured out a better way. Not entirely top-down and not entirely bottom-up, but rather, a judicious mix of both. "On the one hand," Eagleman explains, "she wants to operate as close to the border of the present as possible, but on the other hand, she has to account for the way things get smeared out in time," based on the mechanics of the retina, and the fact that other modalities are processed at different speeds.[10]

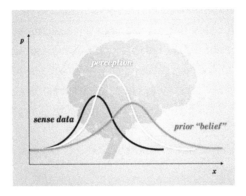

Bearing all these facts in mind, I want you to picture a cousin to one of your ancestors. The kind of cousin who was just a little too amazed by everything he saw that he failed to see the lion lurking in the tall grass.

If we were to seriously consider this stark hypothetical, what do you think the fate of his genetics might've been?

Notes from the Editor:

When you take into account the fact that the brain needs 500 kCal/day just to operate (i.e., 25% of a 2,000 calorie diet). And you simultaneously try and entertain the myth that human beings only use about 10% of their brain, it becomes increasingly clear how an organ as metabolically expensive as that never could have evolved in the first place.

—Rhetorical questions aside, I think it's fairly evident that Nature favors the swift, the efficient, but most importantly, the uneaten. Nature favors the kind of mind that can think fast and act quicker, resulting in a form of consciousness that tends to guess and fill in the blanks, rather than provide an infinitely perfect representation of reality. Remember, consciousness is ultimately about managing information, not overwhelming the operator.

If every time you picked up a piece of fruit you had to shed a tear and have a moment, nothing would ever get done. It's maladaptive to be taken in by everyone and anything. So to circumnavigate this, our brains use internal models and a part of our brain the godfather of LSD research calls "the novelty detector."[11]

Really, it's just a tiny group of cells located in the brainstem called the locus coeruleus, but we can visualize this structure as being like a switch that orients attention and directs focus. In all fairness, we just can't focus and be aware of everything at once, so instead, what we do is, scan our environment for things that don't fit into our model. Then, once we detect some kind of novel stimuli this structure in our brain fires, our eyes saccade, and we start snapping up pictures like the paparazzi. In effect, the brain starts investing more time and energy into perceiving the thing.[12]

Which, as we're about to find out, has an interesting effect on our *experience* of time.

The Oddball Effect

Suppose you were sitting in front of a computer right now. And suppose I started flashing you a series of pictures one after the other. So first it's a penguin, then it's a crane. After that, you're flashed a picture of a rooster, a chicken, a crow, David Hasselhoff, a stork, a parrot, and lastly a pigeon.

What might we expect to happen?

Well, given the conditions of the experiment, we would expect for you to subjectively perceive the photo of David Hasselhoff as having lasted longer. Spinning around my laptop, I could even show you the back-end of the study, where each photo was clearly on the timeline for 0.500 seconds and not a millisecond more. But, seeing as that photo of the Hoff was rather oddly placed amidst a series of birds, your brain would actually devote more attentional resources into trying to perceive it. More specifically, the burst firing of the locus coeruleus would cause you to fixate on the image longer, creating a distorted sense of how long that photo actually appeared on the screen.

This particular phenomenon has been appropriately dubbed *the oddball effect*, and it's just another one of those findings that clues us into the fact that, according to the brain, time isn't just one thing, but is instead composed of separable subcomponents which can be teased apart in the lab.[13,14] Again, just as we saw in David Eagleman's Texas Tower experiment, we see a richer encoding of memory causes an event to appear as if it lasted longer than it actually did.[15]

———

You really want to know why I bothered re-*presenting* that photo when I could have just asked you to flip back to an earlier section? It's because I was trying to make the following point. The first time you saw that photo, you should know that your brain actually spent more time and energy trying to perceive the thing than it did the second time.

So, supposing I had you in an fMRI machine the first time you saw it, I could've shown you a whole lightning storm of activity buzzing through-out your skull as your brain tried to perceive the image. However, if I had you inside the scanner the second time you saw it, or the third time, there would've been a decrease in novelty, such that, with each successive showing, your brain would actually be devoting less and less resources into trying to perceive it.[16] Reason being, you've already seen that picture.

Why waste precious resources trying to perceive something when you could just do what internet browsers do and fetch the file from its cache. Whenever you surf the internet, your internet browser stores up all the images and logos from your most frequently visited sites, so it doesn't have to waste time re-downloading the same files again and again.

Taking all this into account, when you consider what it must be like to perceive a strawberry for the three thousandth, nine hundred and thirty-second time, the question we should really be asking ourselves is—Am I even seeing it at that point?

And the answer seems to be, "Sure, you're seeing pieces and parts." Little idiosyncratic differences of shape and color that violate your brain's predictions. Although, it seems fairly obvious that whatever amount of attention you are investing, it's clearly not as much as you could. Which is also why LSD and other psychedelics are purported to be enjoyable. Their effect on the locus coeruleus alone lets us experience more novelty than usual, reminding us about the *Elasticity of Attention* and the power of that spotlight. Instead of leaning heavily upon internal models (the way it does in normal life) the brain starts to perceive reality more like it did when we were just kids. Back when experiences of ATMs and freshly picked strawberries were still new and fresh.

Needless to say, but this novelty detector is used sparingly under normal conditions. But under abnormal conditions, say a couple hours after you've ingested $400\mu g$ of high purity blotter acid.

Well, in that case...all bets are off.

Chapter 23
Synesthesia

REGENT'S PARK TASTES like malt vinegar, Charing Cross tastes like apple pie, and Lancaster Gate tastes like thin, crispy bacon. Now, if I were to say all these sentences to you would you look at me funny, or start taking notes? Supposing I said, "Russell Square has the taste and texture of celery," or "Knightsbridge tastes of pork jelly," you'd probably think I had some wires crossed, but you wouldn't ask me any follow-up questions.

On the other hand, if you were a neurologist, you'd definitely think I had some wires crossed, and you probably would start asking me follow-up questions. Because as strange and hypothetical as this all sounds, it should be known that almost this exact same scenario has already played itself out. Although, when James Wannerton left the room, doctors didn't just draw tiny circles around their ears with their finger...they actually listened.

Naturally, it took some convincing—and some persistency on James's part—but eventually, they stuck this man inside an fMRI machine, and after carefully reviewing the results, finally started to take him more seriously.

Of course, it's one thing to have someone tell you they get a distinct taste and texture in their mouth whenever they read a word, but when you see the cross-activation for yourself, it's a whole different matter.

Believe it or not, but this moment actually marked a turning point in J.W.'s life, because before this test, synesthesia (or a cross-blending of the senses) wasn't very widely known. In fact, it wasn't even believed to be all that common, until recently. And now, not only is James Wannerton the president of the UK Synesthesia Association, he's also one of the world's most famous synesthetes.[1]

Where does synesthesia come from?

To answer that, perhaps, we should back up a sec and make sure we have all our ducks in a row. Your brain, as we've already stated, isn't all wired together, but when you were just a baby, it was more wired together. As you'll recall, we have about 86 billion neurons in our heads right now, but back when we were just toddlers, that number was closer to 100 billion. At the current moment, each one of your neurons is making around 10,000 connections to all its neighbors, but back when you were about 3 years old, that number was closer to 15,000.[2,3] So what happened? Did we all just lose it because we didn't use it?

In a way, yes, but this wasn't necessarily a bad thing. In fact, we underwent two significant neural pruning events over the course of our development—one which happened around the age of three and the other at about puberty—which had the effect of removing any unused or infrequently used neural pathways. Much like a garden that needed to get pruned back in order to redirect growth, shape, and promote flowering, this overproduction of neurons and synapses actually helped promote a healthy competition of synaptic connections.

The routes that worked best, we kept. The routes that didn't work so well, we lost.

Which brings us to another very important point about the strategy of Mother Nature. She tends to produce endless variants, most of which fail or die, but the ones that do end up surviving, end up surviving for a very particular reason, like efficiency.

This is true of organisms, as 99.9% of all species once living are now extinct.[4] This is true of mutations, as most mutations end up killing the organism. And it's even true of neuronal connections, as your brain ended up tossing out at least forty trillion of these things.

How does synesthesia develop?

To be completely fair, we don't exactly know, but the general consensus is that, we all start out synesthetic when we're born. For most of us, these connections get trimmed back, although for at least 4% of the population, this trimming might go a bit odd leading to a cross-blending or mingling of the senses.[5] Normally, there isn't a lot of neuronal cross-talk going on between the auditory cortex and the visual cortex, however, for someone with sound-to-color synesthesia, there is a lot of cross-activation. So much so, that whenever they hear a sound, they see it as well. Meaning, the sound, word, or musical note will trigger an internal experience of color. Not a hallucination, rather, a difference in perception.

As one synesthete describes it, it's more like the internal experience you might have when you're driving but also thinking of a memory. Or, take as an example, someone who has grapheme-color synesthesia (the most common type). For them, they'll see the alphabet written on a page—know it's written in black ink—but then in their mind, perceive each letter as having its own distinctive color that is fixed and unchangeable. In this regard, if their "5" is green, then it will always remain green throughout their entire life. They never consciously chose to make this association, nor can they just turn it off because, for them, the joined perception appears inside the theater of consciousness unbidden.

For some synesthetes, colors can cause them to hear sounds, or sounds can cause them to hear colors. Basically, any kind of cross-wiring you can think of, chances are someone has it. Hearing to color. Taste to smell. Timbre to shape. And one out of twenty three people has it. Which means, odds are, you probably even know someone who does.

But the question is: Do they know they have it?

For a number of synesthetes, they don't even realize anything about their internal experience is unusual, until they ask a question like, "What color is your Tuesday?" or "What does the word 'bananarama' taste like to you?" Take J.W.'s life for instance. The man was twenty years old before he even heard the word synesthesia.[6] Growing up in London, Wannerton would hear words and taste sounds, but he didn't always know his experience of reality was anything different. And why would he? Matter of fact, why would any of us assume the version of reality we get is fundamentally different than anyone else's?

240

240

How do we know all these people aren't just faking it?

By rigorously studying and testing them. For instance, when they stuck J.W. inside an fMRI machine they couldn't help but notice that his brain was reacting in very different ways in response to sounds than in non-synesthetic controls. So this isn't just a hyperactive imagination and a preponderancy to free-associate—*there's some literal cross-activation going on here.* J.W. didn't choose to make an association between the stop at Snaresbrook and malted milk biscuits, the name Snaresbrook just happened to generate the taste, and he just happened to be the guy who wrote it all down. Just Google the words "Tastes of London Map" and you'll be treated to an entirely different world.

A map of the London Underground.

But then, when you look a bit closer, you'll find that all 270 stations have been cheekily replaced by tastes. An undertaking which took him nearly forty years to complete.

Wannerton would get off at Queensway station, try to figure out the exact taste and texture he was experiencing, and then he'd jot it all down in his notebook: *Tastes like evaporated milk*, or *Tastes like soft-boiled egg*.

To avoid confusion, not everyone with lexical-gustatory synesthesia will have the sensation of chewed matchsticks whenever they get off at the Northwood stop (these associations are highly idiosyncratic). But they will definitely taste something, and they will always taste that same something, because for these people the taste-processing areas of the brain are activated whenever they hear a word.

Why do I bring this up?

I bring it up because: (a) it furthers the point that we can have perceptions that don't come from our sense organs, (b) it furthers the point that our perceptions are internally generated to begin with, (c) it furthers the point about how differences in brain structure can lead to differences in perception, and (d) it opens up a dialogue about how this internal experience can vary person to person.

Again, there's nothing different about the sense organs of a synesthete—their sense organs are sending up data just fine—it's just the way their brain decodes this data that's different.

What does synesthesia feel like?

Funny you should ask.

Koreatown Cont'd

Because synesthetic experiences can be so confusing. And because this subject is already confusing enough. And because this story is already crazy enough; I'm going to attempt to give you a little roadmap to hold onto before we dive back into an experience Erowid Center would surely file under *Train Wrecks & Trip Disasters.*

First Cali and I head to a music venue uptown. We take too much, hail a cab home, and forget everything on the ride home. Somehow we make it up into my apartment, eat some strawberries, lose my keys, and are subsequently locked out of my room. Feeling things quite heavily, we collapse on the floor next to Brian's room, and start banging on his door in hopes that maybe he'll come out and play.

It's 3:20 AM and Brian's dead asleep. But even still, he puts his glasses on and opens up his bedroom door. Peering through the cracks with only one eye open.

Ten minutes after this, and I can't even remember which molecule we've taken.

How could you possibly forget you've taken a psychedelic?

The same way a person might forget they're asleep and having a dream every single night of their entire existence; it's certainly possible to forget you're on something. More to the point, your attention can get so wrapped up in the moment you're in, that you can no longer extract yourself from the present moment, which is precisely what happened. No longer could I imagine a future where the molecule had worn off, nor could I even remember what the word "baseline" even meant. Altered had become my new reality.

And this was the moment things really started to turn sour.

Timelessness leads to time-dilated thought loops. Time-dilated thought loops gives way to ego dissolution, and somewhere in the confusion of 400 mics, Cali decides to throw a steel chair through our living room window (which was definitely closed at the time). The sound of shattering glass must've woken Brian back up because not a moment after that, Brian bursts out of his room and I'm sent spiraling headfirst into a transcendent experience I don't understand. What's worse, I even try to fight it which is another terrible, terrible idea.

In case you hadn't noticed, but everything about this night is a perfect example.

"...of what *not to do*."

Notes from the Editor:

Before the Controlled Substances Act of 1970, psychedelic agents like mescaline and LSD were actually studied as a tool to help facilitate creative problem-solving. The requirements for the particular study in question were: a) the subjects had worked for several months on a problem and failed, b) that they were highly motivated to solve the problem, and c) were evaluated to be psychologically normal subjects with stable life circumstances. There were a total of 27 professionally employed subjects (engineers, mathematicians, physicists, architects, etc.) all of whom were given either 100μg of LSD or 200mg of mescaline; and then provided ample time to work on their problem. The solutions coming out of the session included: "(1) a new approach to the design of a vibratory microtome, (2) a commercial building design accepted by client, (3) space probe experiments devised to measure solar properties, (4) design of a linear electron accelerator beam-steering device, (5) engineering improvement to magnetic tape recorder, (6) a chair design modeled and accepted by manufacturer, (7) a letterhead design approved by customer, (8) a mathematical theorem regarding NOR-gate circuits [chip design], (9) completion of a furniture line design, (10) a new conceptual model of a photon which was found useful, and (11) design of a private dwelling approved by the client."[7]

—Nevertheless, there was a global increase in network communication going on inside my brain, such that disparate regions that are normally highly segregated finally got to have a little chat resulting in massive amounts of synesthetic experiences. In those moments, ideas formed, novel connections are made, and I finally got to experience what neuroscientist Arnie Dietrich has termed a "mental singularity," what Ava and I call "the Possibility Space," and what Buddhists and long-term meditators refer to as nirvana.

But what exactly is this experience?

In one sense, it's an idea. In another, it's an experience. You could say it's the experience of an idea, although I would hasten to add that it's probably the most complex idea you could ever dream of. Evidenced by the fact that when we look at the brain scans of individuals undergoing the psychedelic state, we find that their brains are not only hyperactive, but highly entropic as well. In actual fact, psychedelics like LSD, psilocybin, and low-doses of ketamine will actually increase levels of brain complexity significantly above baseline.

See *Perturbational-Complexity Index*

See *Lempel-Ziv Complexity*

See *Psychedelics as a Treatment for Disorders of Consciousness*

Which, by the way, is not the same thing as saying psychedelics produce "higher levels of consciousness" because that's actually not what I'm saying. What I'm saying is, the Lempel-Ziv Complexity is a robust measure of quantifying conscious level; it reliably goes down whenever we're asleep, in a coma, or under a general anesthetic. But, whenever we give someone a psychedelic, it reliably goes up.[8]

Indicating that there's an increase in the level of neural signal diversity and complexity (something which only seems to happen under this state). Which is a bit strange, don't you think?

In light of these facts, there's certainly a case to be made about the kinds of ideas one *could have* under this state. Again, not that they will happen, only that they could. And furthermore, the only reason I chose to tell you this particular story and not one of the more controlled experiments is because I'd like to be completely upfront about the dangers of these compounds. Sure, LSD and psilocybin might be hard to overdose on, but as we're about to find out, these compounds have the potential of unraveling you in ways you couldn't possibly imagine.

And in case that sounds patronizing, let me just add that before this night I'd already experimented with mescaline, salvia, ecstasy, psilocybin, MDMA, DXM, DOB, nitrous oxide, DMT, GHB, 5-MeO-DMT, 2C-I, 2C-B, 2C-E, 2C-T-2, 2C-T-7, MDA, the toadstool-looking amanita muscaria—and I still didn't see this one coming.

That's how unraveling this experience was.

Chapter 24
Synesthesia Cont'd

TO MY KNOWLEDGE there are only three ways to get synesthesia. One, it runs in your family. Two, you suffer some kind of brain injury which causes the brain to repair itself, resulting in some new connections being formed that may link one or more of your senses. Or three, you accidentally ingest a high dosage of a potent serotonin agonist like lysergic acid diethylamide that ramps up the sensitivity of your cortical cells to the point that they're now "talking to" (exchanging information with) other areas of the brain that are normally highly segregated.[1] The first two are typically permanent; the third is temporary.

Currently, there are two main hypotheses about why synesthesia develops, and seeing as we've already discussed the first one—neuronal trimming goes a bit odd—we'll just briefly run through the other one. But first, I want you to show you an alphabet like you've never seen. One where the letters are represented by shapes.

Below, you'll find two shapes, one they call "bouba" and the other they call "kiki." Before carrying on, can you tell me which is which?

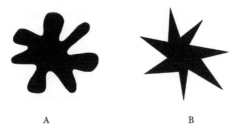

A   B

***Notes from the Editor:*

When trying to remember or explain something, we tend to learn best by making associations, connecting that which we're trying to learn with that which we already know. In this way, those who are gifted at making associations, like synesthetes, possess a natural edge that tends to help with memorization. One synesthete was actually able to memorize the digits of pi out to 22,514 decimal places. He then recited it, out loud, without any mistakes in just over five hours.[2]

—So, if you happen to guess that A. was "bouba," and B. was "kiki," then you just had a minor synesthetic experience. When you first read the words "kiki" and "bouba" the narrator in your head subvocalized the words (causing activation in the auditory cortex). Then, when you looked down at the shapes available, you mapped this rather sharp auditory pattern of "ki-ki" to the shape that appeared most suitable: the jagged one.

Originally documented by psychologist Wolfgang Köhler back in 1929, we should acknowledge that all over the world, the overwhelming majority of people report the more rounded shape as being "bouba" and the sharper shape as being "kiki."[3]

So that's what minor synesthesia looks like and we do this sort of thing every day.

We engage in "light" conversation with those we do not know; we get caught in "heavy" conversations with those we wish not to know. Those who are nurturing and thoughtful, we characterize as being "warm" or *sweet*. Those who are a little difficult to be around, we refer to as being "abrasive" or *rough around the edges*.

When we're at a store, and someone drapes a highlighter-colored tee across their chest, we shake our heads and say, "Too loud."

They offer us a white cheddar cheese, "Too sharp."

So minor synesthesia isn't something that's super foreign to us. It's just that we have certain parts of our brain that have specialized over the course of evolution to deal with specific types of sensory information and not others. Correspondingly, we might say these areas of the brain have a certain 'way' of doing things. Although, when you get data from one sensory modality, like vision, creeping into the auditory cortex, the auditory cortex interprets the data in the only way it knows how (i.e., you hear what you see). Not long ago, I mentioned how when all this sensory data comes into the brain it doesn't come labeled. Which might've been a hard pill to swallow, given that our senses feel so qualitatively different from one another. But now, let's see if we can't sweeten the pill just a bit.

———

From David Eagleman, we were given the hypothesis that it's all about the *pattern of the data coming* in. And to take that point just a bit further here, I'd like for you to imagine a scenario where you're sitting at a cafe inside an international airport. Seated behind you are five different families, sitting at five different tables, each from five different countries.

So one table is speaking French while the other is speaking Mandarin. Then, right next to them, it's a table speaking Arabic, a table speaking German, and on the far end a family speaking Russian.

Now, supposing you were in earshot of all five of these tables—and you weren't looking—would you need some kind of internal labeling system to know which family was speaking German, or might you be able to pick them out simply by paying attention to the types of phonemes used?

Even if you didn't speak French, couldn't even tell me what the phrase *Je ne sais pas* means, wouldn't you be able to distinguish French from Arabic just by paying attention to the melody of the language?

Sure, they might all be using the same medium (chopping up an exhalation of breath into discrete utterances of sound) but it's the 'way' they use their mouths that's different, isn't it? For example, some languages feature guttural sounds or rolled R's, while others, just don't. Thus, there's a musicality to every language which help us distinguish one language from the next, and to me, French appears to be more about the vowels than anything else. While Russian, on the other hand, is just the opposite.

"бодрствовать" vs. "jouaient"

Therefore, to ask a brain, "Why don't sense organs come labeled," might really be as ludicrous as asking you, "Which one of these tables is speaking Russian?"

Admittedly, this is just a hypothesis, but when you consider that anybody is capable of having a synesthetic experience (provided they're given the right stimulus) it's really not that far of a stretch. Which leads us to the second hypothesis about where synesthesia comes from. Instead of it being a literal increase in the wiring of synesthetes, the other hypothesis speaks more about a loss in regulation surrounding the excitation and inhibition of neurons.[4] Meaning, there might be a higher probability that the neurons of one sensory modality *excite* other neurons in another sense, causing a joined perception like colored hearing.

In any case, whether it be from overly excitable neurons or untrimmed connections, really makes no difference for the argument. The fact that it happens, is present in at least 4% of the population, and can be induced by psychedelics is more than enough. Not only that, when you simultaneously start to consider what an idea actually is—from a neurological perspective—you soon find that an idea is really just a network. Hence a new idea is really just a *new* network of neurons firing in sync with one another. A network that's never fired together before.

So after the moment I perceive to be my last…after my brain enters into this peak state of entropy and chaos…some kind of cognitive reset ensues. Essentially, my Narrative Self fractures into a million tiny pieces, and yet, somehow, a part of me remains conscious. Then, while everything starts to boot back up again, this awareness of mine catches hold of the entire production.

In the final analysis, something extremely fundamental to the study of consciousness ends up revealing itself; and it's this idea that I'll be trying to explain over the next few sections.

***Notes from the Editor:*

Below is another example of what minor synesthesia looks like: a song represented in waveform. In the next chapter, the narrator dives headfirst into what major synesthesia might look like, concluding the story about Koreatown night, and how all these revelations are inextricably tied up with a transcendent experience.

Chapter 25
Off the Rails

WATCHING WATER FLOW down a stream long enough and you start to understand how thoughts flow. How people operate. People like to do the easy thing. Follow the path of least resistance, take the easy route, and consciousness isn't all that different. The reason I had such a hard time stopping myself from saying "you know" at the end of every sentence is because, somewhere along the lines, my axons formed a neural pathway that was efficient and worked. Like a stream flowing downhill.

But also like a stream flowing downhill, all it really takes for new pathways to open up is one good storm or flood. One torrential downpour is sometimes all it takes to alter the shape of a river, forcing entire maps to be redrawn, and this is precisely what happened on the night in question. That night, there was such an influx of data coming in that the normal stream just couldn't handle it, so naturally there was some spill-age. "Sensory overload" might be one way to capture the spirit behind a chemically-induced synesthetic experience, but on this particular night, the only way it really makes sense is to call this a Hard Reset. Some kind of massive update to my mind's OS.

Although, before any update can ever be installed, the previous regime of software has to be dismantled. And the way this manifested for me was through a handful of synesthetic experiences, a pilgrimage through the four mental rites of passage—DEMP—and a sad magician's reveal into the darkest parts of my psyche. A moment in which the narrator of my experience ("James") finally got to see all the fossilized, pathological behavior patterns that were holding him back.

Major Synesthesia

To explain major synesthesia, we need only look to music streaming services like Pandora. How it works is, you type in the name of a song you like, and then Pandora's sorting algorithm analyzes the track based on melody, pitch, instruments used, BPM…and then scans its vast library of pre-analyzed tracks to find similar artists. From personal experience, I can say that their Music Genome Project is actually quite good, however, before this algorithm existed online, it should be noted that we were already doing this—*in our heads.*

Simply by playing a song for your friends and judging their reaction, it's rather easy to pick up on a vibe, or notice the lack of one. Then, by asking the right questions, you might even begin to suggest artists or albums they might also enjoy.

How do we do this?

For the moment, let's not worry about the *how* and instead focus on the fact that even non-DJs possess some form of this skill. Implying that somewhere in our brains there exists an ability to extract some kind of musical thumbprint to the songs we hear, and then contrast that against all the songs we've already heard. A mental sorting algorithm, if you like.

How does this all tie in with major synesthesia?

With the following thought experiment. For this one, I want you to imagine that instead of typing in a song to Pandora's search box, you could insert a photo, a memory, or maybe even a dance move. I mean, if a sorting algorithm like Pandora's could be unleashed on a different kind of media, like the memory system…What might that look like?

Patient 11 describes it by saying, "I experienced blocks going into place, things being rearranged in my mind…gold blocks going into black drawers that would illuminate," and that's all well and good, but what if the experience went deeper than that? What if, for argument's sake, this sorting algorithm was able to break down your entire childhood musically? What if it could analyze, condense, and reshape all of your most pivotal memories into a new experience you could re-witness in just a few seconds?

I hesitate calling this experience the "waveform interpretation of your life's choices" on account of this being way more than just an image. Although, calling it the waveform interpretation does get us quite close. And short of seeing this image for yourself, this is part of that 'something' that's powerful enough to break addictions, lift people out of depression, and incite serious change.

Koreatown Cont'd

It's only on experiments with 250µg+ that Ava and I have ever gotten to experience true synesthesia, and being that the majority of these experiences can appear counterintuitive to the normal flow of consciousness, I will now refer to each second of my experience as a "now moment."

For instance, there was a now moment in which Brian disappears, thinking Cali and I were too gone to be harmless, followed by a now moment in which I completely forget which molecule we've taken:

Is this 5-MeO-DMT or is this LSD?

The next now moment, my thoughts start to feel less punctuated. Like there's no space between one thought and the next. One second, I'm amazed it's even possible for a molecule to do this. The next, I'm quite terrified, seeing as I can no longer remember what normal cognition is supposed to feel like.

The next now moment I shiver.

The next now moment I laugh-cry.

The next now moment I'm unlocked from the shackles of a first-person narrative and am somehow above my life looking down on all my memories like they were on a strip of film splayed out in front of me. I saw my birth. I saw my first time on water skis. I saw myself at my best friend's funeral.

All of my most significant milestones were laid out before me, and while I felt each one again intensely, it wasn't like I was re-experiencing them, in real time, one after the other. More like, this was one of those series recaps you might catch before an episode of *Breaking Bad*. The only difference was, I didn't have some comforting narrator saying, "Previously on *Breaking* ▮▮▮▮▮▮," because for me, I could no longer tell the difference between then and now. Here or there. Leaving me stuck in a time-dilated thought loop where I spent the better part of eternity experiencing and re-experiencing the most pivotal moments of my life again and again.

Me, the human-sized hamster, stuck in a giant wheel of karmic reckoning.

The next now moment, I felt like a cat who'd just stuck its nose in a bag of black magic. I remember shaking my head from side to side, trying to "shake the experience off" followed by the realization:

This is going to be the kind of experience you'll never be able to shake off.

And that's when it finally hits you. A memory from back when you were at the club. That bitter, metallic, battery acid taste that filled your entire mouth the second you started sucking on those tabs.

This wasn't LSD, this was 25I-NBOMe; a potentially dangerous research chemical.

Just then, an echo forms in your head.

"If it's bitter, it's a spitter," the voice says, and right then, a series of revelations descends upon you. This wasn't acid. It was N-Bomb. Derek wasn't trustworthy. He lied. And now everything certainly wasn't going to be okay, because overdosing on 25I-NBOMe is *way* too easy. Having it delivered to your doorstep and pawning it off as LSD is even easier. And while you're trying to work out exactly how much you're fucked, Cali's picking up the chair you usually sit in at breakfast. She's grabbing the sides where the steel sled base meets the red plastic seat, and now she's leaning back on one leg like she means to throw the thing.

You don't think it's going to happen, but then it does. You don't think someone's going to sell you N-Bomb, but then they do.

And now Brian's coming out of his room all, "I know I didn't just hear the sound of breaking glass coming from our living room," just for Cali to grab the windows off his face and smash those too.

"Just accept it," she says.

And now you're stuck in the corner pacing back and forth saying, "I know, I know. I know." Over and over, like a broken record.

What exactly do you know?

Only that N-Bomb is mistakenly sold as acid all the time, that it's been making the rounds, and that there were already even a handful of deaths associated with its use.[2-5]

Your average psychonaut, if they know anything, they'll tell you about the slightly bitter taste associated with LSD. But if you ever bite into a tab that tastes like it was dipped in battery acid—and numbs your whole mouth—then the jokes on you because it's not real LSD. "If it's bitter, it's a spitter," the aphorism goes. Quite simply, because certain research chemicals, like the NBOMe series, are not to be trifled with. I knew this. I've even said this. And still, no less than a few hours ago, I was the guy who said, "Let's take more." *Twice.*

The thought, "You irreparably fucked up your brain," passes through my head, followed by the realization. Followed by the physical manifestation of that realization: all the hair on my neck and arms is sticking straight out.

Glancing up from my arm hair, I look over to Brian more terrified than I've ever been, only to find those big brown eyes of doom just staring back at me, as if to say, "This is how it ends for you, remember?"

Only I don't accept it.

I knew I'd fucked up. That of all the times I'd pushed it too far, that I'd really gone and done it this time. But, even still, I don't accept that I'm actually about to die.

So now I look over to Cali hoping she's gonna have some kind of way out of this, only the look on her face says the exact same thing.

You took way too much N-Bomb bro.

I knew I was about to die, and worst of all, I even knew how it was going to happen.

Elevated pulse, vasoconstriction, confusion, delirium—that part already happened. Things I had to look forward to were seizures, tachycardia, hyperthermia, multiple organ failure, and cardiac arrest. Same as all those teenagers I'd recently heard about, I was going to be another sad statistic. About to be handed a Darwin award because I was too lazy to perform a simple Ehrlich test on those tabs.

The next now moment my vision flattens out into these two, two-dimensional sheets that keep drifting apart.

The next now moment my dead best friend flashes in my mind and I realize my entire life has been about this moment: *I've always been about to die, and here it finally was.*

My entire life is shrunken down into a series of jump cuts that keep getting faster and faster, and for the first time in probably ever, there's no egotistical narrator standing on the sidelines helping me to excuse away all the pain; there's just the pain. All the lies and all the trouble I've caused, I see it perfectly encapsulated inside the waveform interpretation of my life's song.

In the second to last moment, I'm utterly ashamed of what I've become. In the last, last moment—when I'm collapsing to the ground—I somehow accept the monster I've become, and let it all go. After that, my head hits the floor. I lose consciousness. And the room I wake up in next is definitely *not* my apartment.

—

Collapsing on the ground, I let go of that last breath. I watch my vision change from focused to gestalt. The kind of dolly zoom, rack focus, vertigo effect filmmakers always use to show a character having an intense realization happens, except when I go to re-focus my eyes, I no longer know where we are.

Looking out across the only plane I can see, I can tell Cali's there, only she doesn't look like Cali. Some of her features have changed. Her eye, for example, is now vaguely reminiscent of a predatorial cat.

This place we're supposed to be in, it might be most readily described as being "outside of time" or inside *the Possibility Space.* But really, it's just the kind of mental summer home you could accidentally stumble into if you only spent about ten years seriously looking.

Here, I'm looking into Cali's right pupil, and inside this pupil I notice a single point. Inside that point, I see every female I've ever interacted with.

She's my mother.

She's my first kiss.

Then she's "Cali" again.

From whatever the hell this place is, I feel the thing Buddhists, Hindus, and meditators spend upwards of ten years trying to experience, and when I do, there's no more fear because here...*I remember.* All the knowledge I've ever had feels like it's playing at the same time—and at the same volume—although, what you might expect to sound like a train wreck, actually sounds like a symphony. There's this grand orchestra of information being played, and because I somehow remain conscious during the whole thing, I'm able to witness all that there is to see.

How conscious works, how it stacks up, how it evolves, *and why.*

Buddhists call this experience "knowing," because in this moment, there is no subject, there's just the knowing.

A state of complete selflessness.

No longer am I just a "poor little me" as the philosopher Alan Watts might say, because in this moment, I'm something larger.[6] No longer is Cali just Cali, because in this moment, she's something larger than me.

We stay here long enough for me to remember just how much I hate it. And then, because I finally have access to a memory of "before," I just start chanting. I just start chanting, "I'll remember. I'll remember. I'll remember," over and over, until my entire visual field is filled with tiny flakes of something red.

Specks of red set against the background of scuffed hardwood.

Had I done it?

Had I actually just willed us back from whatever the hell that place was?

As quickly as I had gotten us unstuck in time, we were suddenly thrust back in a world where time ticks and conscious experiences are embodied. I'm crouched on the ground with hands on both sides of my head, staring at the floor, when I glance over to find Cali.

The way she's blinking her eyes, it's like she's just woken up from a coma. Probably because for the last...however long we were gone...

neither one of us felt like we had a body. So to suddenly be back, in Kore-atown, with a body—and the normal passage of time—I still take to be the greatest gift I've ever received.

Reclaiming my earthly body, I pick myself up off the ground and shout like a cowboy after a train heist.

Cold night air still seeps in from the chair-sized hole in our living room window, but I don't care. There's still glass and crushed pharmaceutical-grade MDMA scattered all over our living room floor, but none of that even matters because, somehow, I just survived my own death and willed us back from whatever the hell that place was. Nodding my head in approval, I do another victory lap and shout again. Never have I ever felt more alive than I do in this moment, and all I really want to do is go over to my notebook and document the momentous occasion: I am alive! I am conscious! And everything about me feels shiny and new.

This was heaven.

Chapter 26

What Really Happened

TURNS OUT THAT wasn't N-Bomb. I didn't actually die. And I didn't actually bring myself back to life. Don't get me wrong, those experiences all happened, it's just the way my first transcendent experience chose to manifest itself was directly after I went through a terrifying little scenelet, in which, I genuinely believed I was about to die.

Because before I could reach "heaven" first I had to go through "hell." Or, put in another way, before I could reach nirvana, first I had to confront/end all of my karma.

To reach transcendence, I genuinely had to *think* I was dying, *feel* I was dying, and then come to terms with who I had become as a person. And Tom, let me just say that Professor Roberts would've never wanted to see what I saw that day. My ugliness without all the filters. My patterns of destruction without all the excuses and justifications the Narrative Self provides. The moment I saw my entire life splayed out in front of me like it was on a reel of film, it should be reminded that this wasn't like your normal first-person experience of consciousness layered with commentary and point of view. Rather, this was just raw patterns of me being a cheater, or me being a liar. Raw instances of me thinking, I was more special than everybody else.

Truth was, I was a hypocrite and I engaged in half-truths all the time. One morsel of truth pressed between between a sandwich of lies.

Kind of like McDonald's, only this shit was coming out of my mouth.

Worst of all, I even felt *justified* during all this. And why? Because I've had a few really lousy experiences happen to me over the course of my life, that's why. The details aren't important, as we all have sob stories we could spin. The difference between you and me is, I used my pain like some kind of a wild card to get, do, and take whatever it is I wanted.

And *why?*

Because, as gross as this next part sounds, but I still felt the universe owed me something. Because my parents got divorced when I was twelve. Because I lost my best friend at age nineteen. Because I lost my faith in women, friends, and humanity at age seventeen. Because of all these things, my Narrative Self concocted the perfect trifecta of pain that I could use, like a wild card, to beat down any argument poor, old Jiminy Cricket ever tried to muster.

Did I realize this is what I was doing?

Not until the night of Koreatown. Up until that night this behavior pattern was completely unconscious to me. Much like a piece of malware that starts up the second you turn on your computer, I had this giant ball of emotional scar tissue that sat on the desktop of my mind affecting every decision I could make.

It kept me closed off. It kept me small. It kept me from having open, honest relationships with people I genuinely cared about.

Now, I'm not going to sit here and tell you some story about how it only took one good look in the metaphorical mirror for me to turn my whole life around. But, after my head hit the floor, it was like the real deep-down me finally got to see the malware wild card for what it actually was. And if I were ever going to use it again, I knew it was going to be my conscious choice. Essentially, there was a dialogue between a part of my mind we're calling "the Awareness" and the Subconscious OS, and that dialogue went something like this:

[*AdminChat*] Now chatting in channel: "*Possibility Space*"
Admin: Look James, this is who you were.
[Cue all the horrible things I'd done].
Admin: This is why you say you did it.
[Cue all the horrible things that've happened].
Admin: And now, I'm about to remove all these painful memories from the desktop of your conscious experience.

[Cue memory_purge.exe]

[Cue defag_recontextualization.bat]

Admin: If you still want to access these memories, you can, but no longer is there going to be the "hundred icons of pain" sitting on your desktop, affecting all of your decisions.

And then I'd say something like, okay.

And then it (my highest self/Subconscious OS) would say:

Admin: Okay so your brain's about to restart, I'm about to show you two final images: The first, of who you were (a high resolution snap-shot of who you actually were).

The second, an image of who you still could be.

[Cue waveform_interpretation.tiff]

And it's just this giant photoshop style image you could keep zooming in on forever and ever...but it's painful.

[Cue potential_james.jpeg]

And after seeing that image, I have chills on the back of my neck.

—The next thing I know I find myself chanting, "I'll remember, I'll remember, I'll remember," with hands on both sides of my head.

The next moment I was "James" again and the dialogue was over.

A breath or two after that, being so impressed that I'd just willed myself back from whatever the hell that place was, I shout "Wooo!" like a cowboy after a train heist. If this was a hard reset to my mind's OS, then I'd just gotten an update:

This is where you get malware.

This is how malware slows down your CPU.

And now you can't say, "I didn't know" anymore, because I literally just told you.

Debriefing & Story B

Too often, at conferences and events, I'll hear colleagues of mine toss around words like ego death and ego dissolution, when the reality is, it can look exactly like that.

Pure chaos and broken glasses.

In all seriousness, unless you're a chemist, LSD is rather hard to overdose on, as you'd literally need thousands of doses.[1] But, in almost the very same breath, if you aren't 100% sure about the source of your compound, does that part even matter?

I mean, if you don't know for certain what it is you're taking, how are you supposed to know the difference between an internal fear narrative and a legitimate concern?

For Cali and me, I take it as a matter of luck that our tabs were overlaid with genuine LSD and not some other research chemical like 25I-NBOMe, because it really could have been otherwise. In actual fact, I was sold DOB as LSD two times before that night, so this concern of mine wasn't completely out of the random. Even so, the real issue at heart is that Cali and I didn't Ehrlich test our tabs, we didn't set a clear intention, and we certainly didn't respect this molecule in the way that we should've.

The result of all this poor planning, we already went through.

The emotional revelations gifted from this experience, we briefly touched on as well (See *Emotional Coding*). Indicating to me, that we can now try and explain the experience/idea one often has the moment *after* they transcend the death of their ego. An experience which can be likened to the following illusion.

Named after the psychologist Edgar Rubin, this is another one of those cleverly drawn images that can be interpreted in one of two ways: *Story A*, where you see the vase, and *Story B*, where you see the two faces. Now normally, the brain would go on alternating between the two interpretations that are possible every few seconds, but here's where I ask you to consider the possibility: *What if it didn't?*

"Rubin's Vase"

What if, for argument's sake, you lived your entire life without ever seeing the two faces, until one night when you and your former partner ingest a few hundred micrograms of some fungus derivative and for the first time in ever you see the two faces.

It's so clear! So obvious!
Why didn't I see this before?

But then, anywhere from a couple seconds to a couple hours later, the orientation suddenly switches back and no matter how hard you try, you can't seem to see the two faces. In a nutshell, that's the transcendent experience. Except, that's the shrunken down billionth scale model about what actually went down.

To scale this analogy back up, you'd have to imagine that your entire conception of life is akin to the vase in that illusion. Basically, you'd have to imagine that there's some kind of standard view about life that most people subscribe to—a *Story A* type lens, let's say—and that, every now and then, someone catches hold of a *Story B* perspective.

A new interpretation about life that completely changes everything.

For instance, Newtonian mechanics was the *Story A* perspective that everyone accepted for over two hundred years. It wasn't perfect, it didn't describe everything about gravity, but this model of the world was 'good enough' until about 1905 when a man working in a Swiss patent office came along and tore it all down. He didn't try to build on top of what Newton did, rather he saw where Newtonian physics broke down and then offered us something better (an interpretation about gravity that was far more comprehensive).

More specifically, it was the idea that gravity wasn't a force, it was the curvature of space-time. In this way, Einstein was looking at the universe—the same universe Newton was looking at—only when Einstein stared up at the sky, he saw something different. Instead of seeing space and time as separate, he saw space and time as conceptually linked.[2] An idea that we accept as gospel now, but at the time, this was equivalent to seeing the faces in that photo.

In essence, Einstein threw out the fundamental assumptions of the time (space and time are separate, time ticks off at the same pace for everyone) and pitched us a *Story B* perspective—and the guy was actually right! Demonstrating, that there's great benefit in tossing out all your preconceived notions from time to time, because like the philosopher J. Krishnamurti once pointed out, "a mind that already knows, cannot possibly learn."[3] Therefore, if we seriously are trying to understand something new about the study of consciousness, perhaps we too could benefit from re-examining our intellectual starting points.

The Birthplace of Story B

Back in the 1970s, Professor Richard Dawkins turned the field of evolutionary biology on its head when he published his international bestseller *The Selfish Gene*. And while this book wasn't necessarily bringing any new facts to the table, it did provide an alternative way of viewing the facts, which at its heart, is really the spirit behind any transcendent experience. Granted, the following summarization is never going to hold a candle to the experience of actually reading this highly celebrated book, but its recapitulation should help advance our conversation immensely.

To set the scene, ever since the neo-Darwinian synthesis of 1930s, there's been a heated debated amongst biologists about the level at which natural selection acts. Or, to put the matter more delicately, *who* adaptations are said to be 'good for.'

Are Darwinian adaptations 'for the good of' the group or the individual? Nit-picky as it may sound but this distinction actually matters. And what Professor Dawkins basically argues is this: We're so used to looking at evolution through the lens of the individual or the species, that we've failed to see something hiding in plain sight. Namely, that adaptations aren't 'for the good of' the individual but for the genes themselves. And by using the word "gene" in this way, Dawkins was, of course, referring to the individual bits of heredity that are contained within the molecule of DNA.

***Notes from the Editor:*

For purposes of clarification, this wasn't an idea that Professor Dawkins came up with entirely on his own, George C. Williams, W.D. Hamilton et al., were also pushing for this point of view. However, it was Dawkins' book that finally tipped the scales in favor of this way of thinking.

—Regardless, it was a well thought out and well executed argument in the form of a book which got people to radically shift their ideas about how evolution works. Through numerous field studies, thought experiments, metaphors, and reason, Dawkins gets the reader to start imagining life from the perspective of a gene. Championing the idea that when you start to consider life from its point of view, you really start to question what's alive and who's in charge. Of course, we like to think that *we're* the ones driving the ship, and that we live a long time, but that's only if you completely ignore life from the perspective of a gene.

A perspective which Dawkins takes a couple hundred pages to really flesh out.

The way he speaks about DNA, he likens them to being like these "immortal coils" (tightly-wrapped crystallized bits of information) that essentially just use lifeforms, like a vehicle, to get into the next generation. Aside from this being scientifically true, it's actually an extremely elegant and philosophical point of view Dawkins espouses. Because, according to this view, it isn't the elephant or the eagle that's in charge— it's the genes themselves. The genes that built the body you're sitting in. From this perspective, both the eagle and the elephant are just temporary vessels, built by the genes, so that they can ultimately survive and get into the next generation.

> Genes are immortal in the sense that the coded information they contain is reproduced with almost total fidelity. Significantly, not absolutely total fidelity, generation after generation after generation. Such that, there are genes that are identical to what they were tens of millions of years ago. Hundreds of millions of years ago in a few cases.
>
> - Richard Dawkins

To be crystal clear, it isn't the literal molecule of DNA that's being passed down, rather the information contained within it. So to use the analogy of DNA being like a blueprint, it isn't the physical blueprint paper that's being passed down for millions of years, although the information written on that paper most certainly is.

At the time when this was published, group selectionism was rather popular, and so, consequently, evolution by natural selection was being interpreted as being 'for the good of' the species or group, rather than the individual or genes. Yet, as Dawkins contends, this is the wrong interpretation. Making the case that it's really the genes that persist down the generations, not the organisms themselves.[4,5]

> The genes that survive are the ones that consistently provide
> slightly longer necks, slightly keener eyes or improved camouflage
> and so help their vehicle to survive and therefore pass those same
> genes on. The 'Survival of the Fittest' really means the survival
> of genes because it is only genes that really survive down through
> many generations.
>
> - Richard Dawkins, *The Fifth Ape*

Once you adopt the gene-centered view of life, he says, what you start to notice is that "an individual is a *survival machine*. A throwaway survival machine for the self-replicating coded information which it contains."[6] If the sun didn't destroy DNA, arguably, that's all there would be: just coils of DNA or proto-DNA floating around making copies.

But, as it turns out, the sun does destroy DNA, which is how skin cancer even develops. Too much exposure to our host star literally breaks apart the code of DNA causing an uncontrolled growth pattern we call melanoma. So, out of necessity, the early replicators figured out a trick; make a membrane; make a body. Make a 'machine' that can preserve its genes intact, such that, genes can replicate successfully and ultimately get into the next generation.[7]

> "A monkey is a machine that preserves genes up trees, a fish is a
> machine that preserves genes in the water; there is even a small
> worm that preserves genes in German beer mats. DNA works in
> mysterious ways."
>
> - Richard Dawkins, *The Selfish Gene*

To really drive home his way of looking at life, here's a little thought experiment designed to help us start reconsidering life at different time scales:

The Magic Camera

Imagine owning the greatest time-lapse camera ever conceived: a camera that was ideal for shooting over extremely long periods of time, at any conceivable frame-rate, and with every lens/filter you could ever dream of. A truly magical camera.

With just a few clicks of a button, you could be shooting in slow motion; with another couple clicks, X-ray vision. Basically, whatever lens you could ever hope for, this camera's got it. It even has a little switch on the side that allows it to perceive nothing but raw genetic code (ATGACGGATCA).

Supposing that were the case, and you had the camera pointed at yourself, that means you could perceive the three billion base pairs that combine to form your unique genetic code. Or if you had it aimed at your brother, you could literally see the slightly varied genetic code that makes up his DNA. Armed with this camera, we can now imagine going back in time and setting it up to record the entire history of our planet, but—and here's the important part—in super fast time-lapse. If we were to then play this footage back at a speed where decades whizzed by faster than seconds, agreeably, the first little while would be a little bit boring. Nevertheless, by the time more complex life started to arise (fish and other sea dwelling creatures) because of this DNA filter, we wouldn't be seeing their bones, scales, and bodies swimming around—we'd be seeing their unique genetic code flitting about. And keeping in mind that this is all happening in time-lapse, instead of seeing larvae hatching out of their eggs, we'd be seeing unique strings of genetic code trying to burst through. Hundreds, perhaps even, thousands of little codes all springing from a central source. Like a colorful bouquet of flowers. Except, instead of varying in color, each sequence would actually carry a small genetic variation that sets it apart from being just a mere copy of its parents or its siblings.

Without a doubt, being able to witness DNA mutate in front of our very eyes would teach us a lot about the nature of time and evolution. But perhaps even more exciting would be our ability to examine the ebb and flow of gene frequencies in gene pools. Using this camera, we could watch how genes that are good at surviving tend to get passed on, while other genes (other lines of code like AACCGCTGTAT) just spring up once and are never to be seen again.

Examining this footage, we'd surely notice certain lines of code which get passed down for hundreds, maybe even thousands of years, while others persist down the line for well over a billion (See *Cdc2 gene* for cell division).[8] Toggling back and forth between the DNA filter and the normal video filter, you could even begin to jot down which sequences belonged to which animals while, at the very same time, observing the way in which an organism's genome differs from parent to child, or child to grandchild. At any rate, you'd finally be able to see the way in which an entire species' genome changes over the generations, either in response to its environment, or in response to all the evolutionary arms races it's currently involved in. Meaning, you could finally observe how genetic improvements on one side literally *force* counter-adaptations to start evolving in the other. Or, if you were watching close enough, you could even track how various environmental factors (climate, source of food, competitors) impose certain selection pressures on one group in a given region, and not another, ultimately leading to divergence and speciation.

So, supposing you had this camera down in southern Africa around 60,000 to 70,000 years ago, you could actually stand back and admire as that first big wave of *Homo sapiens* began marching out of southern Africa, spilling into the Middle East, traversing through India, and eventually making their way into Europe.[9-12] All the while marveling at the incredible lack of genetic diversity that exists between all of us.[13,14] Scratching your head, you might even speed the video up to see how all this overwhelmingly similar *Homo sapiens* DNA ends up spreading itself to almost every end of the globe. With each new settlement, these tightly wrapped coils of information would effectively be creating new gene pools under new environmental conditions, leading to an asymmetry in selection pressure, which ultimately results in these settlements diverging away from one another.

Carefully scrubbing through the footage, you and I could scrutinize over which genes have become more popular in the gene pool of India vs.

the gene pool of Europe. And by studying the accumulation of these tiny differences, we might even begin to tease apart which genetic mutations started causing Population X to start looking different from Population Y. One example of which being, Europe, sometime between 6,000 and 10,000 years ago. If we'd been present during that time, then we'd actually have been able to see the mutations happening along all the genes that code for eye color. Recognizing the mutation, you might even leap out of your chair and shout, "That's the first blue-eyed person!!!" Carefully tracing on a map how this one mutation ended up going viral, scattering copies of itself all across the globe, reaching nearly 10% of the population today.[15,16]

In terms of what it means to be a successful gene, 10% of the population isn't bad. Granted, all the bodies in which these blue-eyed genes sit will all eventually turn over, but it's still possible that the raw information that codes for the blueness in eyes (that string of letters) is potentially *immortal*.

—

The immortality of genes, the transience of individual bodies, this among many other poetic notions in regards to evolution were charmingly made back in 1976. But for the purposes of our discussion, this way of looking at the world (the gene's eye view) is really what we're after. Especially, when you can hold this perspective and simultaneously consider a few other beguiling facts about DNA. Like the fact that all animals, plants, fungi, and bacteria we've ever found all share this exquisitely wrapped set of instructions; or the fact that if you compared your DNA to anyone else's on the planet—even a pygmy from West Africa—the difference would still be less than a tenth of one percent.[17]

Hell, even the difference between you and a chimpanzee is only about 2.1%.

From the opposite perspective, we could say, that if you printed out the entire DNA sequence of a chimpanzee and that of a human being, and then laid them side by side, they would be 97.9% identical.[18] If, however, we were just counting protein-encoding sequences, chimps and humans would be 99.1% identical.[19] Mice and men, 85% identical.[20]

So to bring this all back to what we know about computers, all computers use a two-digit language we call binary. They all use a series of 1s and 0s to transmit, process, and store information. At the same time, it's important to mention that none of this information exists in any one digit, rather, it's located in the relationships between the digits. In other words, it's the pattern of 1s and 0s that make binary information what it is, and the same could really be said of DNA. Only DNA has four letters instead of two. The four base pairs of DNA are adenine, cytosine, thymine, and guanine (abbreviated to A, C, T, and G); and it just so happens that all living creatures use this same code. Meaning, the only real difference between you and a mushroom is the *sequence* of A's, C's, T's, and G's.

Taking all this into account, one view you could adopt about this life is that this planet is just completely riddled with DNA. From the bacteria living in your mouth, to the bacteria living in your gut, all the way down to the cracks and crevices deep in the ocean floor, life has somehow figured out a way to survive. And the way it does this is by altering its blueprint. Not entirely. But by building and sitting in the types of bodies that are best adapted for each environment, this one molecule has come to dominate the entire planet.

———

During Koreatown night, I didn't quite see what the "magic camera" sees, but that's about as close as I can get it for this section. In the simplest of terms, when I looked out at Cali, the filters of memory, belief, expectation, and categorization were no longer functioning properly. Normally, they'd filter and give structure to the incoming data, in much the same way that a search filter like "Open Now" or "Sort by Distance" might constrain the data of a given search query. In this sense, having zero filters meant the sense data coming in was no longer being constrained by any form of categorical structure. It was just pure raw data. And so when I finally peered out at Cali with "untrained" eyes, I no longer saw someone that was all that different from me. Matter of fact, I saw something that was 99.9% the same.

Traditionally, whenever I'd be walking around town before this night, my brain would focus intently on all those 0.1% differences. And if I'm being perfectly honest, that's all it would focus on. It would identify, categorize, and list all the ways that made this person different from that person. And yet, while I existed inside *the Possibility Space* the one thing I noticed was that my brain was finally able to focus on the overwhelming similarity.

Instead of focusing on everything that made us different, it started focusing on everything that made us the same.

The 99.9% view, not 0.1% I'm used to.

The reason I found myself chanting, "I'll remember, I'll remember, I'll remember," is because *Story B* is what you get when you stop focusing on the tiny tenth of a percent, and I was so desperately trying to hold onto the memory. Delysid had unzipped my conscious experience all the way down to its most basal operating layer (the source code; the Subconscious OS; a kind of proto-conscious arena completely free from all the filters of memory, prediction, and expectation). And while I'm down there, a part of me gets to see the most primary visual processing possible. Which isn't all that unusual, as there are numerous anecdotal reports of that happening before. But this unitive experience bit, where people report feeling "one with Nature" or "one with the universe" is a component Ava and I find to be massively underappreciated. Especially, when you take into account that it's one of the hallmarks of a transcendent experience, and that it happens so often, we even have a name for it: oceanic boundlessness.

But still, that's where the conversation usually ends.

Researchers never deny the experienced loss of ego boundaries associated with a transcendent experience, nor do they deny that people commonly use words like "infinity" or "eternity" to describe it.[21-24] Unbelievably, they acknowledge this dimension of experience resembles the so-called mystical-type experiences reported in religious exaltation.[25-28] But even despite all of that, that's *still* where the story usually ends. The difference with Ava and me was, that's where our story began.

Part III
Origins

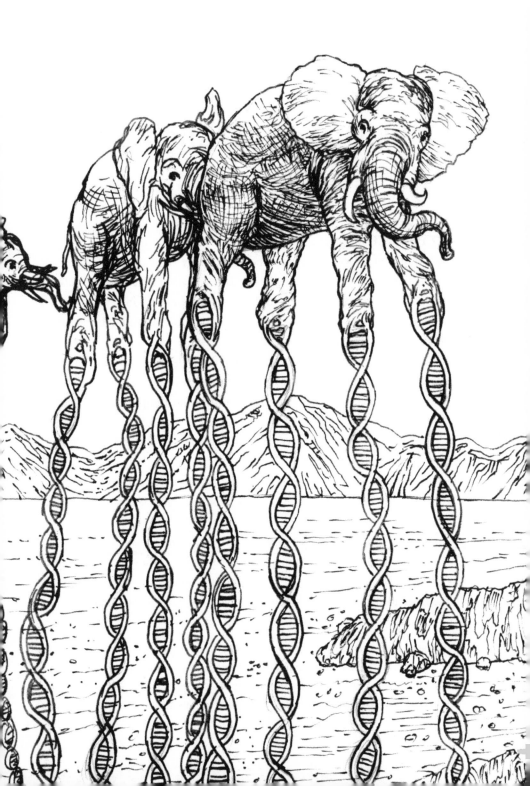

Chapter 27
Gold, Diamonds and CS5
(*The Origin of Everything*)

DO YOU KNOW where gold comes from? How about diamonds? They're both quite interesting stories—and completely relevant to our topic of discussion—although, with movies like *Blood Diamond*, I think it's rather common knowledge that diamonds aren't rare. Matter of fact, we can even make them in the lab with as little as four simple ingredients: carbon, pressure, heat, and time, that's it.

And where do we find conditions like this? Deep down in the Earth's crust; a place geologists call *the mantle*. Down here, with pressures upwards of 725,000 pounds per square inch and temperatures around 2,200 °F, coal gets compacted so hard that its carbon atoms are forced into a lattice-like structure. Therefore, under extremely difficult conditions, carbon becomes diamonds.[1]

On the contrary, gold is actually quite rare, and unlike diamonds, gold does not form here on Earth. Truth be told, gold is actually formed inside stars like our Sun.

Correction: **Stars much bigger than our Sun.*

To explain how stars make gold, we're going to have to back-track a little, but trust me when I tell you that it's worth it. Because once you understand the broad strokes of how the universe created itself, I'll be able to tell you my favorite story of all time.

***Notes from the Editor:**
In this section, the narrator finally begins to marry a Story A type interpretation about life (the standard model of cosmology) with a Story B perspective. For this reason, it is advised to consume this chapter all in one sitting.

A Crash Course in Cosmology

Let's begin with a startling fact: We live in an ever-expanding universe. It isn't static; it isn't fixed; it's progressively getting larger and larger. And the way we started to figure this out was with a man named Edwin Hubble. The year was 1929, and aside from discovering a whole range of new galaxies, Hubble also observed that all these distant galaxies were moving away from one another. A discovery which ultimately led us to the realization that the universe was, in fact, expanding.[2]

How do you picture a universe expanding?

By imagining the universe as being like a big blob of raisin bread dough. As the dough starts heating up in the oven, all the raisins start getting pushed further and further apart, and this is precisely what's going on in our universe. All the galaxies are moving away from one another, uniformly, in this same manner. And because they're all drifting further and further apart, cosmologists were able to reverse-engineer the birth of our universe by first rewinding the tape. Because when you reverse the rate of expansion, eventually, all those galaxies converge to a single point 13.82 billion years ago, intimating that the entire observable universe was once contained inside a space much smaller than an atom.[3]

Sample illustration below.

Looking at this simple depiction of *the Singularity*, right out the gate we should point out that, at this time, there wasn't time. There wasn't space, or mass, or gravity because, at this stage in our story, all the energy and all the matter that exits everywhere "right now" was once all neatly packed and folded in on itself like the most epic IKEA box you ever saw.

Then, somewhere between 10^{-43} and 10^{-36} seconds after the universe first popped into existence, which is a zero followed by a decimal point and forty-three zeros, the force of gravity splits from the grand unified force, causing the universe to start expanding rapidly. Going from the size of a subatomic particle to the size of a golf ball almost instantaneously, expanding faster than the speed of light, until 380,000 years later, when the entire universe actually looked like this:

Cosmic Microwave Background Radiation
Image Credit: ESA

This is the earliest baby photo of the universe that we have. Taken by the Planck satellite, and as you can see by the tiny temperature fluctuations present in this photo, the early universe wasn't completely uniform or smooth. In more specific terms, there was an imbalance of matter present, such that, even back then, the seeds to countless galaxy clusters were already underway.[4,5] The orange and red blobs in this photo, that's where more hydrogen was.

Now, during this stage of our origin story, it's important to add that the only elements that existed in the entire universe were atoms of hydrogen, helium, and trace amounts of lithium...*there were no other elements*.[6] And so, not surprisingly, the universe was more chemically homogenous than it is today. Yet, despite these humble beginnings, these simple ingredients were enough to get things started, because once the universe cooled off enough to allow raw energy to congeal into matter, gravity finally had something to work with.

Flash forward about 200 million years after that and now, atoms of hydrogen are so close to one another, that they're starting to collide, creating friction and heat. Inevitably, the internal temperature of these clouds soars to about 10,000,000 °F, and something magical starts to happen. Protons begin to fuse. And, even though it's probably been a while since your last day of high school chemistry, I'm sure you'll never forget what was at the top left of that table they had hanging on the wall.

Hydrogen!

The simplest element there ever was. And hydrogen, as you'll probably recall, only has one proton and one electron. Because of this, when these clouds of hydrogen finally did reach the threshold temperature of 10,000,000 °F, the resulting fusion of its protons created the second element on the Periodic Table, which is helium. An element composed of two protons and two electrons (element number two).

This process is what's known as nuclear fusion, and this is what our Sun is doing every single day. It's fusing atoms of hydrogen into atoms of helium and then releasing a tremendous amount of energy in the process. The amazing part is, at the center of really big stars, it actually gets so hot that not only do atoms of hydrogen combine to form helium, but they also combine to form more complex elements like carbon, oxygen, and nitrogen. In truth, this process happens all the way down to iron. Although, whenever a star starts fusing elements into iron, that's the death note of a star, given that iron doesn't fuse with anything. Once this starts to happen, the star starts running out of fuel, the core starts to harden and crush in on itself, and *BOOM!*

The star goes supernova.

And when a star goes supernova, that's when atoms of hydrogen can fuse into really interesting combinations like platinum, silver, and gold.[7]

Now, I want to pause for a second and just reflect on how remarkable that is. We haven't known the true origin story of gold until recently, and yet, it's always been a prized commodity.

Which kind of makes a gold ring a little more interesting now, doesn't it? To know that it was forged inside the heart of a dying star *billions* of years ago. That it then shot out from one of the most destructive forces in the entire known universe, traveled across vast stretches of space and time, crash-landed on the third rock from our Sun, and then sat in the ground for billions of years. Eventually, one species of ape thought it might be special, so they mined it from the earth, melted it into a circle, and now that glittering, interstellar traveler is sitting on your finger.[8]

That's the real story of gold, and it's my second favorite story to tell. So, if you're still with me, then let's proceed to my favorite story of all time, which is: *How we got here.*

How Did We Get Here?

Let's recap. For the first 200 million years, the early universe was mostly dark and cooling off from the initial explosion. Then, right around this time (T = 200 million years) our first stars began to turn on, leaving behind all the elements they made during their lifetime.

But then more stars ignited. And more stars ignited. And more stars ignited, making the universe much more chemically diverse than it ever was before.

Flash forward to about 9 billion years after that, when another little star starts to form in a region of the cosmos that would soon come to be called the Milky Way.

Now, what's interesting to note about this particular star is the fact that when it first started to form, there was already such a rich milieu of elements and particles present. Because of this, the Sun whipped up all the dust and debris around it, giving rise to rocks, and asteroids, and planets, and moons; and this is how our solar system was formed about 4.5 billion years ago.[9]

Naturally, the solar system wasn't as neat and orderly back then, but since I'm still dying and I'm still trapped, let's fast-forward another billion years or so, because by about this time, something rather interesting starts to happen on that little ocean planet right there. Thanks to the Sun shining and the Earth being just the right distance from the Sun to allow oceans of liquid water, all these elements, chemicals, and amino acids (i.e., all the building blocks of life as we know it) were free to swirl around and play with one another.

At some point—and we don't exactly know how yet—but these molecules linked up in such a way that an organic replicator was born, and by that I mean *life* came into being.

Irrespective of how it happened this was, arguably, the most primitive life-form of all time. Something like bacteria. But details aside, it was almost certainly some kind of single-celled organism that was alive, and that could make copies of itself. Then, thanks to a copying error in the replication process, this single-celled life-form started spreading out, diversifying, and evolving into every living thing you see around you today: plants, trees, mushrooms, bananas, people, orangutans, everything on this planet—*everything*—came from that one seed of life.

How do we know we all come from the same origin?

Because, again, we all share that beautifully wrapped double-helical template for building bodies we call DNA.[10]

What's more, if we really start to consider what life is—on a fundamental level—ultimately, we're reduced to the idea that life is really just energy and information. Information that is "alive" and information that can make copies of itself (viruses and RNA, we can talk about later).

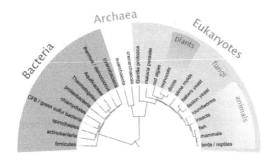

Image Credit: Madeleine Price Ball

Now, when life is rather simple, like bacteria, it just makes copies of itself in a process called *meiotic division*. Basically, it just splits in half, and you've probably seen a million videos of bacteria doing this. The first cell splits in half and then you have two. And then those two split in half and then you have four. And then those four split in half and then you have eight, until pretty soon, bacteria is all over your sandwich.

Just to give you an idea of how fast this process can occur, the bacteria E. Coli can divide itself every twenty minutes.[11,12] That means from one parent cell, E. Coli is capable of creating a million other copies of itself in about seven hours.

No eHarmony, no dating, no sex.

They just swell up and divide, making an error in the replication process about once in every billion rungs of DNA.[13] Which, when you think about it, almost sounds like too low of a mutation rate to even have an effect—that is, until you realize the amount of generations bacteria can get up to in a day. Or 5,000 days? Or 3.5 billion years?

Because once you fully start to appreciate the amount of generations that must've taken place between then and now, complexity starts to become the rule, not the exception. Because once you combine some form of genetics with inheritance and selection, you arrive at the evolutionary algorithm necessary for squeezing design out of chaos.

From this perspective, it seems inevitable that a random mutation would occur which allowed bacteria to develop the sorts of proteins necessary to sense the difference between light and dark, just as it was inevitable that a beneficial mutation like this would get passed on. Thus, however life got started—once it finally got going—it was constantly evolving, re-configuring, and adapting itself to the ever-changing environment. Different forms of life started evolving different tricks to survive, different tricks to survive eventually gave rise to speciation, and so, from one seed of life, this essential life-form began evolving down different evolutionary paths, in much the same way that every branch on a tree grows according to *its individual trajectory*. On top of that, just as many branches and twigs sometimes get blown off by the wind, the tree of life has also seen more than its fair share of destruction, given that 99.9% of all species once living are now extinct.[14,15]

In light of this fact, one way to start conceptualizing the way life has evolved might be to recall that spherical shape-sorter toy we all used to play with as kids. The one where you'd put the plastic star in the star-shaped hole, and the rhombus in the rhombus-shaped hole, and the circle in the circle-shaped hole.

Well, I'd like for you to imagine that the stuff of life (DNA) is really like this malleable, play dough like substance while, at the very same time, envisioning the environment as being like a complex shape-sorter toy.

If we were then to imagine squeezing play dough out from the center of this ball, what might we expect to happen?

Well, we would expect it to start coming out looking like a star on the on the right side, and a rhombus on the left side, and a triangle on the backside—and really, this is the simplistic view of evolution I'd like for you to grasp. The shape-sorter toy being like the harsh, ever-changing environment that's constantly setting up the shapes, and this primordial, malleable stuff of life as being the thing that always seems to 'find a way.'

Sex 101

Believe it or not, but there was a time before you knew sex existed, and crazy as this next part may sound, but there was a time before *life* even knew sex existed too. Because before 1.2 billion years ago, sex didn't even exist. Wasn't even an option. Which also means that prior to 1.2 billion years ago, most forms of reproduction could almost be better conceptualized as forms of cloning or growth.[16,17] We already mentioned how bacteria just swell up and split in half, so now let's talk about how plants do it.

Life-forms which make up 80% of the Earth's biomass.[18]

—

A lot of plants reproduce vegetatively by sending out suckers. What happens is, the parent raspberry bush will send out an underground root, called a shoot, which grows out a few feet from the old raspberry bush and eventually gives rise to a new raspberry bush.

So that's its way of making copies of itself. And if we wanted to, we could actually sever the underground root and give one of these plants to our friend. However, if we never severed the underground root, and just looked underneath the earth, would you still consider this to be reproduction?

 I mean, this just looks like growth more than anything else, right?

Strange as it might seem, but asexual reproduction, of this kind, is the method by which all life on this planet spread until about 1.2 billion years ago. Because right around this time, life finally figures out a new trick for increasing genetic diversity, which is sex.

Males contribute half the genetic material; females contribute the other half.

In our species, once the female's egg has been fertilized, it subsequently grows inside the mother's belly, until it literally becomes too much for her to handle. After that, the mother's water breaks, the baby is delivered, and what's the first thing the doctor does?

He cuts the umbilical cord.

He severs the root.

And *BAM*—you got yourself a new human, right? But I gotta ask you, Tom: Is it really a new human, or is this just a more refined version of growth?

I mean, I'm sure the legal system would finally recognize it as a new person the moment we severed that umbilical cord. But let's suspend that thought for just a moment, as we simultaneously start reconsidering that raspberry bush. If we never severed the umbilical cord—which is kind of a gross thought, but just supposing that were the norm—wouldn't it be much easier to start viewing human reproduction as just a complex form of suckering?

Because if we can accept that asexual reproduction, of the type I just described, is really just a different form of growing and staying alive, then, by the same logic, couldn't we consider human reproduction as being just a more complex version of that? Making sex really just another way for life to shuffle up its genes?

From this perspective, I think it would suffice to say that sexual reproduction is really just *growth*, but a bit more fun.

—

One last example of reproduction and we'll wrap this whole thing up. In Utah, there's this species of tree spanning for over a hundred acres that looks like this:

Now, it used to be that we thought all these trees were separate living trees, until one day, someone looked under the ground and discovered that they weren't all separate trees, but in fact, all shared an intricate and interconnected root system. Meaning, they were actually all one tree.

All fifty thousand.

So this collection of trees, sprawling for 106 acres, all came from one seed, weighs in at about 13 million pounds, and is considered to be the largest living organism ever found.[19]

One organism, they call *Pando*, which is Latin for "I spread."

And just like that raspberry bush, Pando reproduces by suckering too. A clonal colony of trees estimated to be about 80,000 years of age.[20]

So why am I talking to you about trees?

It's because I think we're like Pando, but on the grandest scale. And while we might not be able to see the root system with our eyes, we still feel it.

Sometimes when you're too focused on the details, you can't see the forest for the trees. And it would certainly appear, that amidst all the details of being a human, the vast majority of us have missed the most important point of all, which is that we're all immensely interconnected.

Evolutionary biologists have no problem talking about the Tree of Life as a useful tool to visually grasp that all life on this planet is related to one another (i.e., we're all cousins of one other). But in the final analysis, they always seem to fail at grasping the bigger picture, which is, that the Tree of Life is a living tree we're all still a part of and connected to.

You want it stated in slightly different terms?

The only reason you're alive right now is because you're a part of an unbroken line of ancestry. Not a single person alive today came into this world without a mother and father, and not a single one of their ancestors failed to reproduce. To be an ancestor you have to leave descendants, and so, we sitting here now, are the direct descendants of those ancestors who all had what it took to survive. Against this backdrop, when you trace the line of ancestry back through time, eventually you come to a point where all our phylogenetic trees converge to that last universal common ancestor that probably lived in hydrothermal vents at least 3.5 billion years ago.[21] And this works with anybody.

All humans share a single common ancestor (the so-called "mitochondrial Eve") who lived about 200,000 years ago.[22] So saying that we're all cousins of each other umpteenth removed isn't some kind of charming metaphor; it's literally true. You're cousins with every single human being on this planet, just as you're more distant cousins with every animal, plant, or fungi that's ever lived too. And look, you'd know this was the case if you were looking at any detailed family tree, or if we never severed the umbilical cord because, at that point, the connection would be unmistakably obvious: All life on this planet is fundamentally related, and I can even see the "roots" that connect us all.

But deeper than that, you're also a part of an unbroken line of energy that connects us back to the force that kicked off this whole thing to begin with 14 billion years ago.

The first law of thermodynamics states that energy can neither be created nor destroyed, it can only be transformed from one form to another. So, if that's true, then we are nothing other than the same essential energy of this universe expressing itself as that tree, as that fly, as you, and as me. Or put in another way:

We are not *separate* from this universe, we *are* this universe.

A point that probably flies in the face of whatever the Ego/Narrative Self would have you believe. But once the smallest part of your identity is allowed to die off—the ego narrative that says you're just a poor little Tom—and there's no more ego pollution clouding your view, you come to realize that you really are this universe.

"You did not come *into* this world, you came *out* of this world, just as a plant *flowers* or an apple tree *apples*, the universe *peoples*."
- Alan Watts (philosopher)[23]

Can we see the root system with our own eyes?

Technically, no. But as previously discussed, there are already so many layers of reality we're completely oblivious to that part shouldn't even matter. Tables, for instance, are mostly empty space. So are chairs. The distance between one atom and the next, as you know, is incredibly far. And yet, we still perceive tables and chairs as being solid and opaque. On top of that, the tiny fraction of light that our brains can even pick up on is: red, orange, yellow, green, blue, indigo, and violet. But beyond that there's infrared, microwaves, and radio waves (all forms of light we just can't see them). Birds see ultraviolet light. Pit vipers see infrared.

Bats can even see sounds.

These little synesthetes emit a frequency of sound so high in pitch, to our ears, it doesn't even exist.[24] That withstanding, are we really so biased with our interpretation of reality to say, "That sound doesn't exist," simply because it's not a part of our umwelt?

I mean, your phone's connected to the internet right now, but because you never evolved to see microwave light, you don't perceive that connection, and really, isn't that the whole beauty of science? To see things past our evolved senses? To know things beyond what our biology 'allows' us to know?

Brief Recap

So one last time, let me put it all together for you: When the universe first began, raw energy shattered into the four fundamental forces and congealed to form matter in the form of hydrogen. Thanks to gravity's ability to pull things together, clouds of hydrogen condensed and compressed, until they got hot enough to form stars. Stars are the factories that created all the elements in our entire universe, and as new stars formed, they began whipping up all the dust and debris of all the dead stars that came before it (creating planets and moons); and on **one** average-sized planet, in an average-sized galaxy, amidst a sea of trillions, there existed a planet that was just the right distance from its host star to allow oceans of liquid water. The energy from the Sun then mixed with the growing chemistry of the environment, somehow giving rise to a self-replicating life-form that grew as much as the environment would let it. Then, with the help of random mutations, geological barriers, five mass extinctions, and at least 3.5 billion years of geological time, one species eventually rose up to become the most dominant species on the planet. Then, one of those organisms (who I've renamed to Tom) happened upon a letter from a stranger and got way more than he ever bargained for. Not only did he get a coherent breakdown about the inner workings of his own mind, but he also came to realize that his conception of himself was nothing more than what I just said. A concept. A useful tool for navigating the world, but still, just a controlled perceptual hallucination generated by a network of neurons firing in your brain.

So what's the big idea here?

There are several, but perhaps most pressingly it's this: We humans like to think we're better than our environment, when the truth is: *We are our environment.*

We did not come *into* this world; we came out of this world.

You want it phrased in a more scientific light, then let's consider the five most common elements that exist in the entire universe (*listed in their order of abundance).

1. Hydrogen
2. Helium
3. Oxygen
4. Carbon
5. Nitrogen

Now, let's consider the four most common elements that make up the human body. Elementary knowledge already tells us what the first two elements are, seeing as we are mostly water. Yet, when you rank order the elements we've discovered human beings to be made of we also find that they are a one-to-one with the universe.

1. Hydrogen
2. Oxygen
3. Carbon
4. Nitrogen

The only exception to this rule being helium, on account that helium doesn't bind to anything, so really, we can just set that one aside. As Neil deGrasse Tyson once put it, "Not only do we exist in this universe, but it is the universe itself that exists within us."[25]

It's no coincidence that we are chemically the same as the universe because we *are* this universe. We are not separate from it, we are it.

Of course, Western society has brought us up to believe that we're all individual, organisms born *into* some kind of environment, when really, we're continuous with this universe in all the ways I'm about to show you.

Truth is, we are this universe *personified*, which sounds a little bit better if you say the first two syllables—pause—then say the rest of the word. Or, if you'd prefer the idea in picture form then try re-examining the photo of eyes and nebulas, once more.

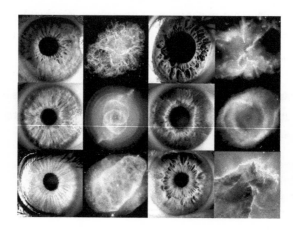

Nebulas, as you may or may not know, are stellar nurseries: giant clouds of hydrogen and dust where young stars are born. Really take a second to pour over this photo and then imagine someone whispering in your ear, "We are the universe looking back at ourselves," because that's what the transcendent experience sounds like.

It's that moment when you not only see the interconnectedness of all things, but you feel it as well. As the comedian Bill Hicks used to say, "We are all one consciousness experiencing itself *subjectively*."[26] As you, as me, as that tiger at the zoo, it's all one continuous process.

So, when you put all this knowledge together, you find that people are really just an extension of this universe just as an eye is merely an extension of your brain. An eye is a tool that the brain uses to see the outside world, and by the same token, a "person" is really just a tool the universe uses to know itself. We (the environment, Earth, you, me) we all come from that same Big Bang energy 14 billion years ago. It isn't that there's a universe…and then there's us living inside it…because that would imply they're separate things.

They're not.

It's all one thing.

One continuous blanket of energy that's unraveling itself throughout all of space and all of time. And once you *know* that, and you *see* that, and you *experience* that—there are no more lines. There's no more boxes, and there are no more problems with identity.

I mean, why would there be...we're all the same thing!

One of the fundamental aspects of every religion is, "Do unto others as you would do unto yourself."[27] It was first introduced to me as the Golden Rule, or the Golden Rule of ethics. But if you walk into the United Nations building in New York city, you'll find a poster hanging on permanent display which shows this same "ethic of reciprocity" rendered in all twelve of the major religions.[28] And why do you think that's so?

Did you ever consider that it might be literally true? That maybe our brain with all its processing power and filtering capabilities just screens this little part out? That maybe all these perceived differences in all of us are only magnified by the game we're all playing.

I won't lie to you Tom, this is definitely one of the biggest reveal moments I could ever hope to conjure up, but it isn't the only one.

And Tom, you really need to see how deep this game goes.

Chapter 28

UNITY

IT'S CALLED A phase transition. Ice becomes water. Water becomes vapor. Vapor becomes ice. You're wondering how it is that we could be all one thing? One energy? Well, the concept of phase changes or phase transitions might be able to help us with that one. Leave ice outside and it's melted by the sun becoming a pool of water. The pool of water eventually evaporates into the air traveling up into the clouds where it subsequently reassembles to form snow.

And this is just what matter does, we say.

Given enough energy, temperature, and pressure, matter is able to transition from one state of being, like a solid, to that of a gas, a liquid, or maybe even a plasma. With phase transitions, the constituent parts all stay the same, but the molecules recombine to form something else entirely. And the reason this happens, generally speaking, is because of external factors like temperature and pressure. Transitioning from solid to liquid, then from liquid to gas, and finally from gas back to solid, it's important to note that at every step of the way this is still H2O. All that's really changed is the simple addition of heat and/or pressure, which causes these molecules to start behaving differently.

See *Hail*

See *Steam*

See *Condensation*

All these forms, they're all still molecules of H2O, it's just the environment (namely, the temperature of the environment) that's providing the necessary pre-conditions for all these different properties of water to become manifest.

To help explain, how about another thought experiment.

For a moment, I want you to pretend that instead of a human being 60 to 70% water, they were 100% water instead. Supposing that were the case, and you still had the same conscious awareness that you do now, could you imagine how difficult it might be for me to convince you that you were also ice?

I mean, we could talk for days about the idea of a phase transition. I might even be able to reach into my bag and start pulling out photograph after photograph of beautiful intricate ice crystals—each shot with a 50 megapixel camera—and point at each needle cluster saying, "This is you...this is you...this is also you." But if you never saw your literal "Water Self" transition from water to snowflake and then back again, would you even believe me, or would you just cast me off as "just another fanatic" yammering about something that can't be seen?

Now, do you see how this is precisely the kind of situation we're in?

At its most fundamental interpretation, you are a living system comprised of energy. Right now, you exist in the form of a human, but that isn't the end-all-be-all of your existence, just as water isn't the end-all-be-all of its existence. Water can be frost. It can be an icicle. It can even be a glacier 250 miles long! Temperature and pressure dictate water's form, just as temperature, pressure, genetics, and so many other factors dictate *our* form. Temperatures below 32 °F result in molecules of water becoming rigid, fixed, and immobile. Temperatures above 212 °F result in the molecules of water starting to wriggle free from one another and become gaseous. So the idea of one thing manifesting under different forms isn't a really foreign concept to us. Matter of fact, the further we've progressed in the field of science, the more unity in Nature we've discovered.

For instance, we used to think there were distinct and separate colors that just existed out in Nature until a man named Isaac Newton came along and began unweaving the rainbow. By holding a prism up to the pinhole in his window shade, Newton was able to recreate the same effect water vapor has when it rises above a cascading waterfall. The prism he held refracted the light shining in, breaking it apart into all its constituent colors, and then producing a tiny rainbow across the room. Newton then placed a second prism in the path of the refracted light and was even able to turn that rainbow back into white light, proving that white light was, in fact, composed of all the colors. The year was 1672, and from this very experiment, Sir Isaac Newton gave us the idea of the visible spectrum.[1]

Then, a couple generations later, the astronomer William Herschel comes along and discovers something below red. He realizes that there's something more to the spectrum than we can even see, and in the year 1800 discovers a form of light that's invisible to our eyes but that our bodies register as heat. Infrared, we call it.[2]

A year after that, someone else discovers another type of light we can't see: ultraviolet light. Then microwave light gets discovered. Then radio waves. Then x-rays.

Flash forward until today's time when we've rounded out the entire electromagnetic spectrum, and realized something rather poetic, which is that gamma rays, x-rays, microwave light, the visible spectrum of light, and so on, are all made up of the same underlying substrate—*light*. It's just that gamma rays have the shortest wavelengths and the highest energies, while radio waves have the lowest energies and the longest wavelengths. But, of course, you and I don't perceive either of these because the photocells at the backs of our eyes can only detect three wavelengths of light (RGB). Out of which, we're able to generate the entire "visible spectrum."

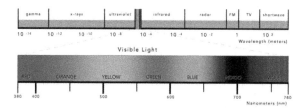

Even upon a closer inspection of that word "electromagnetic spectrum," again we're clued in to another bit of unity. Electricity and magnetism used to be conceptualized as separate and distinct forces until one day in 1820, when Hans Christian Oersted noticed the electric current from the wire he was holding made the needle on his compass move. A very curious phenomenon, he thought. And yet, it was precisely this accidental discovery that ended up revolutionizing our entire understanding about the way the universe works. Because what this observation actually ended up doing is setting scientists down a rabbit-hole of thought which ultimately culminated in our understanding that electricity and magnetism weren't just related, they were fundamentally connected.[3]

Two sides of the same coin.

Or rather, two manifestations of the same force; *electro-magnetism*.

Soon after that, Michael Faraday figures out how to turn this connection between electricity and magnetism into a machine, effectively turning a cute little parlor trick into something we could actually use (electrically powered motors and generators).

Fundamental Assumptions

Look, I only have a couple hours of life left in me, so here's the new plan. In Part I of this thing we discussed all the terms and case studies needed to have an intelligible discussion about how the brain 'appears' to give rise to consciousness; and then we took that about as far as we could possibly go.

In Part II, we began unifying some of our disparate concepts while, at very same time, expanding the breadth of our knowledge to include the peculiarities of NOSC in an effort to expose the entire spectrum of conscious states that are possible.

Part III had us piecing together the grand evolutionary narrative, of which we're all apart, and then, most recently, we even revealed our fundamental position: There really is something to this "expanded sense of self" theme that's so often overlooked.

Now, with what little time we have left, I'd like to really back up the claims we've made so far by tracing the idea of oceanic boundlessness as far back as we possibly can. In doing so, we should arrive on the other side with a much more detailed understanding about how everything we've been discussing thus far hangs together.

Although, before we continue, I'd like to first come out and say that the only reason this idea isn't more widespread, in the West, is because, by and large, we're still carrying around a handful of fundamental assumptions about the nature of reality that we probably don't even realize we've made. And these fundamental assumptions (*keyword being *assumptions*) are clouding our ability to think clearly. Plain and simple. There really is no sense in doubting the obvious culture divide that's cropped up between the East and the West over the millennia, and while the idea of unity certainly has made its appearance in Western thought many times over, it's clearly more popular in the East considering their intellectual starting points.

What do I mean by intellectual starting points?

I mean it in the sense that we all start from a place of assumption, and where you spent your formative years generally dictates the sorts of ideas you'll pay attention to.

You could say, given our cultural conditioning, that our brains have already been *primed* to think about the universe in a certain way based on the most influential thinkers our society holds dear. But what this actually means is that even when we tell someone we're starting from square one, we're actually already coming to the table with a fixed set of ideas about the way the universe is. And that's the problem.

To truly understand what consciousness is we have to be willing to toss out all of our preconceived notions about the way the universe should be and re-consider what is.

The following breakdown, originally introduced to me by Alan Watts, should help elucidate the problem.

———

Essentially, cultures all over the world and throughout all of time have all tried to make sense of a stupefyingly complex universe and in the process have only ever produced three great views about the universe. The most familiar of the bunch, which is mainly a Judeo-Christian and even Islamic interpretation, is that the universe is *something made*, like an artifact. Whether it was crafted by an intelligent designer, like the ultimate carpenter, or whether it was pieced together through some other kind of natural, scientifically explainable process, the universe is viewed more like that of a clock or a machine than anything else. Most crucial to this point of view, however, is that the universe is either expanding according to some fixed universal laws, or that it's unfolding according to some deity's divine plan. In either case, it's important to specify that under this conception of the world, "we" are the ones that are alive…the universe is not.

Next in line, and in stark opposition to the first view, we have the inherently Eastern or Chinese way of looking at the world which sees the universe as if it were an organism; *something that grew.*

And last but not least, there's a model of the universe, which we get from Hindu philosophy, that looks upon the universe as if it were a play or a drama; *something that plays.*[4]

***Notes from the Editor:**

For the sake of transparency, there is no smoking gun type evidence in favor of one way of thinking over the other, it really is just a matter of location. If you happen to have been born in the West, then there's an overwhelming likelihood of you believing in the first myth (the universe as blind, unintelligible force), whereas, if you'd been born in the East you'd likely believe in one of the other two.

—I'm sure it goes without saying but not all these views are capable of being right. And as we'll come to in a future section, one of these models actually has a fatal flaw built straight into it. For the time being, I won't say which model has made such a calamitous error, but rest assured, one of them has. And for the sake of argument, let's just pretend that we're the ones starting from a faulty model.

Supposing that were the case, how well do you think we'd do?

I mean, if we were seriously trying to answer the question, but we were forced to start from a model that effectively shoots itself in the foot, what hopes of solving consciousness could we possibly even have?

—

To cite an example of this, just consider the state of the world before the 17th century, when for an embarrassingly long time, we thought the entire solar system revolved around the Earth. Of course, we can smile and laugh at our forebears for having believed in the geocentric model put forth by Ptolemy circa 150 CE.[5] But as you'll recall, this was the church-sanctioned view of the time, and because the church was such a domineering force, it was the view that persisted for nearly 1400 years!

That is, until a Polish priest named Nicolaus Copernicus came along and made some observations that suggested otherwise, giving us a fresh perspective: *Story B*.

We also used to think the Earth was flat.

That we were the only galaxy.

That the universe was fixed and static.

That time was absolute.

That black holes were just theoretical.

That gravitational waves were just theoretical...and by now I'm sure you've seen where this is headed. We could keep listing off times when we as a society weren't just wrong—we were *impressively wrong*—but that isn't really the point.

The point is, if we were trying to construct a theory of gravity, but we were forced to start from a flawed model, our chances of success would be near impossible. Which is precisely why a great mind like Isaac Newton didn't come along until after the Ptolemaic model of the universe was displaced.

Moreover, if we actually did start to trace the evolution of certain ideas, like gravity, evolution, and relativity, surely we would notice ideas that were fashionable, easy to grasp, and consistent with the current zeitgeist. But then, we'd also notice those pesky revolutionary thinkers whose ideas blew apart the substructure of reality in a way that most people (especially the church) really didn't like.

Notes from the Editor:

Nicolaus Copernicus actually discovered the Heliocentric model in 1515, but didn't publish his work until the year he died, in 1543, out of fear of the church. Scientists like Giordano Bruno, who later taught Copernicus's idea, were burned at the stake.[6]

—Heresy is what they always call it. You say something that goes against the great worldview of those that are in power, and in previous climates, you may become history, but you do not become old.

Notes from the Editor:

In the year 1610, the astronomer Galileo Galilei, validated the observations made by Copernicus a century earlier, and was later convicted on "vehement suspicion of heresy." In the end, they banned his book, and even made him renounce his beliefs under threat of torture and life imprisonment.[7,8]

—Religion certainly has given scientists no shortage of problems over the years, which is why I find it rather ironic that the key to understanding this whole thing only comes to us *after* we shine a light on those fundamental assumptions handed down to us through religious stories. Because after we've had a good romp through the history of Eastern religions like Hinduism, Buddhism, and Taoism, you may finally come to understand why this subject has become so unnecessarily convoluted.

But before we get into any of that, how about one last shot from a more Western perspective?

Chapter 29
Concepts of Infinity

INFINITY. IT'S ONE of those concepts we usually don't like to think about whenever we talk about it, because the truth is, humans aren't very good with big numbers. Take the concept of a trillion for example. Most people have never even seen a billion of something, let alone a thousand times that number, so whenever most people try to imagine what a trillion looks like, their brains just give up. Much like when someone tells you their neighbor's brother's half-sister just found out their landlord had a baby out of wedlock and needs money to help bribe the headmaster of their preschool so that baby can get on the shortlist to being pre-accepted.

At the end of whatever the hell they just said, your brain just offers a shortcut:

Get out your wallet.

It doesn't start by drawing a family tree and then tracing the lines of lineage back to the person whose mouth just stopped moving because that would actually involve mental effort. And it's not as if we don't care, it's just that we're intellectually lazy. It's literally too much for our brains to process, so our brains just shorten it.

Someone you know, knows someone they know, and that person needs help.

It's the same as when you tell someone the age of the Earth is 4.5 billion years old, or that the number of neuronal connections in their head exceeds half a quadrillion! Pay close enough attention and you can literally watch their brain start to stall out. Their eyes get wide, they start shaking their head, and then they just stare at the floor while letting out a breathy laugh, because what else are they going to do? We only live about eighty years, and our mental calculators only go out to about the seventh or eighth decimal place, so what chances of success do we even have?

I mean, are humans even capable of conceptualizing such tremendous numbers? Let's find out.

—

Thanks to radiometric dating, we've calculated the age of the Earth to be just over 4.5 billion years old. The oldest fossils on record, at least 3.5 billion years old.[1] But in order to appreciate what a billion even looks like, I want you to think about your 32nd birthday real quick. Real or imagined, I want you to picture yourself blowing out 32 candles atop a multi-tiered red velvet cake, smothered with cream cheese icing, because in that moment you would be exactly 1 billion 9 million and 152,000 seconds old.

It takes 31.7 years of living just for a billion seconds to pass. And we were talking about years, not seconds. And not just 1 of these ginormous things but 4.5x that number.

This is the problem with humans and math. The real problem being, that we only evolved to perceive reality at a particular level of analysis (Dawkins' "Middle World") and of a particular temporal resolution, which makes comprehending vast scales of geological time, truly difficult.[2]

How about infinity? Can we picture infinity though?

Well, considering that infinity means the unbounded. Something without edges or borders. Something without limits or ways to define it. And because to actually define infinity would be to limit in some way, I really don't think we can. But, we can still talk about it, which is where our attention must turn to next; the eternal, the limitless.

If something like eternity does exist, then it would necessarily follow that it would be something that *has always* and *will always* exist. Not just for 1 billion 9 million and 152 seconds, but forever.

Are infinities actually real, or are they just an improvable concept?

To answer that, we should probably go ahead and call upon some of the greatest thinkers, mathematicians, and philosophers who have ever lived.

Although, right out the gate, I want to go ahead and state my position, which is, that infinite numbers are real and that we even encounter them everyday. Circles, spheres, triangles, deep space…all these things contain infinites. Nevertheless, when it comes to the size of the cosmos, technically, it's an educated guess and not an actual measurement, that says we're living inside an infinite universe. Because in actual fact, we have no way of measuring the infinite.

I mean, how would we? By borrowing our neighbor's infinite tape measurer?

So, admittedly, we don't know for a fact that the universe is infinite, but all signs certainly point to it being so, as it's been expanding ever since time as we know it first began. And even weirder than that, it's actually speeding up.[3] This being the case, we really may never know if the universe is infinite, which is really the best we can ever hope for being an evolved species of ape wielding nuclear weapons.

All that withstanding, you can take any circle you want; divide its circumference by its diameter, and what you'll end up with is an irrational number that goes on forever. An infinite number we call pi. Any soap bubble, any billiard ball, measure its circumference (c), divide that number by the circle's diameter (d), and the answer you'll get is π. An infinite irrational number that just keeps on going and going.

To date, pi has been calculated to 62.8 trillion decimal places; and as much as it's been "solved" we've still never found a repeating pattern.[4] Suggesting, at least to the smartest mathematicians of all time, that the number just keeps on going without ever repeating (a truly infinite number). Not surprisingly, we could probably debate the true existence of infinity until the end of time. But instead of going down that rabbit hole of thought, let's just momentarily agree that infinites probably do exist, even if we can't actually prove or disprove their existence.

The World of the Dream

Later on tonight, after you lay your head down to sleep, a part of your brain is going to flicker back on while the majority of you lies motionless in your bed. You're going to have an experience, but I doubt this experience is going to make a whole lot of sense. Instead, it'll probably be filled with shapeshifting entities, gaping plot holes, and a bizarre law of physics.

For instance, you may find yourself back in the kitchen of your childhood home, making a sandwich, while your mom screams at you for leaving your waffle-copter parked on the grass. In the dream, she might be yelling about how this is the hundredth time she's asked you not to park your waffle-copter on the front lawn, and yet, somehow, amidst all the confusion of the dream, you still might not register the implausibility of the situation. There could be huge red flags indicating you're dreaming, like the fact that your childhood home was sold, or the fact that helicopters made from waffles could never fly, and still, these absurdities might be overlooked. And why?

Quite simply, because we accept the reality we're presented.

Granted, dreams aren't always this ridiculous; after all, there is a component of memory involved. But even knowing this, it still never ceases to amaze me how sufficiently weird things can get before someone begins to notice the abnormalities. In any case, we need to acknowledge two points in the aforementioned hypothetical: (i) that wasn't your actual mother, and (ii) it's highly likely she never said those words to you. Rather, this was a subconscious projection of your mother saying whatever off-the-wall thing your subconscious decided to tell you.

Recently, when I was trying to explain this concept to a friend, he totally just nailed it by saying, "Yeah, because in the dream world, you play all the roles." Only the way he said "you" carried the weight that this was a *subconscious you* playing all the other roles while the conscious you got to be the one experiencing the dream.

This is actually such an important point, I think I'll say it again. In the world of the dream, you play all the roles. Every interaction, every scenario, every building, on some level, you created all of it.

Even if we consider the most lucid moments of a lucid dream—when you know you're dreaming and can control things just a bit—there are still parts of your psyche that are creating the very dream you're experiencing. Which also means that even when you're "in control," you're not really in control, seeing how there's still a whole range of components being synthesized that are outside your sphere of awareness. Elements like the design of the landscape, the people you meet, and of course, the erratic things they say.

Even if you're of the belief that dreams are nothing more than random neural activity, you still have to contend with the fact that the way dreams present themselves is like a story. And so, whether or not the contents of the dream are randomly generated, actually makes no difference to the argument. Because in either case, the Subconscious OS is still the one synthesizing the story.

Taking all this into account, wouldn't it then be reasonable to call the subconscious something like the ultimate architect behind the world of the dream? The functional equivalent of an intelligent designer?

Even now, in this very moment, everything you're seeing, hearing, feeling, and experiencing is ultimately a product of your subconscious. The only difference is, when you're awake there's a ton of sense data coming in to help ground all your brain's predictions to this plane of existence, whereas when you're asleep, you have virtually no sense data coming in.

Just consider the most vivid, rich, detailed dream you've ever had—the one you where you woke up sweating and had to phone a friend—and then realize that even that was 100% a product of your subconscious. Signifying that our brain is completely capable of generating rich, vivid, detailed experiences without any help from the senses.

And as long as *it's* the one generating all the stories, *it's* the one creating all the characters, *it's* the one rendering all the graphics…and *it's the one playing all the roles*…I think it's rather fair to say that the subconscious is pretty much on par with any deity you could think of (at least in the world of the dream).

So much is this the case, that even when someone asks me to speculate about what an image of God might look like, this is how I begin to explain it. By first asking them to consider their subconscious as being nothing other than their highest self; and then by asking them to re-consider the role that their highest self plays in the world of the dream.

Because from the perspective of "Dream Tom" the subconscious really is omniscient, omnipotent, and omnipresent. Especially, when you take into account that it's the one creating the world each night, it's the one spinning all the narratives, and it may even be trying to communicate with us through mysterious ways.

As a matter of fact, that's the one hypothesis about dreaming that seems to make the most sense. Dreams are there to communicate information to us, through some kind of narrative, that the conscious self either hasn't realized, or is refusing to realize.[5] And in the pursuit of helping us to acknowledge this suppressed information, your highest self may have taken on the guise of your mother, your boss, or perhaps even an entire classroom of kids pointing their fingers and laughing at you.

Who made the kids laugh? And who decided to make you the one all vulnerable and exposed? Go stand in front of the mirror one more time because I think you already know what I'm about to tell you.

Don't get me wrong, it may certainly have felt like you were conversing with your mom or your brother in that dream, but in actuality, you were only ever just interacting with parts of yourself. And so, whenever I hear someone say the phrase "God is in everything" this is what I take them to mean: that they have a conception of God, in the real world, which is roughly analogous to the relationship that must exist between an entity like Dream Tom and his highest self. Only, in the dream world, we call it the subconscious, while in the real world, we tend to call it God, the Universe, Allah, Yahweh, Brahman, Elohim…you get the idea.

Infinity Cont'd

Concepts of infinity go back at least as far as Anaximander and Pythagoras (sixth to fifth century BCE), and ever since then we've been trying to make sense of the unbounded.[6,7] Something we can seemingly can do in concept but never in principle.

No finite creature ever could hope to understand infinity, because like pi, infinity just keeps on going and going. To understand infinity, you'd have to be infinity, and even then, there's no guarantee. Be that as it may, there have been some seriously powerful thinkers who've devoted their entire careers to the study of the infinite, and the conclusions they've come up with are truly extraordinary.

Starting at the top of the list, I'd have to put my man Nicholas of Cusa. A German philosopher, theologian, and mathematician living in the 1400s, who made a few poignant points about the nature of God and that which has no limits. His conclusion, that if there were such a thing as infinity, then everything *must be* included within it, arguing that the world itself must be within God.[8]

Or phrased in a slightly different way, that the universe itself must be within God.

Which would include you reading this letter.

Which would include me writing this letter.

It's not that there's a universe...*and then some magical other thing called "us."* There is only the infinite, unbounded universe that we all live in and are constructed from. So you can call it "the Universe" or you can call it "God" just realize that, at the end of the day, all cultures throughout all time have been trying to describe the same exact thing: the infinite, eternal, creation, creating, creator that is infinity.

Being raised in a Judeo-Christian ethic, my first inclination is to call that thing you connect to God, as you certainly tap into a feeling of shared connectedness with everyone and everything around you. But after studying all the major religions of the world, and actually looking into what they believe, the conclusion we've finally come to is that, basically, we're all trying to describe the same thing. We're all trying to say 'what' the Possibility Space is, we're just using different words.

Some call it nirvana, some call it satori, some Buddhist philosophers have even called it the Void (Śūnyatā).[9] Shakespeare once said, "A rose by any other name would smell as sweet," although, if we actually took note of how many wars have been fought over what to call this infinite thing, we'd find that a rose really has to be called a rose, or else there's inquisitions, genocides, holy wars, and a whole lot of murder.

Long story short but Shakespeare was very, very wrong.

Because apparently, the names of things really matter to people. And the symbol systems we've become attached to (aka religions) really, *really* matter to people.

So what's the problem?

The problem is, we never should have tried to define the infinite thing, but we did, and we've been paying in blood ever since. The problem is, no culture ever should've tried to attach a name or a gender to a thing that is above genders and beyond names, but still, we did that too, and now we're living in the aftermath of that decision. The reason I like Nicholas of Cusa so much is because he tried to use the universal language of mathematics to describe the God-world relation. My favorite line of his being, a quote about God's shape that goes something like this.[10]

> "God is an intelligible sphere—a sphere known to the mind, not to the senses—whose center is everywhere and whose circumference is nowhere."
>
> > - *Book of the Twenty-four Philosophers,* et al.

For Nicholas of Cusa, humans were the creatures that stood at the boundary between the eternal and the finite, but nothing was *outside* of God. And once you understand that, and you know that, and you feel that, you start to remember who you really are, not just who you pretend to be.

Philosophers Cont'd

A couple hundred years after Nicholas of Cusa, the famous Dutch philosopher Baruch de Spinoza (born in 1632) also argued for an interpretation of this kind. Stating that if God was infinite, then God is *the one substance.*[11] Meaning, there can be no other substances besides this one. And if there are no other substances besides this one, and you are that substance, then technically speaking—you're God too.

To paraphrase Spinoza for just a bit, his argument follows something like this: God has to be eternal to count as a god, because by definition, that's what a god is; an "absolutely infinite being." Spinoza then reasoned that a being couldn't be truly infinite if there existed something outside of it, so for Spinoza, God and Nature were the same thing.

The body and the mind being just two of God's infinite attributes.

There wasn't Nature…*and then people*…who somehow stood above that Nature. Spinoza said, "Deus sive natura" (God, that is Nature). Because for him, God was Nature, and Nature was God. If there was a highest ethic to be achieved, Spinoza tells us, then the highest ethic would be for us live in accordance with the laws of Nature.[12]

Which sounds reasonable enough, but what does science have to say on the matter?

Science Cont'd

I think it's rather undeniable that we're all just concentrated forms of energy, after all, that's what matter even is from a definitional standpoint; concentrated energy.[13]

But more to the point, matter and energy are also interchangeable (completely governed by the equation $E = mc^2$), which is also why when you split an atom of enriched uranium you release a tremendous amount of energy. So matter really is just energy that's been slowed to a particular vibration. Or, if you prefer, matter is really just energy after it's been through several phase transitions. And yet, despite all of our knowledge about what matter *does*, science has failed to teach us anything about 'what' matter actually is, for the plain and simple reason that the majority of scientists are still being unconsciously influenced by 17th century cartesian dualism.

***Notes from the Editor:**

In the 17th century, French philosopher René Descartes famously divided the universe into two substances: *res extensa* (the stuff of the material world, that things and bodies are made of), and then on the other side he had *res cogitans* (the stuff of thought, the non-substance substance that minds and souls are made of) popularizing the philosophy of dualism.[14]

So what am I proposing?

I'm re-proposing that we are not something separate from that primordial explosion cosmologists are so keen on describing, but have instead just gone through so many different phase transitions that we don't even recognize our own baby photo (See *Cosmic Microwave Background*

Radiation). I'm proposing that we're all still the same energy that kicked this whole thing off in the first place, only this energy has been steadily increasing in complexity and unpacking itself through the dimension of time, such that it's even able to ask questions about itself like, "Where did I come from?" or "What am I made of?"

Sir Julian Huxley once said, "Evolution has become conscious of itself," which is a lovely quote even when it's read with a *Story A* type lens. But when you start to realize that evolution is something the whole universe is doing (that the entire universe has been evolving since the thing first went 'bang'); then you simultaneously start to appreciate that we are the process. We are not a result of the process—somehow commenting on it—we are the process itself. Evolution incarnate.

It isn't that there's a universe…and then some special place inside it where life just happened to start evolving…the proposition is that the entire universe has been evolving from the second things "began" 13.82 billion years ago.

Why do I say "began" in quotations?

Because, as you already know, there is no beginning to infinity; infinity always was and always will be. And in case that sounds confusing, it is. And look, I agree with you.

I seriously don't know anyone who refutes the facts or the predictions of the Big Bang, which is why I'll be quick to add that these two views aren't mutually exclusive. The Big Bang model doesn't stipulate that the universe definitively had a beginning, nor does it even specify 'what' it was that first banged in the first place. All it really says is: there was a time when the universe was highly ordered, extremely small, massively dense, and contained a very low amount of entropy. Thus, Big Bang cosmology isn't a description of how the universe began as much as it's a description of how the universe *evolved*.[15] And so this perspective I'm trying to share with you doesn't negate any evidence from evolution or the Big Bang, but rather, provides a new way of understanding the facts. It very well could be that the Big Bang only happened once and here we are living inside it, just as easily as it could be that the Big Bang was just one 'bang' amidst a series of infinite bangs. At the present moment, science couldn't tell you the answer to this riddle, but in case you're still interested, here's how it looks to me...

***Notes from the Editor:**

In some ways, this is where the story ends. As of today, the scientific community still cannot say what consciousness is with any semblance of authority because, as we're about to find out, the scientific community is still shaky on the question: *What is life?* Believe it or not, but this question is actually more fundamental than the original formulation, and for this reason, deserves to be attended to first. In the final portion of this book, aptly titled "Interpretations," the narrator attempts to provide a new answer to these questions by first holding up a mirror to our most cherished ideas about existence and then showing us where they fall short.

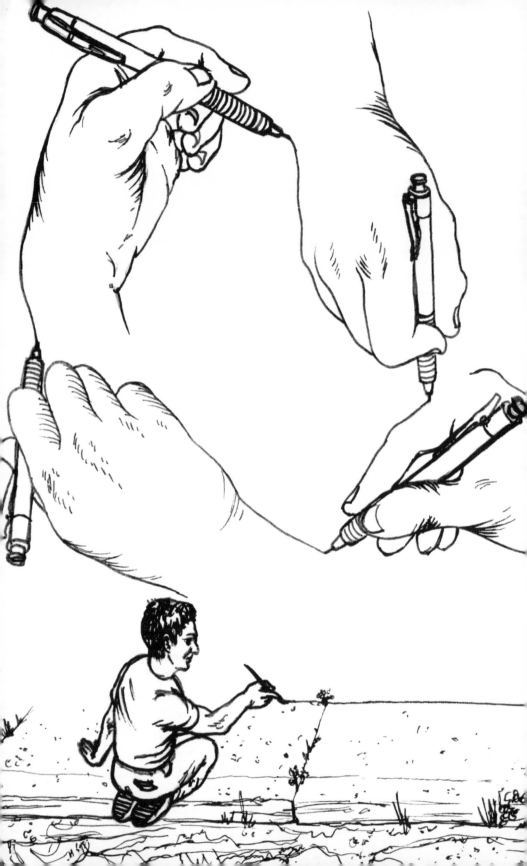

Part IV

Interpretations

Chapter 30
Piecing it all Together II

LOOK, IF YOU were to ask me a question like, "What are we?" I'd have two responses for you: a short one and a long one. The short answer would've been, you and me are nothing more than the primordial energy of the universe, *plus* time, whereas, the longer answer would've been that anthology. An anthology of epic proportions that eventually circled back to the exact same conclusion. You know, in hindsight, maybe that's what I should've done (taken this thing apart more slowly). But seeing as I have neither the time nor the energy to go back and provide you with that level of detail, you're going to have to just settle for me bumbling through one of the most complex conundrums humanity's ever faced, half-asleep, and with only one eye open.

So what's really going on here?

Consider the following analogy: It's as if there is only one tree in existence and on that tree exists various types of leaves. Of course, if we wanted to, we could imagine that each leaf had its own life-force, but in actuality, there is only one living entity under consideration, and that entity is the tree. Now, I'm sure if you spoke Tree, I bet you could probably ask each leaf what its name was and where it was from. Hell, they might even be able to spin you a story about what life's been like from *its perspective*. But, what the leaves don't know, can't see, or are refusing to notice, is that they all have a stem, and that they're all connected to the tree that birthed them.

This being the case, each leaf might be able to tell you about the twigs that birthed them (their parents), and the branches that birthed them (their grandparents), but that's usually where the narrative ends—with a *Story A* type interpretation. *Story B* only comes into play when you stare deep into our ancestral past and realize that we're all just manifestations of the one tree, each trying to maximize our place underneath the sun.

Why does each leaf on the tree believe itself to be an isolated entity?

In short, because all the leaves on this tree have a local memory, and this fundamental truth about us all being connected is buried under years and years of *local memories*.

For example, if you burn a leaf with a magnifying glass, a brown spot will remain, so there's a form of memory there. But if I were to ask you: Who did you burn, the tree or the leaf? You could say, "Technically, both," and technically, you'd be right. Although, what really happened was, you burned a part of the tree which we call a leaf, indicating that the enduring enigma of consciousness is really just a problem in perspective.

It's a problem of magnification.

If we were instead to zoom out and imagine the archetypal image of a lonely tree sitting atop a hill, from that zoomed out perspective it'd be much easier to embrace the idea that you're only looking at one entity. But once you start zooming in and examining any one particular leaf, perhaps burning it with a magnifying glass and then recording the results, you may start to see things differently.

"This leaf appears to have its own life," you might say.

And so it is with humans.

We have *local* memories, with *local* origin stories, and an attentional-based form of consciousness that only selects out pieces of reality to focus on, in much the same way that a telescope only frames up *parts* of the night's sky. No one here's doubting the utility of using such a device when trying to observe objects that are certain distance away, but when you're trying to look at yourself in the metaphorical mirror, a high-powered telescope is not the instrument you should be reaching for. And so, the connection here is this: From close up, we all appear to be self-contained, isolated entities, but upon adopting a more gestalt or zoomed out perspective, we notice that we're all really just modes of the one.

Leaves on a tree, not the self-isolated systems we pretend to be.

Naturally, some of the leaves on this tree have caught sight of their connection. Somehow or another, they've seen the stem that connects them to everything else, and they may have even tried to share this knowledge with others. But without the proper set-up or vocabulary, the message is rarely received, creating a tendency for leaves to just believe whatever they've been 'brought up' to believe.

Instead of them actually looking into the matter themselves and seeing the larger evolutionary story they're all a part of, they choose to stay small and scared. They choose to believe that this life, this season... this *leaf*...is all they're going to get.

So what happens next?

They cling and they shake. They live a life filled with anxiety and regret, until eventually, like all leaves, they fall off the tree and die, failing to realize the big secret of Being, which is:

You were never 'just a leaf' because you were always the whole tree.

But again, because each leaf only has access to a local memory, and a small localized body in space—replete with a stem, veins, and a midrib—they mistakenly think that's all there is. They see themselves as an isolated island of life, while remaining completely ignorant of the vast interdependent Tree of Life which gave rise to them in the first place.

"You are something the whole universe is doing, in the same way that a wave is something that the whole ocean is doing."

- Alan Watts (philosopher)[1]

Three Worldviews Cont'd

As previously mentioned, there really are only three great models of the universe ever put forth, and the first is that the universe is "something made," like an artifact. The second, that the universe is an organism. And the third, that the universe is a drama.

Now, on a first pass interpretation, the first view sounds like it's the most scientific, however, upon closer examination, you may have noticed that it actually shoots itself in the foot by presupposing that we know what life is, when the stark reality is…we don't.

We genuinely don't have a good definition for what life is.[2-4]

And because of this one undeniable fact, it seems rather foolish to even consider starting from a model that already presupposes such limitations. For this reason, we think it wise to go ahead and dispense with the first model as we simultaneously start opening ourselves back up to the possibility of the other two. Even if they seem implausible, or utterly hopeless, please try and keep an open mind here, because once we burn through some of these older points of view, the real answer to the question, "What is life?" may finally start to make sense.

According to the Hindus, there is only one thing, and that thing is the Self. It's the thing that always was and always will be, and they call it *the Atman* or *Brahman*. For sake of simplicity, we'll just stick to calling it the Atman for now.

Notes from the Editor:
Out of the 12 major religions that exist today, practitioners of Hinduism make up about 16 to 17% of the population (about 1.3 billion people). Just behind Christianity and Islam, Hinduism is the third-largest religion, and is also considered to be the world's oldest religion.[5,6]

—So there's this eternal self that hides behind each and every one of us (a divine essence or spark they call *Atman*), and this Self likes to play games with itself. It likes to play hide and seek with itself, because really, that's the fundamental game of the universe: chases and escapes. On that basis, from the standpoint of a Hindu, the entire universe is essentially a never-ending drama in which the eternal self (or Atman) hiding behind each of us gets lost in stories and personas, like Tom and Ava, ultimately forgetting *who* it is. Then, after a sufficient period of time, the Self gets tired of playing "hide" (the forgetting portion) and starts on the path to "waking up" (the seeking portion) where it eventually realizes *what* it truly is, which is, the Great Self or Brahman.[7]

*** Notes from the Editor:

In a way, the Atman could be thought of as being like a soul, only according to Hinduism, this spark of divinity is the same in everybody. While Christianity believes souls are individual and open to being judged by God, Hinduism sees everyone as being fundamentally the same. It's the superficial layers 'stacked on top' that are different.

—Back on Koreatown night, when all the external layers of my psyche began melting away (memories, VMs, personas), the awareness of who I actually was started to shift as well. Because down in the Possibility Space—or the level of the Atman—that's all that there even was: *a proto-conscious state of awareness where I finally caught sight of my essential self*. Something a non-religious person, such as myself, has taken to calling the Awareness, while the Hindus have decided to use the Sanskrit word for "inner soul" or "breath" (pronounced aht-muhn). In this regard, if we were to imagine all the complex systems in our brains like image parsing, edges, shadows, depth perception, and facial recognition all had their own layer to the conscious experience, then this was like watching all those layers getting unticked one by one.

In Photoshop, clicking the eye next to each layer either hides it from view or makes it visible again. And by experimentally toggling layers on and off, we're able to see not just what each layer does, but how each layer contributes to the whole. However, during the chaotic ego dissolution experience that was Koreatown night, it's as if I got to see all the layers of my conscious experience getting unticked at random.

First to go was my periphery, followed by my ability to detect edges, followed by my ability to stitch together three-dimensional images...all the way down, until my Narrative Self got off-lined and I fell to the ground "dead," genuinely believing that's all that I was. But when my head hit the floor and a piece of my psyche somehow transcended the death of my ego, I caught sight of Cali's essential self which ultimately reframed my entire life's existence.

But what exactly is it that transcends the death of my ego?

In our shared terminology, it's the Awareness (pure intellect unmarred by the confines of identity), whereas, for the Hindus, it's simply called the Atman. From a neurological perspective, it's whatever type of consciousness you have left after the orchestrator of your thoughts gets obliterated, although from a subjective perspective, the way it feels is nothing short of nirvana. Infinite bliss. Like you've somehow entered into the gates of heaven, only the gates of heaven are in the here and now.

"How could this be?" you think to yourself. But then you realize how we don't live in the present moment; we live in the slightly distant past. About 500 milliseconds, to be more precise.

And so it's only in the absence of all our categorical structures that we come to see this world as perfect. Because once your brain is no longer grasping or clinging (holding onto fear or desire), you come to appreciate reality for what it actually is, not what you hoped it to be. For me personally, it was seeing Cali's essential self that first reminded me about my own essential self—and once I'd caught sight of that—I quickly realized that this was the ultimate "Background Layer" of experience.

How does the feeling of nirvana or moksha come into play?

The feeling of liberation or release comes into play when an individual realizes that Atman is, in fact, Brahman (the force underlying all things). Or, to put the matter a bit more delicately, that your essential self is one with all of creation and everything that is possible. And the way this reveals itself to you, generally speaking, is through a feeling and an intuition.

A feeling that you are one with the universe, not something separate from it.

An intuition that you are part of Nature, not something standing outside of it.

And a perception that you are one with whatever God archetype you happen to subscribe to, not something separate from it.

So, if you once believed there was some kind of difference between *the maker* and the *made thing* (first model), then this fundamental assumption you'd been operating under would soon dissolve along with all the external layers of your perceived identity.

Bearing this in mind, you can take this essential "Background Layer" of experience and pile on as many superficial descriptors as you'd like, but at the end of the day, we all have a spark of divinity, which is Atman, and this divine self likes to get lost in the game of life by pretending it's 'just a Tom' or 'just an Ava.'

Tracing the Idea through Time

From the Upanishads (ancient Hindu texts written between the eighth and sixth century BCE) there's this dialogue between a teacher and a student which helps get us at the connection between Atman and Brahman.

> As the rivers flowing east and west
> Merge in the sea and become one with it,
> Forgetting they were separate rivers,
> So do all creatures lose their separateness
> When they merge at last into pure Being.
> There is nothing that does not come from him.
> Of everything he is the inmost Self.
> He is the truth; he is the Self supreme.
> You are that, Shvetaketu; you are that.

Written at least 8,000 years ago, and it seems we even knew it back then.[8,9] There is an infinite, eternal being. That thing we call the Self, with a capital "S", and we should never try to classify or name this thing...but we do...and therein lies the problem. We named it; we gendered it; we even started using narratives to try and pin the thing down.

And what was the result?

A divergence in stories, a speciation of religions, and a distrust in our fellow man.

—

From our perspective, it really does seem as if all human beings are capable of having this experience, where they catch sight of their own divinity and realize that they're one with everything (oceanic boundlessness). But other than the fact that she and I have been gifted this experience dozens and dozens of times, there really is nothing special about her or me. She and I just happen to be at the right place at the right time. Nonetheless, the crucial point under consideration is that this same essential experience keeps happening to people. And not just to her or me, but to countless individuals all across the globe.

***Notes from the Editor:*

"We're wired to have these kinds of experiences," remarks Dr. Roland Griffiths of Johns Hopkins University when asked to reflect on the last twenty years of psychedelic research.[10]

—And the reason this keeps happening, we would argue, is because, deep down, all these experiences are speaking to the same metaphysical realization: *we really are all one.*

Make no mistake, we all have a sensation that tells us we're little egos, piloting around these individual meat skeletons. But whenever that sensation gets off-lined—even for just a little—people keep inexplicably waking up to this same idea. Which wouldn't be a problem if people just took the message and went about their day, but they end up trying to make a religion out of it, which is really the whole crux of the problem.

This is where the Monk would wag a long index finger at me and shake his head. "Once you've started *any* religion," he'd say, "you've missed the point." Because the Monk already knows the true nature of thought and religion.

Thought cuts and religion divides.

On the face of it, religions sound like a good thing. They sound like they unite people under a common cause and lift up communities. But as we're all quite painfully aware, that isn't what happens at all. What happens is, they unite people under a particular "brand" of infinity, let's say, but in the process, end up cutting themselves off from everybody else.

What's the problem with one group believing something different than another group?

In simple terms, it's that whenever a group becomes attached to *the form of God* (their particular culture's interpretation about infinity) they lose sight of the real message. In essence, they begin to worship the metaphor, not the divine itself, losing sight of God's true perfection.

For example, something that is supposed to be beyond name and form, suddenly becomes a human being living in a particular time, carrying a specific set of genitals, and particular shade of skin that just happens to be different from the neighboring culture's interpretation of the exact same energy. In other words, by failing to recognize that all religious stories are *just a reference* to the transcendent, not the transcendent itself, we become "brand loyal" to our particular culture's interpretation about infinity, and start casting down judgement upon all other interpretations as *less than*. After that, we start organizing ourselves into special groups that meet on Saturdays or Sundays. What's supposed to be a personal relationship to the divine suddenly morphs into a group's relationship to the divine, and once that happens, all hell breaks loose. In effect, God becomes "God, Inc.," and while the genesis of most every religion might have been paved with good intentions, the unfortunate and pernicious side effect of creating any system of beliefs is that it necessarily produces in-groups and out-groups. Believers and non-believers. Right people and wrong people.

"The second you start drawing lines in the sand," the Monk would repeat, "you're on your way to chaos."

Because what always starts as a just a quick division, swiftly devolves into one group feeling more superior than another group. After all, "right people" can't exist on both sides of the fence, can they? And so, with this one simple demarcation, "We believe in so and so," come all the problems we know and hate: war, murder, genocide, missions, crusades, inquisitions, Holy land disputes, scandals, secrets, lies, corruption, fundamentalists, propaganda, more war…the cycle is endless as long as there are still people making in-groups and out-groups.

So how do we stop it?

By not using the small identities we're handed as children. You're a boy, and you're a girl. You're an American, and you're Chinese. You're a methodist, and you're a football player. Instead of placing all these metaphorical boxes around ourselves, and subsequently limiting our intellect, we re-identify with the largest part of our identity, which is: *that we are life and that we are human.* Which doesn't give us dominion over the animals, or the right to destroy the environment. All it really means is, that of all the creatures living on this planet, we just happen to have the most responsibility.

Besides, from this perspective, the external world isn't even something truly external anyway. As a matter of fact, from the standpoint of a Hindu, which is fundamentally the same as Nicholas of Cusa or Baruch de Spinoza, there is only one substance and we are that (monism).

In this connection, if you were to ask me what I see when I look out a lecture hall and there's someone sitting there, holding their hand up, I'll tell you what I don't see. I don't see someone with a particular color of skin wearing a funny hat. I mean I do, but that isn't what I choose to see.

I choose to see *life*, temporarily manifesting itself as a human.

The 99.9% view, not the 0.1% view.

And really, shouldn't that be the ultimate agreement we all have before interacting with one another? A kind of mutual respect for all shades of life and that which supports it?

The Monk's Final Lesson

Years ago, after one of our final sessions together, the Monk and I were walking through a park near my apartment complex, when my teacher starts to ask me about some of my solo meditation sessions. He was going to be traveling to Indonesia next, and he wanted to make sure I had a firm understanding about the alternative path I'd been asking for.

The Monk and me, we're just meandering along this long windy sidewalk, in no real rush to get anywhere, when from almost every direction the eye can see there's dogs running around without leashes, bicyclists riding by ringing their little tiny bells, and young couples walking side by side, swinging clasped hands.

With eyes fixed straight ahead, the Monk's asking if I've started to notice any themes to the million and one thoughts I keep sending away with my mantra, and when he asks this, I really have to pause and go inwards. As the Sun starts to move behind a giant wall of clouds, I plunge myself deep within the abyss of ceaseless mind-chatter and re-emerge about a half-second later with three important realizations.

"I'm angry," I start to say, but then have to bite my lip. "I'm judgmental," I add, staring at all the cracks on our path. "And scared," I finally admit, avoiding my teacher's eyes.

Going over the response in my head, I can't stop nodding. Angry, judgmental, and scared: that pretty much sums up my entire life's existence. Just then, another jogger pushes past, bumping into me. The jogger says nothing. He just rams those earbuds into his ears even farther and starts punching at ghosts with taped hands.

Disgustedly wiping another man's sweat off my arm, I can feel myself start to get mad. I even go so far as to turn towards the Monk to say something, only when I finally look over, this man's just smiling.

There's a family flipping burgers on one of those park grills, and he's watching that, while a guy with really long hair is about catch the perfectly thrown frisbee—and now he's watching that.

Over to our left, there's these two dogs playing chase without a care in the world and I can't help but envy their innocence.

I used to be like that, I think to myself.

I used to be excited for life and want to play with others…but then something unforgivable happened, and now I can't seem to walk around the block without getting re-stimulated. It's thoughts like these that always used to make me spiral out of control, except when I feel them come on this time, I finally start to counter them with something productive.

"How are you able to be so patient with me?" I ask the Monk.

Because by this point he'd already taught me how to do yoga, how to meditate, how to listen better, how to eat better—he even got me to start questioning why I felt the need to keep punishing people with my actions.

But why?

How could one person be so compassionate, patient, and giving, while I was still so quick to judge, so desperate to be right, and so unbelievably petty. It just didn't make any sense. So the Monk responds, but as is often the case, the response comes in the form of a question.

Turning his face towards the Sun, the Monk goes, "Do you know why we always bow to each other in India?"

Respect, I say, trying to let another jogger past. Except when I go to step off the sidewalk, the Monk moves in front of me and turns to face me head on. At first, it looks like we might be about to spar, until his palms kiss together and he starts bending at the waist. Lowering his head in reverence to me, he keeps his eyes closed and his back entirely straight. Then, about ten to fifteen seconds later, he utters the words, "Namaste," and raises his head to find my eyes once again.

"It means, I bow to the divine in you," he says.

We bow to one another to honor the Great Self that's gotten lost within us.

We bow to one another because "self" and "other" aren't really separate things.

As he says this, a glimmer of sunlight reflects back off his glasses and I can't help but notice the patchy pink clouds hanging in the sky.

The whole rest of the walk, I didn't say a damn thing.

Chapter 31
Speciation Events

LOOK, IF AVA took you to her house to meet her parents and all five of her brothers, by about the fourth handshake, I'm sure you'd be able to pick out some of the quintessential Adams features. Like most families, everyone in Ava's family doesn't look identical to one another, but there still exists enough similarities between them for you to realize they're all related. For example, not a single one of her brothers is under six foot, which was a little intimidating, to say the least. But as we leafed through embarrassing photo albums and yearbooks that Thanksgiving, I couldn't help but notice my ability to pick up on these slight similarities became more and more astute. Whatever features her family tended to express (freckles, fair skin, reddish hair) after getting to meet her entire family (and then getting to meet her entire extended family) I finally understood what her Mom meant when she said, "that Adams chin" or "that Adams nose."

Essentially, there existed an entire range of common denominators between Ava and everyone else praying around that turkey; and because I had gotten to see so many different variations of it, my brain had gotten rather good at deducing which members were related by blood and which ones were related by marriage.

I guess you could say I became somewhat good at spotting the resemblance.

Which isn't really something to brag about as it's mostly an unconscious process. Even still, it was ultimately this same process which led Charles Darwin and Alfred Russel Wallace to synthesize their theory of evolution. Except, instead of being able to spot the resemblance between one family, they were able to notice the resemblance between all families.

Not to take anything away from one of the most celebrated geniuses of our time, but Charles Darwin wasn't really all that special; he just happened to be in the right place at the right time.

And then, of course, there was Alfred Russell Wallace, the long forgotten co-discoverer of evolution, who also just happened to be at the right place at the right time.

Both were naturalists.

Both were keen observers.

Both were given the chance to sail around the world, visit archipelagos, and see the world in a way that so many people of their day just couldn't. And so, not surprisingly, they both ended up synthesizing a theory to explain how and why animals speciate. The reason I say Darwin and Wallace weren't all that special is largely because the seeds to this idea were already in existence long before either of them were even conceived. And the fact that this theory was formulated, *independently*, almost the same year, by two naturalists working in completely different countries only furthers the point.

This was an idea that had to come out.

In actual fact, Darwin and Wallace were even correspondents, which is also why, in 1858, Alfred Russel Wallace sent Darwin something he wrote in the midst of a fever dream.[1] A theory about how and why animals speciate written in an essay entitled "On the Tendency of Varieties to Depart Indefinitely from the Original Type."

Later that year, astonished by the similarities between his own work and his colleague's, Darwin and Wallace jointly publish a paper to the Linnean Society. The following fall, in 1859, Darwin publishes *On the Origin of Species*, and the entire world starts to change.[2] His conclusion, that all animals, plants, and reptiles are, in fact, related. And that islands, like the Galápagos archipelago for Darwin or the Malay archipelago for Wallace, were some of the best ways for a person to catch evolution in action. And why? Because islands just happen to provide reproductive isolation, which just happens to provide the pre-conditions necessary for animals to speciate.

Darwin's Finches

For five years, Darwin got to sail around the world aboard the HMS *Beagle*, collecting and cataloging more than 1500 species of animals. And after examining his notebooks with a fine-tooth comb, we know that Darwin had already worked out most of the ideas behind his theory of evolution by the year 1839.[3] Without a doubt, his trip to the Galápagos was clearly the most important part of the trip, as it was here that he first began to take notice of Nature's uncanny ability to produce variation.

In sum, there were many different variations of plant, bird, and animal species present that appeared to be completely unique to the Galápagos but also seemed to be mysteriously related to the mainland species. In other words, as Darwin was traveling from island to island, he couldn't help but *spot the resemblance* between all these finches. Where ornithologist John Gould saw twelve new species of finch, Darwin was beginning to see one ancestral population of finch "that had been taken and modified for different ends."[4,5]

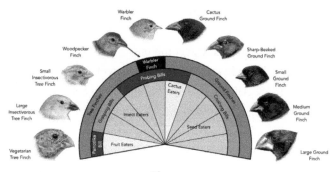

Fig 1.1

In all sincerity, Darwin had no idea about the importance of the finches he'd collected, or what this evidence would eventually mean. But after visiting a dozen or so islands, each with its own distinctive type of finch, each with a beak that seemed specifically designed for the exact type of food available on that island, like a good detective, Darwin could no longer ignore the coincidence. Contrary to popular belief (that all species were fixed and unchanging) Darwin was beginning to suspect that all species were evolving and continuing to evolve according the environmental niche they found themselves in. And if we were to ask ourselves what Darwin saw when he looked out at those finches, we may venture a guess that it was something like this (See Fig. 1.2).

But how exactly is it that animals speciate?

To explain that, let us revisit the concept of islands.

Islands

To make a new species, all you really need to do is have a source population (like a group of mosquitos) and then you need to introduce some kind of geographical barrier which sections off one bit of the population from the other. So, for the next thought experiment, we can imagine taking a subset of these mosquitos and then releasing them on an island far enough away from their original source population that there's virtually no gene flow occurring between the two populations.

After that, all we have to do is wait.

That's it.

In biology, you'll often hear terms like "islands" or "islands of life" being tossed around, but this doesn't always mean a bit of land surrounded by water on all sides. Islands could be literal islands, like the Galápagos archipelago or they could be metaphorical islands, in which, the term "island" is really referring to any kind of geographical barrier that prevents or restricts gene flow.

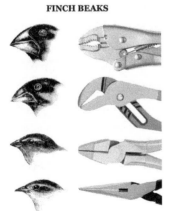

FINCH BEAKS

To cite an example of this, where I grew up, there was a chain of lakes that existed not too far from our house. Each of the six lakes were connected by a little canal you could get your boat through, and so consequently, if you happen to be a largemouth bass, this meant you could potentially swim and breed in all six of the lakes. Your habitat, and in turn, your gene pool, wouldn't be restricted to just

Fig 1.2

one lake because, in effect, you'd have six to choose from. Although, it's also important to note that if a serious drought ever occurred, there'd be no way of getting your boat through. You'd be trapped in whatever lake you started in and so would any fish.

At one time, a largemouth bass living in the first lake could potentially swim and breed in all six of the lakes in question. But during the midst of an extended drought, this chain of lakes would have transformed itself into a reverse archipelago. Signifying that its gene pool, which used to be one large gene pool, has now divided itself into six. Therefore, from the perspective of a largemouth bass, it's like he's now stuck on an isolated "island of life" completely cut off from exchanging genes with any of the other five populations.

By the same token, from the mind of Richard Dawkins we're gifted the idea of "virtual islands."[6] Again, not an island in the traditional sense, but a kind of virtual reproductive isolation put in place by pedigree breeders who wish to control the mating opportunities of their pups. Pedigree breeders will usually go to great lengths to ensure their prize-winning Dane only mates with other Great Danes of a certain genetic purity because they're not trying to sully their Great Dane's genetics.

In light of this, if we now return to those hypothetical mosquitoes being trapped on an island far enough away from their original source population, what might we expect to happen?

Well, given enough time, we would expect these two populations to begin diverging away from one another. Being that no two environments are ever the same, there would be an asymmetry in selection pressure, such that each population of mosquito would start to become populated with different frequencies of adaptations. Just like the thirteen or so geographically isolated populations of finch, the two gene pools of mosquito would inevitably start to become more specialized for their particular environment (mainland or island). After that, if conditions of the geographical barrier are maintained—hundreds, perhaps even thousands of generations later—there would be so many genetic differences present that if the two populations were ever brought back together they could no longer mate and produce viable offspring. In short, a difference in environment almost always comes with a different set of selection pressures, which eventually gives rise to certain adaptations being favored over others; and it's precisely this asymmetry in selection pressure which causes two geographically isolated populations to begin the process of speciation.

How long does it take for a new population to speciate?

Of course, it all depends on the organism and how long its generational time is, but since Darwin published *The Origin of Species* until the year 1999, appears to be long enough. Because this hypothetical situation, where one group of mosquitos suddenly became reproductively isolated from another group of mosquitos, has already happened.[7]

Mosquitos in the Underground

Either in 1863 when the London Underground first opened for public use, or perhaps a bit earlier, there happened to be a group of mosquitos who got trapped in the tunnels and were forced to adapt to life underneath the streets of London.

Flash forward until the year 1999, when two geneticists named Katherine Byrne and Richard Nichols started testing these underground populations, and the results they found were the very definition of remarkable. These populations of mosquito were now so different they could no longer interbreed with the mosquitos they found at the surface. So about the place where Burnt Potato starts to merge with flavors of Mince Pies and Marmite (the Shepherd's Bush tube stop) they collected a sample of underground females, but when they tried to cross them with the males collected from the surface, *no offspring were produced.*[8]

Just to be clear, the underground mosquitos are perfectly fine breeding with other underground populations of mosquito, just not with their ancestral population living at the surface. Indicating, at least from a definitional standpoint, that we may have a new species on our hands. A species is a group of individuals who either regularly interbreed together, or could. And this population they've found living underneath the surface, no longer can. To put this in perspective, just consider the Great Dane and the Chihuahua. Two breeds of dog that couldn't look any different from one another if they tried, and yet, both of these breeds belong to the same species *Canis lupus*. Which also means they're both capable of mating with one another and producing viable offspring.

As you might expect, their mating couldn't be successful without a little physical assistance from us, but the fact of the matter is, their genetics are still similar enough that a mating between the two can still produce offspring that can then have offspring of their own. Even breeding a dog with a grey wolf, also *Canis lupus,* is still possible because, as we now know, all dogs are descended from wolves.[9] Admittedly, there's still an ongoing debate as to whether or not dogs deserve to be called a separate subspecies like *Canis lupus familiaris.* But irrespective of this, and despite tens of thousands of years of artificial selection, there still exists enough similarities between a wolf and dog, that a pairing between the two produces a wolfdog. Not a sterile animal that can't have pups of its own, rather a half wolf half dog that's perfectly capable of interbreeding with either wolves or dogs.

Notes from the Editor:

To put this in a slightly larger perspective, we know that zebras, donkeys, and horses are all closely related. They all evolved from a common ancestor that lived somewhere between 4 to 4.5 million years ago. However, if you tried to mate a horse with a donkey the result of this endeavor would be what they call a *hinny* or a *mule.* Animals which are almost always incapable of having offspring of their own. A true speciation event.[10,11]

If we evolved from chimps then why are there still chimps?

This is a question you might hear asked from time to time, and the answer is, we didn't evolve 'from' chimps but we did share a common ancestor with them that lived about six to eight million years ago.[12-14] So if we're looking at the following phylogenetic tree (Fig. 1.3), where Branch V meets Branch X is where we shared a common ancestor.

Correspondingly, to ask why are there still chimps is really the equivalent of looking at a tree with a branch that forks in two and then asking the question, "Why does a tree need this branch if it already has that branch?" In a way, the question presupposes that evolution was always aiming at a particular goal, like a human, which it never was. Remember, evolution is a dance, not a goal-directed enterprise.

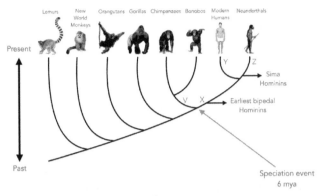

Fig 1.3

And while we're at it, let's also keep in mind that trees never bet all their leaves on one branch; they diversify their efforts. Which is also why the whole Tree of Life idea is such a powerful and beautiful concept (the shape of the tree just happens to match its strategy). The same exact strategy, I might add, that Mother Nature seemingly has put in place everywhere: *Produce endless variants on creation, like branches, twigs, and leaves, and then scatter them about in every direction.*

Of course, not every branch, leaf, or twig is destined to survive the chaos of existence, but by diversifying your efforts, you will have maximized your chances of survival.

Against this background, branches are certainly seen as important, but at the end of the day, they're just like arms on the tree, of which there are several. There can be no doubt that losing a limb is detrimental to an organism, but when you have the ability to grow new limbs, losing a couple branches here or there doesn't jeopardize the whole tree. In this respect, the branch that eventually did give rise to the chimps and the bonobos is seen as just another way of being alive and catching the sun. Again, it's not that we've been evolving longer, or that every other species could be us if only they could "get their act together." Basically, there is just the stuff of life (DNA), which like water, takes its cues from the environment. Water can assume whatever shape of container you put it in, and in much the same way, whenever you subject the stuff of life to the sorts of selection pressures our ancestors endured over the last six million years, you get a human. Whereas, if you were to roll the clocks back and subject our ancestors to the exact same environmental conditions that Branch V saw, then we'd be the chimps and bonobos, it really is that simple.

It's a mistake to think there's some kind of inherent special stuff that we possess, unless by "special stuff" you're referring to the environmental conditions that sculpted our evolutionary story over the last six million years, because if anything is special—it's that.

It's the deaths, the wars, the parasites, the genocides, the predators, and the diseases that have shaped our species into what it is, because really, human beings are just *expressions of DNA* like everything else. Nothing more, nothing less.

So to answer the question, we are not descended 'from' chimps; they are our cousins.

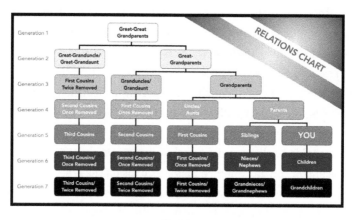

Fig 1.4

Just like you and your first cousins share a common ancestor (your grandfather), which is only two generations away from you, we are literal cousins with the chimpanzees and the bonobos, only it's six million years' worth of generations instead of only two. So, if you were take the following chart and turn it on its head, so that it looked more like one of our phylogenetic trees (Fig 1.3), effectively, you'd be staring at the last few twigs on the Tree of Life. If, however, you wanted a more refined picture about how your own family tree fits in with the grand Tree of Life, then all you'd have to do is start filling in details about your parents, your grandparents, your great-grandparents, your great-great-grandparents...and so long as you kept doing that for the last 300,000 generations, unavoidably, you'd come to an ancestor which was destined to leave at least two descendants. One, which would eventually give rise to the entire *Pan* lineage (the chimpanzees and bonobos); and the other, which would eventually give rise to the entire *Homo* lineage (humans, Neanderthals, Denisovans, etc.).

That's the true meaning of the phrase, "all life is connected by descent," because if you draw your trees with enough detail, inevitably it all comes back to that last universal common ancestor which probably lived in hydrothermal vents deep in the ocean floor 3.5 to 4 billion years ago.[15,16]

***Notes from the Editor:*
Considering that the last universal common ancestor or LUCA, is sometimes confused as being the first organism to have lived, it seems necessary to provide some clarification here. LUCA is not a fossil, it's a theoretical construct. The origin of life is still shrouded in mystery, although the last common ancestor of everything alive today (archaea, bacteria, and eukaryotes) is something they call LUCA, whose origin story keeps getting pushed further and further back in time.[17]

—As you might expect, talking about the exact moment when a species becomes another species can be rather problematic. Much the same as trying to determine the exact second when a cube of ice has become water, a speciation event just can't be pinned down to a specific point in time.

Same as when dialects of a certain language keep diverging away from one another. Eventually, they drift so far apart that they produce a separate language. But to ask *what year* this one regional dialect of Latin officially became Spanish, is not really a question that can be answered readily. It isn't the case that one day everyone was speaking the English of Chaucer's age and the next it was the posh RP dialect you'd hear on the BBC, because the evolution from Old English to its modern day incarnation happened gradually.

Artificial Selection

Before Darwin and Wallace taught us what evolution was and how to talk about it, it should be known that we were already doing it; we just didn't know that's what we were doing. Pigeon fanciers, sheep herders, dog breeders, and crop farmers all seemingly knew the basic precepts of selection and inheritance, they just didn't realize that it was exactly small changes of this type—when stacked end-to-end over the course of billions of years—that result in massive changes like a fish to an orangutan.

What's more, if we now consider the story of the modern day canine one last time, we find that the first domestication event appears to have occurred somewhere between 20,000 and 40,000 years ago.[18] And within that time, one species of animal has taken another species of animal and molded it into the 350-something officially recognized breeds you see around you today.[19] From a St. Bernard to a Bichon Frise, one piece of the universe has sculpted another piece of the universe into meeting its every whim and fancy. Purse dogs, lap dogs, guard dogs, sled dogs, hunting dogs—you name it.

In a similar fashion, we've taken one species of plant, the wild cabbage, *Brassica oleracea,* and simply by breeding for desired outcomes (just like we did with the wolves), we've transformed it into vegetables as distinct as kale, kohlrabi, cauliflower, broccoli, collard greens, and Brussels sprouts.[20] Not as impressive as what we did with the wolves, but the sheer plasticity and transformability of organisms is beyond impressive.

And that's just artificial selection.

The real genius behind Darwin's work was to get people to fully appreciate and understand the power of artificial selection, and then to take that one step further by realizing that we don't always need a human hand doing the selecting because, as it turns out, Nature does it for you. Natural selection, he then argued, was just like artificial selection, only it wasn't the whims of some pigeon fancier doing the selecting, it was the non-random selection of Mother Nature.[21]

Why is it said to be non-random selection?

Because while the mutations may arise randomly, the selection process that Mother Nature imposes is by no means random. Like Darwin first showed us, there is a struggle for existence, such that, only the strongest and fittest individuals ever survive to reproduce. For this reason, the genes that do end up aiding in the survival and reproduction of that organism are more likely to become frequent in a given gene pool than the mutations that don't, consistently providing natural selection with the opportunity to build and accumulate wisdom. The mutations which provide some type of survival advantage subsequently allow that organism to leave more descendants, whereas, the mutations that don't provide some kind of survival advantage ultimately, leave less descendants, until eventually, those lines of descent peter out. Thus, the whole Tree of Life can be viewed as a living tree that is constantly branching and re-branching, sending out exploratory vines under the guise of mutation, and then reproducing the template that works best.

Clearly, not every twig or branch ends up being a fruitful endeavor as 99.9% of all species once living are now extinct. But to all those branches not shown in this picture, we owe a special kind of respect. For, without them once being here, we could not be here now. We depend on them having been here, however long or however short their time in the sun actually was.

334

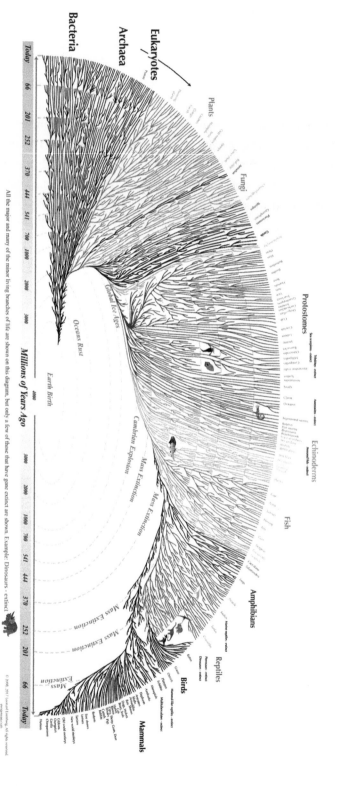

All the major and many of the minor living branches of life are shown on this diagram, but only a few of those that have gone extinct are shown. Example: Dinosaurs - extinct!

Bacteria

Archaea

Eukaryotes

Plants

Fungi

Protostomes

Echinoderms

Fish

Amphibians

Reptiles

Birds

Mammals

Millions of Years Ago

Oceans Rust

Earth Birth

Global Ice Ages

Cambrian Explosion

Mass Extinction

Mass Extinction

Mass Extinction

Mass Extinction

Today 66 201 252 370 444 541 700 1000 2000 3000 4000 3000 2000 1000 700 541 444 370 252 201 66 Today

Chapter 32

Interdependence

AT WHAT POINT does a cake become a cake? Once all the ingredients are sitting on the kitchen counter, or does the oven need to be turned on first? How about when the ingredients are all mixed up, the oven's been preheated, but the batter is still in the bowl, can we call it a cake then? Or does it need to be cooked first?

How about after it's been baked, iced, sprinkled, *and cut?* Can we call it a cake then? Or does it need to be served on a plate too?

At some point in the last ten seconds, I hope you picked up on how ridiculous and arbitrary this whole line of questioning was, because for Ava and me, the question about when life "began" is tantamount to asking, "When does a cake become a cake?" or "When does a seed become a tree?" The question itself is somewhat nonsensical because from the second the universe first went "bang" it was already pre-seeded, so to speak, which is really just another way of saying all the ingredients were already on their way to being on the metaphorical counter.

Nevertheless, seeing as the vast majority of thinkers are still operating out of some genuinely flawed fundamental assumptions (believing that consciousness is something that emerges out of complex systems) that means there's still a great deal of people running around trying to figure out the exact moment when you can call a cake, a cake.

The reality is, it was always going to be a cake.

Consciousness was always going to happen, in some form or another, because consciousness is what there is. It's a fundamental property of the universe. Which means that everything—you, me, rocks, iron, plants, brains, air, space—are necessary for consciousness to be.

The confusion with all this seems to only arise when you believe, as I used to believe, that consciousness is some kind of mental phenomenon only capable of occurring from the neck up. And it's totally understandable to feel that way, just as it's totally understandable to believe in the world of VR when you have one of those virtual reality headsets sitting on your face. But once you've burned through as many examples of us living inside an illusion as we have, perhaps you're beginning to see that there really is only the one consciousness, and everyone and *every* thing has gotten a slice.

Or, to put it plainly, consciousness isn't just a product of the brain; it's a symptom of the universe. The entire universe is conscious because consciousness is what's baked into the very fabric of existence.

The Problem with Emergence

To show how fundamentally problematic the emergent idea is, let's reconsider the cake question from a slightly different angle. Under the emergent idea about consciousness, where consciousness isn't fundamental to existence but emerges *out of* complex systems like brains, you're always going to run into problems figuring out when the lights came on. Much like the difficulty in pinning down a speciation event, there just isn't a moment you could ever point to that wasn't arbitrarily chosen.

Cosmology tells us that the universe is about 13.8 billion years old. Fossils tell us that life is at least 3.5 billion years old.[1] Reasoning from these facts, the reductive materialist would probably tell you that there was no meaning, no happiness, and no conscious experiences for at least the first 10.3 billion years of our universe's existence.

Of course, that's assuming they even think the earliest replicators were conscious.

If, however, they were to assume that these early microorganisms were not conscious, due to lack of complexity or something, then you really start running into issues about when you can call a cake, a cake.

For example, are elephants—who bury their dead, use tools, play music, form strong social bonds, and pass the self-awareness test—conscious? Or are they just giant, mindless robots? How about chimps? Chimps can laugh, use tools, understand the concept of money, pass the self-awareness test, and learn sign language! So, do they get to have conscious experiences? Or are we really trying to conceive of a universe that had zero conscious agents in it, until somewhere around six million years ago, when our ancestors split from the chimps? And if that's the case (which I don't believe it is) can you also tell me at what point during our long evolutionary history that we shed almost all of our instincts and made the leap from soulless, mindless, automatons to fully fledged, fully conscious, soul-carrying, entities with a perceived sense of free will and dominion over the animals?

—

You see, Tom, there never was a magical moment when all the ingredients on the counter suddenly made the switch to being a cake because all those arbitrary limitations *depend* on our perspective. And that perspective is ultimately defined by the depth of our understanding (i.e., where we've decided to stop thinking).

The general point was once summed up by the twentieth century philosopher Ludwig Wittgenstein.

In a remark to his friend, Wittgenstein once asked, "Why do people always say it was natural for man to assume that the Sun went 'round the Earth, rather than that the Earth was rotating?"

His friend replied, "Well, obviously because it just looks as though the Sun is going round the Earth."

Wittgenstein then replied, "Well, what would it have looked like if it had *looked* as though the Earth was rotating?"[2]

A very astute remark, especially when you consider that it cuts right to the heart of the problem we currently face today. Sure, the details might've changed, but the same essential argument still holds. And just like the once held geocentric model of the universe, the only thing the Western worldview really has going for it is the fact that so many people have just blindly accepted it on faith. Even worse, they probably never even knew there was such an elegant alternative to replace it with.

This being the case, that means we now have tens of thousands of people running around trying to figure out the exact moment when "consciousness" suddenly got injected into our animal hardware, when the sad truth is, they're never going to find it. Like trying to make Newtonian physics work under the Ptolemaic model of the solar system, it's just never going to happen.

What you want to say is, "I admire your tenacity, but your starting point is all wrong."

There was never a point when life suddenly got 'injected into' cold, lifeless, matter or emerged 'out of' dead, mindless matter because consciousness isn't emergent—it's fundamental.

Intelligence, life, consciousness, the universe, these are all really just different terms for the same thing. Which is partly a semantic problem, but it's also a serious scientific oversight, due to the fact that our definitions about 'what' life even is are too short-sighted to begin with. Just take viruses as an example. Viruses aren't considered to be alive. They don't have heartbeats or brains, and yet, they certainly behave as if they have desires and intentions, don't they?

The way they adapt to our existence and we to theirs.

Sure, they might be parasitic on living, breathing hosts for replication and dissemination, but to just pass viruses off as non-living entities, simply because they don't fit into the standard model of reproduction is a serious error. After all, let's not forget about all the ingenious ways viruses do manage to get themselves passed on.

When a dog first gets infected with the lyssavirus (aka rabies), the virus *makes* them start salivating and foaming at the mouth. Two things which really increase the odds of it getting passed on. But in addition to this, it also makes the hosts more aggressive, highly reactive, more likely to bite, more likely to roam, and completely fearless of human beings.[3]

"But it's not alive," is still what most biologists will tell you.

And to that I always respond, "Are we sure about that?" Or is it the case that we actually still don't have a good understanding about what life is and all the bizarre ways it manifests?

Notes from the Editor:

By and large, it would certainly appear as if the majority of biologists are still operating under the idea that viruses are better conceptualized as clever bits of chemistry, rather than being some kind of organism. Although, in his latest book *What is Life*, the Nobel-prizewinning biologist Sir Paul Nurse has actually argued in favor of viruses being alive. After making the case that all forms of life exhibit some degree of dependency upon other life-forms, and after acknowledging that viruses are completely dependent upon other living entities for survival, Sir Paul remarks:

> ...so you can see there's a graded spectrum of living organisms, from the viruses through to plants with a wide range in between, and these all have varied dependencies on other life-forms. In the case of the virus the dependency is very strong, whilst in other organisms that dependency can be weak. But I argue that these different life-forms are all alive because they all can evolve by natural selection. If you don't like that explanation, I'll give you another one for viruses: They're dead when they're outside the cell, and they're alive when they're inside the cell.
>
> - Sir Paul Nurse (biologist)[4]

Eastern Philosophy

If there's one thing the East can be said to have gotten right, it's their appreciation of backgrounds and relationships. While we in the West might have gotten a few models for conceptualizing the physics of our universe right, we've completely failed at teaching our society the vital significance of ecological awareness. Darwin himself made a point to stress the ecological interdependence of life by employing terms like the "struggle for existence" and by painting us a picture about life's "entangled bank" in the final chapter of his masterpiece.[5] So for Darwin, the term "struggle for existence" wasn't just referring to an organism's independent fight to survive the tragedies of existence because he was also referring to each individual organism's collective dependency on every other organism in the vastly interconnected *web of life*.*

To give you an idea of this, let's re-examine sunflowers for just a moment.

The existence of a sunflower—and all flowers for that matter—*depends* on there having been some kind of pollinator like bees, birds, or insects. And why is this the case? Quite simply, because flowers and pollinators go together. Predators and prey go together. And, of course, parasites and hosts go together. Then, from great Chinese thinkers like Lao Tzu (the alleged founder of Taoism and credited writer of the *Tao Te Ching*) we're gifted the idea of *mutual arising*. To be and not to be mutually come into existence, and opposites are revealed to share a hidden connection.[6]

Because everything in Nature exists on a dipole, and because you can only know something in relation to its opposite, it is often stressed inside the rubric of Eastern philosophy that these two processes are intimately related.

They're not two coins; they're two sides of the same coin.

In other words, we can only know what "up" is sitting here now because we already have a notion of what "down" is. Likewise, north is only north in relation to a direction that is south, and similarly, positive is only positive in relation to something that is negative. These things not only imply one another, they *go with* one another, just like motion implies stillness, hot implies cold, or self implies other.

Against this backdrop, if positive only makes sense *in relation* to something that is negative—and you can't know one without the other—an Eastern philosopher would tell you, they're both a part of some higher-order process, like electricity. They wouldn't doubt the differences between positive and negative; however, they would probably tell you that if you could only view these forces from a zoomed out perspective, you'd soon find that they're actually a part of some higher-order system:

The predator-prey system.

The parasite-host system.

The "Organism-Environment" to borrow another one of my teacher's descriptions.

> You see that, for example, in the science of ecology. One learns that a human being is not an organism *in* an environment but is an organism-environment. That is to say, a unified field of behavior. If you describe, carefully, the behavior of any organism, you cannot do so without at the same time describing the behavior of the environment. And by that you know that you've got a new entity of study.
>
> - Alan Watts (philosopher)[7]

Or to get at the matter from an entirely different direction, you'll just never find a human floating around in deep space because deep space is one of the most inhospitable environments imaginable. Matter of fact, the overwhelming majority of our universe is rather inhospitable to human life, which is precisely why we evolved on this planet and not anywhere else. In accordance with this, when we take this idea to its extreme, we find that a human being (just the sheer fact of one) *implies* its entire environment.

In the same way that flowers imply pollinators, the complexity of a human implies the whole cosmos. We didn't just spring into existence, we co-evolved with the planet. Human beings are embedded within a deep, interconnected, evolutionary story analogous to a giant house of cards. Currently, it feels as if we're at the apex, and who knows, maybe we are. But we couldn't be here now—in this exact way—without every evolutionary dead-end, mistake, and mutation that's ever come before us. Take any one card away and the whole house comes crumbling down.

Redefining the Environment

Look, the more you go into ecology or evolutionary biology, the more you start to realize that everything depends on everything else, and that nothing, save space, exists in a vacuum.

To put the matter more concretely, our planet has changed immensely since it first coalesced a little over 4.5 billion years ago, and so has its progeny. If we could see the planet as it was 4.5 billion years ago, we would scarcely recognize it. Asteroids, volcanoes, continental drift, ice ages, floods, oxygen, fungi, plants—*and five mass extinctions*—all helped shape our environment into what it is today. And seeing how our evolutionary story began at least 3.5 billion years ago, that means our ancestors have been co-evolving with the planet for at least that long too. As it's changed, we've changed (evolution). And as we've changed, it's changed (global warming).

Earlier we spoke about how reality is really whatever your senses tell you it is, given that our internal experience of reality (aka our internal model) is largely dependent upon the sense organs we're bringing to the table.

But where exactly did we get these sense organs?

The short answer is, of course, evolution. However, the more refined answer is, the environment forced it out of us. So, we might then say: *Reality is shaped by our senses and our senses were shaped by the environment.*

It's often misunderstood that random mutations are the driving force of evolution, when realistically, that's only part of the story. The other part is that the organism is shaped by the environment in which it lives, even further by the selection pressure put in place by the females of that species (sexual selection), even further by the evolutionary arms races that species is currently involved in, and even further still by all the competitions/alliances it has to maintain in order to survive. It isn't the case that toothed whales, bats, and two species of cave swiftlets all independently evolved echolocation...*and then*...sought out a nice dark environment in which to use it, because in reality, the environment came first.

——

If we ever wish to speculate about what life on other planets might look like, it would certainly behoove us to start by looking at all the various ways life has tried to express itself on this planet first, for it's only after studying the *independent evolutions* of things that we begin to unearth some of Nature's archetypes. The eye, for example, is one of Nature's most enduring creations on account that she's evolved it no less than forty separate times.[8]

For the avoidance of confusion, this does not mean that only forty species on this planet have eyes, more like, the blind processes of evolution have stumbled upon some kind of a plan to make an eye at least forty separate times (compound eyes, camera eyes, etc.).

From the looks of things, we might then say, given that eyes are so common on this planet, there's a good chance that life on other planets will probably have eyes too. Echolocation, on the other hand, has only ever evolved four times, independently, which definitely speaks to the utility of being able to see with your ears, but taking into account that this sense is more rare, we can at least infer something about its practicality.[9]

If we then return to our example of the bat, we should remind ourselves about two important points: (i) that the ancestors of the bat were only ever trying to do what we're all trying to do: survive and reproduce. And (ii), that the conditions of the environment in which they called home were wet, dark, and cavernous, providing them with the exact kind of directional selection needed for echolocation to develop. It's a mistake to think that bats evolved echolocation first, and then screeched their way into a dark cave where they could finally use it. Rather, the environment reached into the genetic grab bag and pulled out the echo-locating capabilities nascent in all these creatures.

It's no coincidence that polar bears possess a strong sense of smell *and live in an environment where everything looks the same*. Better to say, the polar bear's nose works so well because they live in a place where everything looks the same. In this respect, polar bears depend on their nose for hunting just as much as pit vipers depend on their heat-seeking pit organs to strike prey in total darkness.[10]

To borrow another Richard Dawkins' metaphor, "It is as if Nature has the lock and the creature has the key."[11] Only the lock is always changing. As a consequence of this, in order to keep on surviving, creatures must always be changing too. And so, with each new environment our ancestors either moved to (or were forced to move to) there existed a different set of selection pressures to contend with. In essence, the locks kept changing, but so did we, and in the process of perpetually having the locks changed on us, we evolved into the most adaptable kind of key there is.

The kind of key that cuts itself.

So far, we're the only species that's ever managed to nip and tuck its own genome (See *CRISPR*); and in many senses of the word, conscious and unconscious selection has played a part in our evolutionary trajectory for quite some time now.

Notes from the Editor:

The real beauty behind Darwin's great idea was in its universal applicability. Not only did he unearth Nature's algorithm by showing us how "she" selects, but he underlined the basic precepts of evolution (variation, selection, and heredity) which could then be applied to almost any field. From tech, to religions, languages to business models, all things evolve.

—I'm reminded of a story I once heard from David Foster Wallace about these two young fish who were busy swimming around when an older fish swimming in the other direction, passes by them, tips his hat, and says, "Morning, boys...How's the water?"

Neither of the two fish says anything. They each just swim around a bit before one of them finally turns to the other and goes, "What's water?"

Notes from the Editor:
This little anecdote, which is meant to spark thought in the minds of the reader, makes us laugh at the two young fish for not having noticed something so fundamental, but really, haven't we all done the same thing?

—So then, we need to start thinking about Nature in much a similar way. Not simply that we've taken our environment for granted but that even our concepts about 'what' Nature even is are too shortsighted to begin with. The environment isn't some stale, dead, lifeless, rotating aquarium of which we happen to have woken up on. Rather, the environment: air, oceans, rocks, minerals, plants, herbivores, weather, the magnetosphere, and the ozone are all a part of one interrelated, interdependent, field of behavior.

> "Through our eyes, the universe is perceiving itself. Through
> our ears, the universe is listening to its harmonies. We are the
> witnesses through which the universe becomes conscious of its
> glory, of its magnificence."
>
> - Alan Watts (philosopher)[12]

Time-lapse

At this point, all I really have to say left is, it'd much easier to notice that human beings go with their environment by reminding ourselves about all those little environmental necessities we consistently depend on for survival. Including, but not limited to: the temperature and pressure of the air, the fact that we have gravity, the fact that the Earth spins on an axis, the fact that we have a moon, the fact that we just happen to be the perfect distance away from a suitably-sized host star to allow oceans of liquid water, and so much more.

As an illustration of this, perhaps the best course of action might be for us to imagine another time-lapse, similar to the one we did before. Only this time, I'd like for you to imagine we were watching the goings on of a human for about 24 to 48 hours. If that were the case, the first thing we might notice is an inflowing of environmental particles (air, food, water, etc.) coming into the body and altering its physical shape. And then, not too long after that, we'd notice an outflowing of environmental particles (sweat, urine, fecal matter, and carbon dioxide) returning to the environment in the form of excrement. CO_2 might be a waste product for us but it's breathable air for a plant. Which seems to work out nicely given that whenever a plant converts CO_2 and H_2O into carbohydrates, they release O_2 as a waste product.

Imagine that: they need our waste product and we need theirs. And just like we saw in the Magic Camera thought experiment, our ability to shift and change our perspective offers us the ability to notice something foundational to our existence. Namely, that we go with and depend on our environment for survival.

Where our skin touches the air, we say this is where my body ends. But then when you consider what a body even is, the most logical conclusion you can come to is that the body is really just bits of the planet you've converted into body.

To make a tree, a sapling doesn't need go around gobbling up bits of the planet and turning that into body because a tree can just use water, carbon dioxide, and the energy from the Sun to make almost everything it needs. Humans, by contrast, cannot photosynthesize light or make all the vitamins, proteins, and amino acids we need to survive, which means that we have to either eat plants (or animals that eat plants) and then use their body to make ours. So where do we draw the line with all this? Where does your body stop and Nature begin?

The answer, of course, they're all one process. Nature is an organism and we're just a piece of it.

Or to paraphrase Spinoza once again, God is the mind of Nature, and Nature is the body of God.[13,14]

Hell, even Joseph Campbell, the celebrated author of *The Hero with a Thousand Faces* once said in an interview, quite unambiguously, "We *are* the Earth...we are the consciousness of the Earth." Then motioning towards his face he adds, "These are the eyes of the Earth. And this is the voice of the Earth."[15]

Over the centuries, taxonomists, and scientists have been fairly judicious about naming, classifying, and placing each species in its own little box. But apparently, the one thing we keep failing to recognize is that by over-fragmenting the world, we render it impossible to perceive Nature as a unified, functioning whole. And because we've been too busy focusing on the parts instead of widening our depth of field, we've also failed to realize that all these seemingly "lifeless" processes, like viruses, rocks—even the laws of physics themselves—are actually a part of a new entity of consideration.

A self-regulating, self-organizing, and self-governing system.

The kind of cake that'll assemble itself—*if you give it enough time.*

Chapter 33
Cuckoos III & Story B

AT LONG LAST, we must return to where this whole *Story B* interpretation began: the Vase-Face illusion. When we finally stop focusing on the vase, which is really a symbol for ourselves, the two faces set against the backdrop immediately start to shift into focus. Faces you didn't even know were there. And so it is with nirvana, satori, moksha, transcendence, the feeling of unity, the beatific vision, ecological awareness, samadhi, or whatever else you wish to call it. Because when it finally happens, it's the background of things that suddenly starts to become important.

Instead of consciousness consistently focusing on objects in the foreground, it finally starts to appreciate the background. Instead of constantly screening the background out, it finally starts to comprehend the necessity of negative space.

To give you a concrete example of this, I want you to picture your favorite song. The kind of song that, no matter where you hear it, *always* makes you want to sing the words. Then, once you have that, I want you to imagine what it would sound like if we took out all the dead space.

If we removed all the time between the drum kicks, took out all the space between the words, pulled out all the pauses when the chorus went silent—would it even be a song at that point? Or would it just be unintelligible noise? On this basis, it seems reasonable to conclude that backgrounds, pauses, intros, rests, spaces between the words, time between the repetitions, and outros are just as important as the notes you do play.

348

Notes from the Editor:

Buddhism, which is really just an offshoot of Hinduism, is not a religion in the traditional sense. They don't have a God or specific deity they pray to, not even Gautama Siddhartha the man they call the Buddha. In accordance with this, Buddhism never requires you to believe that Guatama was a God, just that he was a man who woke up. In this respect, practitioners of Buddhism aren't worshipping the Buddha as much as they're following him. Following his teachings so that they too might reach a state of nirvana.

—Go into the study of Eastern philosophy for more than a few minutes and you're bound to find this symbol crop up again and again. Only now, I hope you can try and look upon this image with fresh eyes.

In America, this is more commonly referred to as the Yin-Yang sign, whereas in the East, it's simply called the Tao (pronounced "dow"). And the Tao means "the way of things" or "the course of Nature."

From here on out, I'd like to purposefully refer to this symbol using the original singular word "Tao" versus the hyphenated designation "Yin-Yang" for the pure and simple reason that the true meaning behind this symbol is, in fact, harmony, balance, and unity—not duality. Thus, the two seemingly opposed forces in this symbol (dark and light) join together to create a new entity of consideration the Taoists call *Tao*.

You needn't be a theologian or a historian to spot the numerous overlapping concepts in Hinduism, Buddhism, and Taoism—heretofore spelled Daoism—although, it certainly doesn't hurt to have someone else there to show you where the dots are. Turns out, Hinduism is the world's oldest religion being founded in India around 2200 BCE, while Buddhism doesn't enter into the picture until much later (around 500 BCE). Then, right around the same time, only in China, Daoism starts entering into the Chinese culture with its supposed founder Lao Tzu.[1-3]

India being right next door to China, it probably comes as no surprise that some of these ancient Indian concepts might have seeped their way into the Chinese way of life, and that's exactly what happened. There was a cross-pollination of ideas, such that, when all the dust finally cleared these two 'philosophical religions' ended up co-influencing one another.[4] In the end, Daoism and Buddhism wound up giving birth to an offshoot called *Zen* or *Zen Buddhism*. However, in an effort to streamline the conversation we can just leave Zen out of the picture, and instead, spend a little time focusing on the Daoists.

Because for the Daoists, their entire concept of the universe really does begin and end with this symbol. The Tao, being the archetypal representation of the basic underlying principle upon which all harmony, in the entire universe, is based.[4,5]

But what is this symbol really?

In many ways, it's the ultimate teacher, but in its most simple formulation, it is the very nature of things. Not just a potent reminder of balance, harmony, unity, mutual interdependency, reciprocity, entanglement, the necessity of opposites, and the importance of relationship, but a guide for how to conduct yourself in the world: *To live in harmony with the underlying force of the universe, which is Tao.*

Kind of sounds like what Baruch Spinoza said, doesn't it?

Although, if I really had to put a fine point on it, I'd probably say that the real intelligence behind this symbol lies in its ambiguity. It's binary. Same as the language of a computer. Meaning, the Tao could really be representative of anything, and it will be representative of everything, eventually, because that's what infinity implies. Everythingness and nothingness. Positive and negative. Masculine and feminine. White and black. Self and other. Organism and environment.

What they call a "coincidence of opposites."

***Notes from the Editor:

The teachings of Daoism suggest that all creatures should try and live in a state of harmony with the entirety of the universe and all the energy contained there-within. To live in harmony, the Daoists suggest following "the way," which is most readily conceptualized as walking the tight-rope balance between the two eternal forces that comprise the Tao; though the Tao itself is not seen as a god. In this regard, Daoism is not a religion as much as it's a spiritual practice or way of being.

The Void

For what it's worth—and just know that this is coming from two rather adventurous psychonauts with a predilection for finding their edge—but every time Ava and I have tested the limits of perception and reached the infinity spot (aka the Possibility Space) this symbol is exactly what we see, feel, hear, and experience.

It isn't a God—it's a fundamental pattern of the universe. And when it's 'teaching' you, all it's really doing is showing you all the places where you've fallen out of alignment with this pattern.

Granted, in order to experience this unity you have to use some kind of tool or spiritual practice to off-line the Narrative Self so there's no more ego pollution blocking your ability to perceive. But once you've done all that, you soon realize that all life really is just a play of energy.

A dance between life and death. Right and wrong. Day and night. Up and down. Chaos and order.

Or perhaps you prefer the terms entropy and evolution?

In any case, whatever two forces you choose to put inside the Tao end up canceling each other out, because the forces themselves are supposed to be polar opposites. Equal and opposing forces. Which is also why the Buddhist philosopher Nāgārjuna (c. 150–250 CE) tried to encapsulate this wisdom in his *Doctrine of Emptiness:* Śūnyatā (pronounced shoon-ya-ta) which posits that the world is essentially void or emptiness.[6]

After attaining enlightenment, the Buddha realized that nothing in life was permanent (all things were in flux). And so, in keeping with this, most Buddhists are rather reluctant to ever conceptualize reality and give it a form. Mainly because they don't believe reality is something that can be pinned down. But apart from that, it's because Buddhism is more concerned with setting up the preconditions necessary for their students to have the experience themselves.

For Buddhists, the experience of nirvana was and is the most important thing, because once you get it, it's like everything in your life undergoes a massive re-contextualization program. The value hierarchy of your mind finally gets adjudicated, and your psyche is exposed to the fundamental pattern of reality, which is Tao. In this moment, absolute truth manifests itself in whatever form your psychophysical organism demands, and after basking in the infinite well of eternity you begin to 'remember' how there really is only one substance—*and you're that.*

Apart from this, you realize—just as the Buddha realized, just as the Greek philosopher Heraclitus realized, and just as the great German polymath Alexander von Humboldt once realized—*That all life is in flux, and that the very nature of life itself is change.*[7-9]

Piecing It All Together III

Pretend you're in a lecture hall, sitting in the front row, while I just happen to be at the front of the room pulling down a silk screen with one hand and then flipping a switch with the other. Just then, the lights dim down low and the projector hanging above your seat starts to turn on. Taking a few steps back, I slip a remote from my pocket and point it at the screen. At first, nothing happens. But then, as I hang off to the side, a video starts playing. Some kind of compilation. A highlight reel pieced together from some of the greatest Nature docs ever produced. First it's a snippet of a Burmese python grabbing a rat, then it's a crocodile taking down a zebra, followed by aerial shots of wolves hunting in a pack.

Alone in that auditorium, it's just you and me watching footage of mosquitos slurping up blood, stag beetles battling for females, antelopes evading lions, bears chomping up salmon, orcas taking down a great white, cuckoos chucking out eggs, great whites snatching up seals. And while all this is happening, I point at the screen and say, "This is the predator-prey relationship, right?"

And you go, "Yeah."

And then I say something like, "Everywhere we look on this planet we find this same essential game, this same *struggle for existence* playing itself out." Sometimes it's predators vs. prey, sometimes it's organisms battling their environment, sometimes it's parasites vs. hosts, but it's always the same essential game playing itself out, right?"

And you go, "Yeah."

So I walk over to the whiteboard and draw a sideways "8" on the board, followed by an equals sign, followed by a picture of the Tao. Putting

352

the cap back on the marker, I turn around and go, "Everywhere we look we see this same essential game, predator vs. prey, playing itself out."

"Mm-hmm," you nod, letting me know you're still following.

"And what I'm saying to you is this: the predator-prey relationship goes all the way up and all the way down."

But since this flies in the face of everything you've been taught to believe, you shake your head and say, "Wait, what?"

Pointing at the ceiling I go, "The predator-prey relationship goes all the way up," then my index finger changes direction, "and all the way down." That's infinity. That's you. That's me. That's everything. Then I take my hands and clap them together really hard.

The sound startling even me.

I press them together trying to animate the only two forces in existence—chaos and order—pressing them up against one another in forced opposition, then I look back up at you to make sure you're still paying attention.

With wrists parallel to the floor, I'm pressing both hands into one another so hard, they're actually starting to shake. And when you can tell that each finger is perfectly matched to do battle with its opponent, that's when the game starts to switch.

One of the hands starts to slip.

Then it falls.

The hand with the scar on it was about to fall to the ground dead but at the last second the left hand stuck its neck out and saved it, making you jerk forward in your chair.

After that, the left hand pulls the right hand up and the two go again. This time, the mirrors of infinity are pressing up against one another—the same as they were before—only when the wrists go parallel to the floor, and one of the hands starts to slip, the fingers begin to interlock propping each other up. They grip each other tight, highlighting their interdependence and showcasing their inability to exist without one another. And because both sides finally understand the true meaning of the word interdependence, some kind of respect starts to develop and the hand with the scar on it begins to relax. Sooner or later, it stops trying to play the losing game of survival and instead realizes the eternal, never-ending situation of it all, which is: the Prisoner's Dilemma.

Predator and prey are locked in a never-ending death dance until the end of time, and by the way I'm holding my hands it's clear to you there's no way of getting out of this fundamental relationship. We can't rewrite the way the universe is structured, but we can rewrite the strategies we use to play this game, so that's what we do instead. We develop the ultimate strategy for dealing with our eternal situation, and we call it *the ethic of reciprocity*. Everyone is equal no matter what.

Once this point is understood, I let go of my hands and try to say this next part with just my eyes.

What I emote to you in that auditorium, we could never put into words, quite simply, because emotions aren't words...*they're something more*. There's some kind of emotional data transfer that happens between your brain and mine, and when you've finally had enough, you have to look away and shut your eyes.

Then, pulling the marker out of my pocket I mumble something at the floor. Letting out one of those elongated, "Yeeaaahhhs" like there's some big story to tell.

"That's it really," I say, somewhat defeated. Because sadly, that's the way these stories always end. Not necessarily with a "you had to be there" type mentality, but surely with the recognition that language just has its limits.

But since you still look confused, I take two steps backwards and then start jogging up towards the podium. "This struggle," I start to say pointing at the screen, "this 'war' that's being waged between predators and prey, parasites and hosts, cuckoos and reed warblers, organisms and environments...this is *an eternal game*." It's an eternal struggle where either side doesn't really want to win, because if either side ever did win the entire game of life would soon be over.

Finally landing back over near the lecture stand, I begin to think out loud.

"If, on Monday," I posit, "all the predators somehow managed to eat up all the prey, what might Tuesday look like? I mean, without any prey animals to snack on, what would predators have left to eat, aside from other predators? And how long can that go on?"

So predators need prey and prey need predators.

Prey need predators to (a) keep their numbers in check, (b) drive their evolution and make them sharper, (c) keep their species from being out-competed for food by a similar species, and (d) prevent the surrounding ecosystem from being destroyed by unrestricted breeding, unrestricted grazing, so on and so forth.

At this point, you have to change positions in your chair and let out a "Hmm." Your hand comes up to cover your mouth while your eyes get squinty. You're on the verge of catching sight of it, I can tell, so I run back over to the whiteboard featuring the symbol of the Tao and tap on the board with my marker. Then, with my other hand, I point at the projector running all of Nature's games and cross my arms to show the connection.

Crossing them again, I add, "The entire universe is that."

It's a series of interactions between chaos and order, matter and time, self and other, viruses and hosts, organisms and environments, yin and yang, positive and negative, masculine and feminine—*all at the same time*—signifying, that we are what happens between the on and the off. Between life and death.

You're nodding now, possibly just as a courtesy, but quite possibly because you're on the verge of having a breakthrough. So I run back over to my laptop and swipe open up another presentation. This time, a power-point. Inside it, tons of high-res photos of cuckoo eggs and all their ingenious means of trickery. After tapping through several, we finally land on a photo of the dunnock and the common cuckoo where I suggest a likely series of events.

"Those cuckoos," I start to say, "it's like they're all involved in one giant evolutionary arms race playing itself out everywhere—only in vari-ous stages of development." So nested inside the grand game of Predators vs. Prey, we have Parasites vs. Hosts, but at a layer even deeper than that, it's the specific, individualized evolutionary arms race that each gens is involved in with their respective host, like Cuckoos vs. Redstarts or Cuckoos vs. Reed Warblers.

***Notes from the Editor:*

In the interest of clarity, it is important to remember that evolution has produced something like different "races" of cuckoos which generally only specialize on hosts of a specific type. This specialization, in turn, allows cuckoos to lay mimetic eggs that are less likely to be chucked out by the hosts in question.

—Scrolling through various photos of the poorly matched cuckoo egg sitting right next to the blue eggs of the dunnock, I harken back to some of the ideas previously presented. Namely, that cuckoos *can* employ egg mimicry, but that it's only in response to host defense.

Fig 1.1 Image Credit: Nick Davies

Even the world's authority on cuckoos, Nick Davies, has argued that the starting place for an evolutionary arms race of this kind is likely with no host defense whatsoever, just as it is with the dunnocks.

Remember, dunnocks don't chuck out foreign eggs; they readily accept the grey, speckled cuckoo egg all the time. And so, we can begin to imagine that in the grand evolutionary arms race, it's as if the dunnocks are still in the first quarter of their game. Their evolutionary arms race being so new, it's like the dunnocks don't even have a goalie yet. They don't discriminate eggs, and so, not surprisingly, the cuckoos who do parasitize on dunnocks have no need to employ egg mimicry.[10]

—Then, right over in Wicken Fen, you have the Cuckoos vs. the Reed Warblers, and judging by the strength of their mimicry, it would certainly appear as if this evolutionary arms race was a bit more sophisticated.

"Quarter 2," I say, displaying a photo of their ability (Fig. 1.1).

"Here the hosts have advanced to the stage where they'll peck or toss out any eggs that look suspect...they'll even warn their neighbors if they spot a cuckoo working in the area," I profess.

—Then, over in Africa, by all accounts, it's an all-out blood bath. We're in the third quarter of an absolutely incredible game between the Cuckoo Finch and the Tawny-flanked Prinia (Fig. 1.2); and the tawnies are not making it easy.

Their ability to produce intricate egg signatures is truly sublime, and as you can see from the photo in question, there's even a high amount of egg signature variability amongst tawny females. What's more, the tawnies have seemingly got the upper-hand here, as some of them have even been able to produce an egg of an olive green hue that the cuckoo finch has been unable to forge...so far.[11]

tawny-flanked prinia *Prinia subflava*   respective cuckoo finch gens

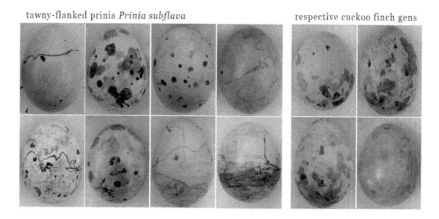

Fig 1.2 Image Credit: Claire Spottiswoode

"But the game is far from over," remarks the coach of the Cuckoo Finches, "seeing as, the cuckoo finch is now laying multiple eggs in one nest to confuse and offset the tawnies."[12]

—Then over in Australia, after a grueling fourth quarter showdown, it's like the Gerygones are stuck in sudden-death overtime with the Little

Bronze-Cuckoos. Although, things are a little bit different down here, seeing as the Gerygones have evolved a new means of host defense:

Chick rejection.

Gerygones have now been observed ejecting cuckoo chicks from their nests.[13]

And how have the Little Bronze-Cuckoos decided to respond?

By evolving chick-mimicry, of course (See Fig. 1.3). Now, the cuckoos have to look like their hosts in order to survive too.

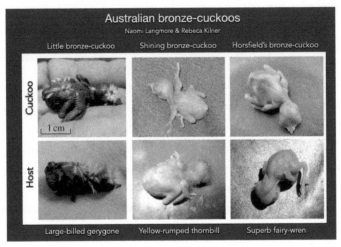

Fig 1.3

"This is an older arms race in Australia," reports Nick Davies, "so maybe there's simply been more time for defense and trickery to escalate to the chick stage."[14]

Looking up from behind the screen of my laptop, I take a swig of water and then close one eye, "Now does it make more sense?"

"Honestly," you start to say, "not really."

Exhaustion makes me chuckle for just a second until I motion at the silk screen behind me with my thumb. Projected on the screen, it's more examples of cuckoo trickery we haven't even begun to talk about yet, and I'm quick to mention that we've only just went through four hypothetical quarters of a meta-evolutionary arms race that's happening between all cuckoos and all hosts. "I got that part," you say, "but what does that have to do with infinity?"

Walking over to the desk near the door, I prop up a medium-sized wooden box about the size of a medicine cabinet, and then spin it around like a magician. With my hands, I make sure to showcase the little gold handle and the hinges on the side.

"Imagine," I say, lifting my shoulders, "if we could see that game in its ten millionth iteration or quarter." If we could somehow let the fundamental game, *Predator vs. Prey*, play itself out for about 14 billion years, what kind of complexity might it produce? The ultimate predator-prey skirmish? The ultimate predator-prey game? "The real answer," I say, "is sitting inside that box."

"Go ahead, open it...*see* what's inside."

"I'd rather not," you say, looking down at your lap, and I don't presume to say anything more either. Instead, I just pack up my things.

Exiting the lecture hall, I take one final look back over my shoulder and notice you still sitting in the front row. I watch your hands come together like mine did before, and when those fingers begin to interlock, I watch the expression on your face start to change.

To myself and no one else, I whisper the words, "I see you," and shut the door.

Chapter 34
A Parting Note on Opposition

ON KOREATOWN NIGHT, after all my categorical structures were stripped from me and I finally came back to baseline, there were four things I knew in my bones: (i) that we were one organism experiencing itself, (ii) that we were infinite, (iii) that we were playing a game, and (iv) that the predator-prey skirmish was just one of our infinite games. But before I even ascended to the mental place of heightened connection where all those realizations were even possible (*the Possibility Space*), my brain condensed my entire life's story, including every decision I ever made, into one giant synesthetic image for my Awareness to experience—and you know what that image was?

It was this:

A symmetry I'd never known before, and I absolutely could not believe it.

Correction: *didn't want to believe it.*

Because to truly take this idea seriously was to realize how many times I'd traded away my humanity for just the smallest modicum of attention. Because before this night, I wasn't exactly what you might call a "good" person. I wasn't necessarily a bad person, but I definitely didn't want to believe we could all be one energy experiencing itself because that would've implied I had a lot of growing up to do.

Why give up the belief that I could do anything I wanted?
Why give up the belief that life was cruel, meaningless, and unfair?

But then it kept happening. And it kept happening. And it kept happening. If it only happened once, that one time (all those years ago) we never could've pieced it all together. But seeing as it's happened to Ava and me well over four dozen times, with and without the aid of pharmacology—and it's happened to who knows how many tens of thousands of people throughout all of time, just with different details—we could no longer deny the evidence. We had to get to the bottom of what this experience was and how it was able to change people. Not long after that, I started speaking about my experience publicly and someone asks if I've ever heard of the philosopher Alan Watts.

"You say a lot of the same things," this stranger says.

Just an off-the-cuff remark said at just the right moment in my re-developing adult brain was enough to send me down the right path. And after five years of continuously pulling the thread, the realization we've finally come to is this:

In the process of waking up, our species has produced a handful of ideas about the nature of reality, like evolution and heliocentrism, that were bound to come out given the fact that all things do evolve and we actually do orbit the Sun—but, for whatever reason, took a fair bit of time to catch on. Evolution, for example, is a concept that dates back at least as far back as Anaximander, a Greek philosopher working in the 6th and 5th centuries BCE, but it never achieved mass resonance until a proper mechanism was worked out by a couple naturalists working in the mid 1800s.[1,2] The theory that Earth went 'round the Sun was pieced together by a few Greek philosophers working from the 5th to 3rd centuries BCE, but also didn't garner much attention until a mathematical model was put forth nineteen centuries later.[3,4] And now, she and I would argue that humanity is standing on the same precipice, only this time the idea concerns, unity, infinity, and ecological awareness.

Nearly a decade ago, I was crouched on the floor staring at flecks of red set against the background of hardwood, thinking *I must've discovered something new*. But, as it turns out, three men named Alexander von Humboldt, Friedrich Schiller, and Johann Wolfgang von Goethe already cobbled together the general pieces of this theory back in the early 1800s.

Schiller was a philosopher; Humboldt, the most famous scientist of his era; and Goethe wrote a little play called *Faust*. A play about a guy who sells his soul for infinite knowledge.

And where did Goethe get inspiration for his main character Dr. Faustus?

From an insatiably curious friend of his named Alexander von Humboldt. A man whose name has been mostly lost to history, but whose ideas are still as relevant as they were two hundred years ago.[5]

So who was Alexander von Humboldt?

Aside from being a world-renowned naturalist and explorer, he was the guy who gave us the very concept of Nature that we still use today: that life is like an interconnected web.[6]

By many scientist's accounts, he was a "pre-Darwinian Darwinist" as he was describing the relationships between plants, climate, and geography in a way that no one else had thought to do before. And his thirty-something books on the matter described his explorations, findings, and measurements with an elegance and flair that inspired so many great minds to pursue the same end, like John Muir, Carl Sagan, Alfred Russell Wallace, and Charles Darwin, only to name a few. Darwin even said if it weren't for reading Humboldt, he never would have boarded the HMS *Beagle* or been able to conceive of *The Origin of Species*. By Darwin's own account, Alexander von Humboldt was "the greatest naturalist in the world."[7-9] And what sorts of ideas was Darwin's hero writing about?

His five year expedition to Latin America, "the gradual transformations of species" (aka evolution), the curious distribution of plant and animal species (aka ecology), the call to abolish slavery...oh and, of course, that time he climbed the highest known peak in the world and had a vision.

Get this, Humboldt's standing on the edge of an inactive volcano, 19,413 ft above sea level, when a combination of exhilaration, excitement, and awe washes over him. He and his team are barely able to stand from lack of oxygen, but when the fog obstructing their view of the peak finally lifts, Humboldt experiences a non-ordinary state of consciousness that changes his life forever. All his knowledge finally comes together in "a single glance," and while he's standing on the edge of that mountain top, Humboldt gets the idea that the entire Earth is really just one giant organism, and that everything was connected.[10]

The year was 1802, and from this one experience, Humboldt grasps a concept of Nature that he would spend the next 57 years of his life trying to explain.

"Nature is a living whole," Humboldt said, not a "dead aggregate" he later remarked.[11] And just to give you a sense of how revered this man was, there are more places, animals, and things named after him than any other person in the world. More than 100 animals, and almost 300 species of plant, counts the historian Andrea Wulf, and that's saying nothing of the towns, rivers, counties, currents, and mountain ranges.[12]

Basically, what I'm trying to tell you is: this guy was the most famous and influential scientist of his era until the year he died in 1859. Coincidentally, the same year Charles Darwin published *The Origin of Species*. After that, people remembered Darwin and his great idea, but they forgot about the man who inspired him and countless others.

Perhaps the world just wasn't ready for an idea of this magnitude.

Perhaps we needed to have a Darwin first, and then chew on his ideas for next 160 years, before we'd be ready to take the next intellectual leap and learn from the guy that spoon-fed him.

Infinity

Look, if you go into the study of infinity for any length of time, eventually you're going to run into images like this:

Circles, fractals, yin-yang fractals, a figure eight, or a snake eating itself; and this, I submit, is the universal tragedy of existence.

Life eats itself.

Life destroys itself, which is basically why the Buddha said that all life is essentially *Dukkha* or suffering. Being enlightened, the Buddha knew that suffering was the basic background of Being, that all life was transient, that Being was becoming, and that there really wasn't anything you could ever hold onto. Suffering, he then said, is due to *Trishna* or clinging (resisting life when it changes).[13,14]

Without suffering, the Buddha knew we could never know pleasure. Without fear, he knew we could never know trust. And, without death, he knew we could never know life. These opposites inform one another, they go with one another, and so, as these ideas later developed in the Buddhist tradition, Nāgārjuna gifted us the idea of the world as Śūnyatā; void, or nothingness. But, seeing as empty space isn't really empty, what manifests between the two fundamental forces in the Tao is actually immense creativity. The universe doesn't appear to us as being perfectly symmetrical, but that's only because we've unconsciously chosen not to hold that perspective.

And why?

Because that's the game of it. The game is to see how many times this eternal self, *the Atman*, the Tao, God, the universe, your highest self, or whatever pet name you wish to call infinity, can divide itself up and then trick itself into believing it's just an ego locked away in a bag of skin.

Why don't we feel like we're all the same person?

Probably for the same reason identical twins don't feel like they're the same person. Fraternal twins are what you get when two eggs descend into the fallopian tubes and get fertilized at once: Two sperms meet two eggs, making two babies who subsequently grow inside the same womb and have two distinct consciousnesses. Identical twins, on the other hand, are what happens when one sperm pairs with one egg—*and then the egg splits*. The result of which being, what are termed monozygotic twins whom share 100% the exact same DNA. Now usually, one sperm-egg pairing only produces one baby and one consciousness. However, in the case of identical twins, one sperm-egg pairing produces two babies, who both feel they have a distinct and separate consciousnesses.

And why is that the case?

Quite simply, because their brain tells them that it's so.

Why don't we experience this unity in our everyday life?

Probably for the exact same reason we don't always notice our nose, or we don't always notice how sore we are until someone touches our shoulders. Most of the time, the brain just screens this sort of thing out, because again, that's the nature of an attentional-based consciousness. We only notice what stands out. We only pay attention to that which is important or that which changes. In this way, we only notice that which violates our internal predictions.

What's the unfortunate side effect of all this habituation?

That we screen out that which is fundamental, background, or foundational. So, if we think back to simplified Vase-Face analogy (where transcendence is likened to finally noticing the two faces) we could say the following:

The problem of consciousness is a problem of habituation. You already know the truth but you've buried this truth under years and years of local memories. To truly understand consciousness, you need to stop looking at the Vase-Face illusion and asking, "Which 'part' of this diagram is conscious?" and realize that the entire thing is conscious. The vase couldn't exist without the two faces, and the two faces couldn't exist without the perfectly placed vase. They depend on each other. Much like day depends on night, hot depends on cold, and matter depends on antimatter.

Tao Cont'd

The Daoists believe this is what there is: Two energetic forces, like chaos and order, predator and prey, masculine and feminine, locked in a never-ending death dance until the end of time. Where one person might look out at a fox chasing a rabbit and see the struggle for life, I now see one organism and a game being played.

A snake eating itself.

A dog chasing its tail.

The left hand writes while the right hand erases.

Instead of viewing life as this one-way thing that sometimes manifests as prey and sometimes manifests as predators, I see the essential predator-prey dynamic (which is Tao) playing itself out *everywhere*, only in various stages of development, and in ways we still don't quite recognize. That is, unless you just happen to reach transcendence and realize that we're all the same substance.

One giant organism, which creates itself and eats itself.

Same as the way your brain creates and perceives the dreams you have each night, life creates and consumes itself in much the same process. In this regard, life is a play of energy between the two fundamentally opposed forces, we may title order vs. chaos, known vs. unknown, left brain vs. right brain, and so on.

That withstanding, I should want to clarify that this "vs." symbol sitting in between all these forces doesn't necessarily mean it's a battle or a competition. All it really means is that these forces are necessarily juxtaposed. The way we've treated our environment (especially in the last two hundred years) has certainly been antagonistic, but it doesn't have to be that way. Remember, as we've already discussed, that opposition on one level, is really unity at another; and so we should never strive to dominate our environment, or foolishly attempt to beat the chaos of life into submission, because: a) it'll never work, and b) we need chaos to survive.

We couldn't exist without death, darkness, sadness, pain, suffering, and disease, nor would we really want to if we could because pauses, breaks, rainy days, insects, and catastrophes are on equal footing with their counterparts. Without them, life could not be, and this is the essential lesson of the Tao. Everything stands *in relation to* something else, or else it has no meaning.

Brief Recap

So one last time, let me put it all together for you: When you first go into the study of consciousness, it almost seems as if neuroscience can account for all the goings-on of a conscious experience, but sooner or later, you realize that there's way more to the subject of consciousness than studying neurological correlates. Of course, there do exist correlations in the brain that match up quite nicely to conditions outside the brain, and this results in the personal subjective experience we colloquially refer to as "consciousness," but really, consciousness—in the grandest sense of the word—is something that the entire universe is doing.

> [Referencing the Big Bang] It's like you took a bottle of ink and you threw it at a wall. Smash! And all that ink spread. And in the middle, it's dense, isn't it? And as it gets out on the edge, the little droplets get finer and finer and make more complicated patterns, see? So in the same way, there was a Big Bang at the beginning of things and it spread. And you and I, sitting here in this room, as complicated human beings, are way, way out on the fringe of that bang. We are the complicated little patterns on the end of it. Very interesting. But so we define ourselves as being only that. If you think that you are only inside your skin, you define yourself as one very complicated little curlicue, way out on the edge of that explosion. Way out in space, and way out in time. Billions of years ago, you were a big bang, but now you're a complicated human being. And then we cut ourselves off, and don't feel that we're still the Big Bang...but you are. Depends how you define yourself... you're not something that's a result of the Big Bang. You're not something that is a sort of puppet on the end of the process. You are still the process. You are the Big Bang, the original force of the universe, coming on as whoever you are.
>
> - Alan Watts (philosopher)[15]

And what's happening right now?

Well, right now, consciousness is trying to discover itself. It's trying to shine a light on itself, and to know itself. The physicist Max Tegmark puts it in the following way, "Consciousness is the way information *feels* when it's being processed."[16]

And what is the universe?

Information.

And what do humans do?

We process/eat latent information and then turn that into body (a form of memory); or else, we turn it into our experience bank (another form of memory).

If what we do in this life benefits us in the struggle for existence, then we pass along that information in the form crystallized information we call DNA. But because not all creatures have 'what it takes' to leave descendants, the DNA that does end up persisting down the generations tends to accumulate wisdom about the kinds of body plans, survival tricks, and behavior strategies necessary to keep it all going. In this way, DNA can be thought as being like the wisdom of all of life's experience rendered into physical form.

A kind of negative imprint about the environment, if you will.

Eventually, evolution gives rise to humans, which are composed of multiple competing sub-personalities (the most boisterous of which we refer to as the Narrative/Ego Self). And this Narrative Self is often mistaken as being one's true identity, when in actuality, it's just a tool of the brain. What's more, the Narrative Self also just happens to be operating from a severely constrained and biased point-of-view, making it very prone to misinterpreting the complexities of reality, ultimately giving rise to a form of "ego pollution" that systematically hinders the scope of your true identity. Nonetheless, once you train yourself to off-line this Narrative Self, either through many hours of focused meditation, some kind of pharmacology, or some other kind of state-changing technology, true knowledge about one's identity can be attained. Then, if you go into this experience again and again, you're likely to realize that all events are really just one event, and that this universe you once appeared stuck in is really just an extension of yourself. Then, if you go into the study of unity for any length of time you're going to slam straight into concepts of infinity, oneness, interdependence, symbiosis, gnosticism, the one substance, dialectical monism, pantheism, the Guiya hypothesis, and panpsychism.

Then, if you go into the study of those fields even further, you'll quickly realize that all religions, at their core, are trying to say the same thing but failing miserably.

Then, if you go into the study of what it is they're actually trying to say and extract out the common denominator, you'll soon realize that all religions probably got their start *after* some influential person or persons had one of these "impossible-to-describe" experiences and tried to share it with others. These experiences, which we've come to call by names as varied as awakening, liberation, nirvana, transcendence, the unitive experience, enlightenment, waking up, the beatific vision…are all fundamentally the same experience too, it's just that as the wisdom got passed down from generation to generation the essential message morphed and changed fracturing into the twelve major religions we see around us today.[17]

Then, if you make your eyes go fuzzy one last time and start reading into what all these disparate religions actually believe, you soon find that cultures all over the world seem to think and experience that there's this connection we all have to one another, and that the creator of it just had to be infinite. The infinite thing, we've tried to brand and repackage again and again, probably for pure reasons at the outset, but like most religions, once they gained any sort of political power they swiftly degenerated into something tyrannical. What always began as 'just an interpretation' soon morphed into God's infallible word, and as we started putting up walls and dividing ourselves into special groups, our actions began to drift further and further away from the original message. The end result of this process being, a population consumed by tribalism, polarization, judgment, and massive amounts of conflict.

Then, if you go into the study of philosophy and comparative mythology, and really start paying attention to some of the oldest religions, you find a few ideas that all seem to line up a little too perfectly:

If there is a God, he must be infinite to count as one (Spinoza).

The infinite is that which has no limits, therefore everything must be included within the infinite (Cusa).

There is only the one substance (Spinoza).

There is only the Great Self (Atman or Brahman).

What's one way to conceptualize infinity? The sideways figure eight. What's another way to conceptualize infinity? The Tao. And then, when you really start to take the ideas of Daoism seriously, you realize that everything that can be explained—or *will be explained*—exists inside this one tiny symbol.

You know, when people usually draw the early universe (the infinitely hot, infinitely dense, highly ordered state the universe was in before its expansion) they tend to render it as a small tiny sphere, while we, somehow, sit on the outside of that universe, observing it from some kind of an outside-the-universe perspective. Disregarding the impossibility of this vantage point, if we could somehow zoom in on what this primordial state actually looked like, we would submit that you would find the exact same symmetry found inside the Tao.

Basically, there would be two infinite eternal forces, *like* dark and light, where each force had an eye that was seeded with the opposing force, indicative of the idea that everything contains the seed of its opposite.

What's more, we would also submit that if you could somehow take that same impossible outside-the-universe perspective on your own life, you'd notice that even your own existence bears the exact same symmetry.

Matter of fact, that's what nirvana even is: it's catching sight of this grand symmetry, and then realizing that you're not something separate from it.

It's noticing the great symmetry of Being and then realizing that this universe really is fair and just, if you can maintain the "Right View" prescribed by the Buddha.[18]

What's the problem with creating religions again?

Aside from creating in-groups and out-groups, the problem with religions is that they are generally very insistent that their interpretation of the great divine energy is God's infallible word when they are, at best, just interpretations pulled down from some kind of higher realm.

What's the problem with pulling down interpretations from a higher realm?

Perhaps this one was best described in Carl Sagan's original 1980's TV series *Cosmos*, which just happened to be inspired by Alexander Von Humboldt's most famous book, *Cosmos: A Sketch of a Physical Description of the Universe* published back in 1845.

In the TV show, Carl's sitting in an office of sorts. Seated at one of those great big drafting desks, when he begins to muse about higher dimensions and lower dimensions.

Pulling out a plexiglass cube, about six inches wide, six inches deep, and six inches long, he holds it in such a way that this perfectly clear box casts a shadow on his desk. Then, tracing the lines with his finger, he says, "That shadow we recognize. It's ordinarily drawn in third grade classrooms as two squares with their vertices connected."

But, of course, we already know it by the name Necker Cube.

This is actually quite an amazing and poetic image because, with his right hand, Carl holds a cube that is perfectly represented in three dimensions, while his left hand is busy pointing out all the ways in which the projection of that cube has fallen short.

"In this case," he says, referencing the shadow, "not all the lines appear equal; not all the angles are right angles." Then, by touching the cube in his hands again, he's sure to restate the problem, "The three-dimensional object has not been perfectly represented in its projection in two dimensions."[19]

So what's the point?

The point is: infinity or God, or whatever you wish to call it, is without a doubt a higher-dimensional thing. Right now, you and I exist as three-dimensional creatures embedded within a fourth dimension of time. Although, if we wanted to simplify our situation a bit and pretend we were just two-dimensional creatures instead. Then to us, the idea of God would be synonymous with *the idea of a cube*. In this respect, God would be the perfect cube thing existing in some kind of higher-dimensional realm, while we, sitting here now, would be the imperfectly represented shadows of that higher-dimensional object, "Made in its image." In effect, we'd each be confused little Necker Cubes walking around our two-dimensional reality spinning stories to one another about the last time one of us caught sight of that 'perfect cube entity' that is the creator of all things.

At first, we might try and imagine him as being like us, only without any faults. Except, whenever we tried to pull down some of this three-dimensional wisdom into our two-dimensional reality, the higher-dimensional thing would almost always be subject to some kind of transformation. Whether that's from a nameless thing into something we call Jehovah, a genderless thing into something we call a "he", or an ageless thing into some kind of old man with a fluffy white beard. In short, one culture would draw their imperfectly represented shadows to the right, another culture would draw their imperfectly represented shadows to the left, and instead of celebrating the overwhelming similarities, we'd wage war over the tiniest of differences.

Worst of all, because each culture had improperly assumed that their interpretation was the real deal—and not just some interpretation—they might even feel justified in destroying the other man's culture.

This is the problem with groups and religions, I echo back.

Labels create groups and groups create war.

People are the source of their mystical experience (that much seems obvious to me by this point), but the second we try to pull down some of these higher-dimensional concepts, like infinity, and then transcribe them into some kind of narrative, cultural and temporal inflections about the *idea of God* become more important than the message itself.

What does the Tao look like when it's rendered in four-dimensions and not two?

Tom...you're looking at it, because that *is* what we are. A coincidence of opposites. An eternal Self playing that it's not. An infinite, intelligible sphere that's fractured itself a trillion times over and then projected itself down to this plane of existence, just so it could get lost and forget that it's all that ever was and all that ever will be.

Chapter 35
The Ava Ending

THIS ONE NIGHT, Ava and I are walking on the boardwalk of some beach, on a mini-vacation of sorts, when Ava tells me she loves me. One of those "I love you's" that just falls out of the sky, you know? The kind of "I love you" that just breaks your heart because, deep down, you know you don't deserve it. This beach we're walking on, there's no one for a couple miles. Possibly because it's freezing. Possibly because it's 3 AM. But as we're tramping through the sand, on a little midnight stroll from our beachfront hotel, Ava tells me she wants to go where the ocean meets the sand. So we walk and we walk and we walk, but in the back of my mind, all I can think about is:

How could a girl like this could actually love a person like me?

We've been together for three and half years and she still doesn't even know who I am yet. And she really deserves to know if we're going to have any kind of future together, so about halfway to the water, we take a break from walking and sit our butts on the cold damp sand. Where the water is, it's still a good two football fields away from us, because this beach we're on is epic. And aside from us being right next to the pier, the amount of sand you have to cross to get the ocean is exceptionally far, even if you weren't "peaking" (which we most certainly were).

This was supposed to be another one of our "reset" weekends—we just didn't realize how much of a reset it'd actually be. So we stop and take a break from walking and because I'm feeling things quite heavily, I stick my palms into the cold damp sand and try to grab the earth. This next part, I didn't plan to happen, it just happened. This was our first real vacation together in eight or nine months, as we'd both been seriously busy with work, so when Ava turns to me underneath that moonlit sky, something old bubbles to the surface. She's looking at me the same way a dog might look at something it's never seen before, when Ava notices something.

I didn't have to say what it was, because on some level, she already knew.

Everything Ava needed to know was already written on my face, but still, I gave it a name. I confirmed her worst fears and spilled my guts about the three most selfish things I've ever done in this life. One of which being, something that happened within the first two months of us being together.

You wanna know why Ava's gone and why we're not together anymore? It's because a long time ago I made a choice. One single choice that would come to define my entire adult life.

This was a couple months after that writers' group, the first time I could say I really met Ava, when I felt myself falling hard. I was falling so hard, Tom, I could barely even stand it.

The idea of not being in control.

The idea of handing away my power to another girl.

I saw Ava had the power to destroy me, much worse than my first real girlfriend did back when I was seventeen. But instead of being a responsible adult and realizing, that even back then I was the one at fault, I clung to the story that made me look the best and played the victim card one last time.

A girl crushed me once, so now I get carte blanche to do whatever I want.

So what did I do?

I did to her what I did to almost every girl before her: I defected. Which is really just a fancy word for saying I cheated. It only took one choice for me to secure my own demise, but I guess I figured it was better to die on your own sword than live under somebody else's.

Of course, it didn't have to be that way. I mean I could've just extended my hand in trust, the way she had done to me, but instead of doing that—I chose to be right.

This girl's gonna screw me over because that's just what girls do.

So I better do it first, right? Because if you were never "all in" to begin with, how could someone else possibly break you?

Was this a conscious thought I had?

Not exactly. But when you're trapped inside a room without doors or windows, stuck in a place where there's literally nothing left to do but contemplate all the worst decisions you've made in this life, you start to piece together the patterns that got you here.

Of course, now I know she didn't. That she never would have. But as the story always goes, the lesson came about ten minutes too late.

—

What seems like forever ago now, she and I are sitting on the beach of infinity, peaking on delysid, trying to see through the fabric of reality one last time, when everything just starts spilling out of my face.

Not everything though, right?

Yes, everything. The sad fact was, I couldn't stand the thought of living a lie for even one more second. So sitting on that beach, for the first time in I don't know how long, I was 100% honest about who and what I actually was. Who knows how long I talked for, but when it was all said and done, Ava just looks at me with those eyes. Those piercing bluish-grey eyes, and at first, she can't say anything. All she can do is shake her head and wipe tears. She's looking at me like no one's ever looked at me before when she says, "You're not a boyfriend. You're not a friend. What are you?"

Then she screams, "What are you!?" so loud my entire visual field shakes. One second, I can't shut the hell up, and the next all I can say is, I'm so stupid, and I'm so sorry. Because in that moment, that's all that I had left: stupidity and apologies.

The future we'd built together…gone.

The book we'd written together…gone.

That's how fast your entire life can change.

—

Later, we're back at my Koreatown apartment, just trying to figure out the next steps, when Ava stops crying, wipes the tears from her face, and says rather dryly, "You know you ruined everything, right?" No emotion. It wasn't up for debate. *This* was a statement of fact. I'd destroyed everything, and what's worse is...I even forgot that it happened.

Because we'd broken up and gotten back together a few times, the most convincing storyteller of all time (my Narrative Self) actually convinced me that I was free.

"That was years ago," I told myself. "And the break-up should've wiped the slate clean...so it's like it never even happened, right?"

These are the lies we tell our selves [sic]. The kind of bullshit technicality games the Narrative/Ego Self always tries to spin.

Sitting on the floor of my bedroom, I'm still trying work out how I actually could've bought my own bullshit—again—when Ava tosses out another painful realization.

"There's no way you'll ever respect me if I take you back."

And with that, she lifts herself off the bed and starts looking for her sling backpack.

That's not true, I try and explain. But Ava just crosses her arms and says, "Yes, it is," so matter-of-factly, it's like she just stepped out of a time machine. So I try to make her believe that I'll respect her, that I'd finally learned my lesson with whole cheating thing, except Ava doesn't budge. Not even a little. She just throws my hand down and goes, "Fine, then I wouldn't respect myself," completely shutting me down. The next moment, I'm sinking back into the carpet of my old college apartment wishing I could trade places with my old best friend. At least he got to go out with some dignity.

It's bad enough knowing you're the one at fault, but when you add into the mix that you were already supposedly "awake" when it happened, the punishment is more than I can possibly bear.

How does this make me feel?

Like I'm somehow less than human. Like I don't even deserve to have life, and Ava can tell. So she rolls her eyes and picks me up off the floor. She hugs me tight like I'm headed overseas, and then leans in to tell me one last thing. It sounds like, "Don't ever call me again," only she can't seem to get the words out. Ava doesn't want to believe this is happening any more than I do. Up until this point, I'd still been holding onto the belief that somehow we might be able to work things out. But the second Ava tells me not to call, not to text, and not to email, I knew she wasn't joking. I'd seen that look on her face once before. But still, I don't accept it.

You can't just let someone like this walk away from you, so when she first tries to peel away, I ask her to really think about what she's doing. To remember all those perfect moments we shared, but before I can even finish the thought, Ava shuts me down again.

"I am thinking about them, James."

Only she says my real name.

Followed by, "And you *still* don't get it, do you?"

That's when I sit her back down and try to figure this whole thing out. What did I miss? What did I overlook? Staring at Ava in the eyes, I try and imagine what it might be like to never wake up next to these eyes ever again, and the next minute, I can't even breathe.

"I don't want to imagine a future without Ava," is what my inner-narrator says.

But when I finally stop thinking about what it is that I want, I notice the only thing the love of my life wants is what she's staring at. So I trace her eyeline to the door and I realize what I have to do.

Oh, I say.

An epoch of silence passes between us, after which I tell her I won't call, but I don't really mean it. I want to have the strength to mean it, if that's what she really wants, but when I try and anchor myself to that decision, one of the clearest thoughts I've ever had descends upon me.

If you really love this girl…then you'll never talk to her again.

And with that, my face splits in two. It's painful enough just to have a thought like that, but when it's bookended by another painful realization like:

Hasn't this girl suffered enough?

Your entire world comes crashing down on top of you. You cry because there's nothing else better to do, and while you're trying to find the strength to move, or stand, or breathe, you simultaneously discover something that hurts worse than any window pane or steering wheel you've ever put your hand through:

You have to say goodbye to another best friend.

So I pull myself together just long enough to say goodbye. I motion towards the door because I know Ava really wants to go, but when I lift my head to take one last look at something good, for some reason, Ava throws me a bone.

On an out breath, she goes, "Look, you can call me in two years." That first, she has to get over me. And without even thinking about it, I backslide.

"Two years?!"

Before even registering the kind of gift she's given me, I shout, "Two years!!" as if it were some kind of life sentence, when Ava hits me back with, "Probably more like three."

Squinting through her eyelashes, she goes, "Definitely more like three." And that's the moment I finally decide to stop talking. Then, as calmly as I possibly can, I manage to tell her I'll call her in three years, but after that it's just silence.

Thunderous silence.

The judge had already hit the gavel. The sentencing was already over. Ava and I were officially done. Except Ava doesn't leave. It's like she's waiting for me to say something, but what that something is, I haven't got a clue. All I know is, everything I say only seems to make things worse, so instead of saying anything, I just sit on my hands and blink. The sound of Ava's fingers tap-tap-tapping against my white bedspread gets louder and louder, but I don't say shit. Finally fed up with the silence, Ava gets up off the bed and walks rather playfully past my desk. She starts to walk through my bedroom door, out into the hallway, but at the last second she puts her hands out and stops herself. Those fingers now tap-tap-tapping against the frame of the door, she does one of those half turns in my direction and then smiles.

"Do you want to play a game?" she says.

On the other side of my bedroom door is the hallway, the same hallway that leads to my apartment door. But when Ava asks me this question, I'm too stupid to work out what she's up to. I know I definitely don't want to play any games, but I also don't want the love of my life to just walk away, so I just stare at the floor and say, "Yeah."

It's the sound of a click that first makes me look up. Followed by the sound of hurried footsteps tearing down the hallway. I'm such an idiot, at first, I'm thinking she wants me to chase her—that is, until my brain finally registers what that sound was.

Glancing over at my desk, I catch the sight of the power cord to my laptop just dangling off to the side, tap-tap-tapping against the wall. So I leap off the floor, charge down the hallway as fast as I possibly can, only to find Ava framed in a different doorway. This time a window.

And now, she's just sitting there, waiting for me to watch what's about to happen next. With one leg inside my apartment and the other hanging out the window, Ava extends her arm out like a traffic cop and tells me to stop. She tells me to stop right there or she'll drop it.

Frozen behind the enormous, shedding couch, I can see the silver laptop glistening in her left hand, when for the second time, she warns me, "Don't move!"

Then, placing her hand against the top of the glass, she leans out the window even farther saying, "Would you rather call me in three years… or after you finish *your* book?"

Only the way she says "your" makes me want to puke.

In my head, I know there's at least two or three more years left on this thing (if we were working together) so my inner hostage negotiator comes out and I ask her to come back inside. Proving her power—my lack of control—Ava goes to lean out the window even farther. At this point, she's so far outside the window I'm legitimately starting to fear for her life, so I just blurt out the first thing that comes to my lips.

"Finish the book," I say. Just hoping she'll come back inside. But really, I'm figuring a way to buckle down and pull eighty-hour work weeks until the damn thing is done.

That's when another painful realization descends upon me:

You probably should've just said three years.

But when Ava stops to ask if I'm sure, since I'm not capable of doing complex math, I just say yes instead. Nodding my head, I'm trying to get her to just come back inside, but before Ava even moves, she looks me dead in the eyes and says, "Are-you-*sure?*" pausing between every word to make sure I really understood the work I'd be committing myself to, and I agree, whatever it is, I'll figure it out.

"Just come back inside, Ava, *please.*"

So she tucks the laptop underneath her armpit and pulls herself back into my apartment, where we both take our first real breath in minutes. Watching Ava step back inside, my entire body sweats and comes alive in an instant.

The same way your skin feels right after almost getting sideswiped by an eighteen-wheeler, that's the way my whole body feels right now—but, of course, Ava has no idea why I'm so scared. I mean, apart from the obvious (me being genuinely concerned about her life) Ava had no idea that the only copy to that anthology was on that MacBook.

Correction: *our anthology*

The back-up copies, the tattered, old notes, those were all at her place. The place she hardly ever slept in. Then, with both hands criss-crossed around my entire life's work, she lets out a small nervous laugh that we both share, until quite abruptly, she just stops laughing and I watch the corners of her mouth turn down.

Because of the pain, she can't look me in the eyes when she hands the laptop back. So when she kicks herself off that windowsill and takes a giant step towards me, for some reason, she decides to stop and pull the laptop in close. One second she's leaning back on one leg like she might be having second thoughts, the next, she's all coiled up like she means to throw the thing. The second after that it's just silence. Followed by the sound of a crash. Followed by the sound of metal and glass tumbling end over end until it finally hits the other side of the street and stops.

The way Ava looks during all of this is happy. The same way you'd look if you were admiring a sandcastle that took you hours to build.

Staring at my laptop's destruction, she has a good sigh, delights in her creation for another minute or two, before finally glancing back in my direction to see what I might have to say on the matter.

And what can I do other than tell her the truth?

"That laptop," I start to say, "that wasn't just my laptop you threw out the window...that was my only copy."

Correcting myself, I say "our" only copy.

"I know," she says, turning her face back towards the window.

She knows she's given me an impossible task. Re-write something that would take two people nearly a decade to write (if they worked non-stop).

Hearing this, I immediately have to slam my face in my hands. I try and wake myself up from this horrible, horrible dream, only when I go to reopen my eyes she's still just standing there staring out the window.

"You want me," Ava says, taking in a mess of metal and glass, "then you know where to find me." And with that she bends down to scoop up her sling backpack. Tossing it over one shoulder she walks right up to me and starts looking me in the eyes.

First it's my right eye, then it's my left, then back to my right, almost like she's studying me. At first, I think, maybe she's just trying to memorize what a cheater's face looks like, until I realize this is the first time Ava's actually seeing the real me.

No persona. No mask. Just a boy trapped in a man's clothing.

It's the sound of a zipper that first snaps me back to reality, after that it's the sound of Ava's hands pushing around the inside of that backpack.

"You lost something," she says grabbing the palm of my hand. Then, flipping it over, she presses something cold and metallic into the center. I don't have to open my eyes to know it's a key, but when I do, she closes my fist around the key and says, "Goodbye, James," and turns to leave. The next thing I know it's just footsteps. Followed by the sound of a door opening. Followed by the sound of a door closing. And that was the last time I ever saw her.

—

I wish it didn't have to end like this, Tom. I wish I could just tell you we got back together, worked things out, had a bunch of kinds, and lived happily ever after, but that's just not how the world works. After Ava left my apartment that morning, I had no motivation to speak, to write, or to do anything other than just exist in a state of catatonia. With sheets pulled up to my neck, all I could do was lie in my bed, and replay every last scene over and over. With detached amusement, I watched as my Prospective Self would desperately try to figure a way out of this hole, only for my Experiencing Self to *Ctrl-Alt-Delete* any plans that were ever made. There was no getting out of this one. I'd really messed things up, and for once, I just needed to feel this pain. So, for what seemed like the better part of eternity, I just stared at the ceiling fan go 'round and 'round, replaying the love of my life throwing everything I had out a window two to three floors above the street.

I remember how bittersweet it felt finally letting someone else see the ugliness that I was. And then, after I consciously made the decision to never make someone else feel that way ever again…a curious thought comes into my brain.

What if?

What if this wasn't an anthology but a road-map instead?

What if I could turn this book, this thing, into something else entirely? Not your usual run-of-the-mill non-fiction monstrosity, but reimagine it as something else entirely.

Of course, the idea doesn't come to me fully formed but I catch the tail end a fleeting thought that pops my head off the pillow and makes me think:

What if?

What if I just wrote an anonymous letter?

Not to her, but to someone else and just explained everything that happened. Because in my head, if there were ever a story powerful enough to wake someone out of the illusion that we're all separate—it'd be this one. And, seeing as I might be one of the dumbest people on the planet, I start believing in the idea more and more.

"If a story like this could wake an idiot like me up," I begin to think, "then maybe it could do the same for someone who's normal?" That's when all these *What ifs* really start running through my brain.

What if I just locked myself in myself in a room and didn't leave until it was finished?

What if I just gave myself a deadline? Six to eight hours, or else.

What if I started right now, before the pain starts to fade? Right now. Just write it right now. "Right now!" the voice says, "Be brutally honest about everything that happened and just tell 'em: 'Look, the night I finally got my act together was just yesterday. The night Ava threw entire my life's work out the window was just last night.'"

The moment she said, "Goodbye James," and walked out of my life forever, that was only eight hours ago.

Who I was before I started this letter is officially dead...gone.

The relationship Ava and I use to have is officially done...over.

But the man I still have the chance of becoming isn't quite through yet. Earlier, I mentioned how if I wasn't dead in six to eight hours, I'd wish I was, and I really meant it at the time, because whoever I was when we started this thing needed to die off. All his habits. All those stories. Those were all the emotionally charged events James still needed to reconcile before he could finally become something more. And in writing this letter to you, for the first time in my entire adult life, I feel like I've actually found something worth living for.

Creating, not destroying.

But as far as Ava and me are concerned, I just can't be the one who sends her this. To be perfectly honest, I don't even feel right about asking you to send it to her either. Although, I must confess, that was my original intention. You see, what seemed like forever ago, I thought if I told you everything I knew about the subject of consciousness that maybe you'd do me a favor and let me ask you a question. But now, I'm wondering if that was just another one of my lame attempts at trying to control the situation—which I can't.

So I guess I'll be letting that go too, as I'm seriously done trying to be that guy.

If, by some force of magic, this letter does happen to wind up on her doorstep, I certainly couldn't stop you. But that would ultimately have to be your decision, not mine. Because for me, I just need to learn to let go and trust.

Mailing Address: ~~Eva Adams~~
                     ███████████
                     ~~Berkeley, Ca 94704~~

*Attach a little note on the front that says: "This is life?" She'll know what it means.

—If you do end up forwarding this letter on, and she actually reads it, then who knows maybe it'll end up getting turned into whatever she wants it to be. A lecture…or maybe a book? Or, who knows, maybe she'll just take her big purple pen through the whole thing and chuck it in the fire like in the before times.

Whatever ends up happening, it'll be her decision, not mine.

Because if there's one thing I've learned over the past thirty-six hours, it's that love isn't about being in control—*it's about being out of control.* It's about handing your heart over to someone else and saying, "Here, take it," without expecting anything in return.

Notes from the Editor:
The only reason people like me have a job is because writers tend to write too much and too hastily. The end result of which being, first drafts that drone on for hundreds and hundreds of pages. "James" always did have a good eye for story, although sometimes, big purple X marks the size of each page are the only way to get through to people.

—Things don't always work out the way you expect, but hey…that's life. The best you can do is just try to live as honestly as you can, which is something I will forever be learning how to do.

Take writing this letter, for example.

Even now, part of me wonders if this wasn't just an elaborate form of cheating. A sly way of me getting out of the five or six years it probably would've taken me to re-write the other one. And who knows maybe that's what she really wanted.

Maybe, what Ava really wanted was to give me a task so difficult, that by the end, we'd be different people. Being this sleep deprived, I honestly can't know for sure, which is why I can't be the one who sends this. I will, however, finally take you up on that question I've been meaning to ask you, and that question is this:

Now that you know what consciousness is and all the ways that it manifests, what are you going to do with it?

Because, for me—this moment right here—this is why *I'm here*. It's to wake at least one more person up before they go and hurt someone else, which is really to say, before they go and hurt themselves. Because, deep down, that's all there is anyway. An eternal self playing that it's not. An omniscient, omnipotent, omnipresent being that likes to get lost in the game of life by pretending it's just a tiny atom of awareness.

So, one last time, you ask me, what consciousness is? What are we even doing here anyway?

And I'd say this: We are conscious entities, assembled from stardust, riding out infinity the only way we know how.

See *Stories*

See *The Beginning*

See *The End.*

Notes from the Editor:

In the original letter there was a return address listed, so Tom could actually write back with his response, although that old California address is no longer valid. The new address can be found hidden within the pages of this text. Hidden in plain sight, I'm told. If only James had given you some kind of tool to help you de**code** this message.

Acknowledgements

My career in writing wouldn't have been possible without the help and support of my mother, Suzanne Nelson. Even from a very young age, she insisted that I always go the extra mile, and while I certainly cursed her at the time, I am now more thankful than ever that she instilled in me the patience to shape an idea, and the will to see it through. For this, I am incredibly grateful. She is without a doubt the most creative person I know. For helping me develop the idea—and for supporting me the entire time—I thank my beautiful wife, Lindy. But for showing me how to think scientifically or write in the first place, I must thank a few secondary fathers, like Chuck Palahniuk, Richard Dawkins, Alan Watts, Jordan Peterson, Neil deGrasse Tyson, David Eagleman, and Joseph Campbell. Each of them has written something that has changed my perspective on life and made learning fun again, and for that I am incredibly thankful.

To Emmanuel, thank you for showing me what truth is. To Dr. David Nichols, thank you for continuously sending me research papers; I've enjoyed our conversations more than you'll ever know. And to Justin Harris, I am extremely grateful for all the useful feedback and assistance you've provided over the years. You truly are a good friend.

Finally, I would like to thank my father, John Nelson, my sister, Lauren, and my extended family for always being there to listen, or help in some way. We really are nothing without our families.

Endnotes/Bibliography

In order to make the story of consciousness as accessible as possible while, at the same time, maintaining a sense of scientific integrity, the majority of qualifying statements, footnotes, and clarifications were pushed to this section. Below, not only will you find the sources for all the facts and findings, but you will also find a parallel argument happening on a more nuanced level.

Introductions:
1. Fultz, N. E., Bonmassar, G., Setsompop, K., Stickgold, R. A., Rosen, B. R., Polimeni, J. R., & Lewis, L. D. (2019). Coupled electrophysiological, hemodynamic, and cerebrospinal fluid oscillations in human sleep. *Science, 366*(6465), 628–631. https://doi.org/10.1126/science.aax5440. See Also: Wamsley, E. J. (2014). Dreaming and Offline Memory Consolidation. *Current Neurology and Neuroscience Reports, 14*(3), 433. https://doi.org/10.1007/s11910-013-0433-5
2. Conselice, C. J., Wilkinson, A., Duncan, K., & Mortlock, A. (2016). The evolution of galaxy number density at z < 8 and its implications. *The Astrophysical Journal, 830*(2), 83. https://doi.org/10.3847/0004-637X/830/2/83
3. Handwerk, B. (2021, February 2). An Evolutionary Timeline of Homo Sapiens. *Smithsonian Magazine.* https://www.smithsonianmag.com/science-nature/essential-timeline-understanding-evolution-homo-sapiens-180976807/
4. Baumgartner, R. J., Van Kranendonk, M. J., Wacey, D., Fiorentini, M. L., Saunders, M., Caruso, S., Pages, A., Homann, M., & Guagliardo, P. (2019). Nano−porous pyrite and organic matter in 3.5-billion-year-old stromatolites record primordial life. *Geology, 47*(11), 1039–1043. https://doi.org/10.1130/G46365.1
5. *Richard Dawkins versus Rowan Williams: Humanity's ultimate origins.* (2012, February 1). [Video File]. University of Oxford. https://youtu.be/zruhc7XqSxo (10 minutes into the video). The quote is also found elsewhere too, see Julian Huxley's essay entitled "The New Divinity" in: Huxley, J. (1964). *Essays of a Humanist* (1st ed.). Harper & Row. Or, Huxley, J. (1968). Transhumanism. *Journal of Humanistic Psychology,* 8(1), 73–76. https://doi.org/10.1177/002216786800800107
6. National Nuclear Securtity Administration. (2018, October 17). Visible Light: Eye-opening research at NNSA [NNSA Official Website]. *Web Articles.* https://www.energy.gov/nnsa/articles/visible-light-eye-opening-research-nnsa
7. David Eagleman. (2014, June 3). *The Umwelt* [Author's Website]. David Eagleman. https://eagleman.com/latest/umwelt/

Chapter 1: Trapped

1. Davies, N. (2015). *Cuckoo: Cheating by nature.* Bloomsbury. (p. 23)
2. Memoirs of Professor Richard Dawkins. (2013, November 19). *Richard Dawkins—Cuckoos and a History of Life* [Lecture]. https://www.youtube.com/watch?v=USdjFRqVI7E (4:20 in).
3. The majority of these facts concerning cuckoos were either derived from Nick Davies' book *Cuckoo: Cheating by Nature* (item 1), or from the lecture he gave at the Royal Society entitled: Davies, N. (2015, May 14). *Cuckoos and their victims: An evolutionary arms race* [Royal Society Lecture]. https://youtu.be/nOO6S4hDDfE
4. Goslings will follow human beings they've imprinted on whether they were wearing striped boots, zigzag boots, or polka dot boots. Leslie Nielsen; National Geographic Society. (1975). *Konrad Lorenz—Science of Animal Behavior (1975)* [Documentary]. Jack Kaufman Pictures, Inc. https://www.youtube.com/watch?v=IysBMqaSAC8
5. Lorenz also found that Goslings would imprint on a box when he placed it atop a model train. See: T.L. Brink. (2018). *Psychology: A student friendly approach.* San Bernadino Community College District. https://www.researchgate.net/publication/335128923_Psychology_a_student_friendly_approach (p. 268).
6. Lorenz, K. (1981). *The foundations of ethology.* Springer verlag. (p. 5).
7. Lorenz, K. (1937). On the formation of the concept of instinct. *Die Naturwissenschaften, 25*(19), 289–300. https://doi.org/10.1007/BF01492648
8. Davies, N. (2015). *Cuckoo: Cheating by nature.* Bloomsbury. (p. 14).
9. Most neuroscientist's cite around 98 to 99% of our cognition being unconscious. George Lakoff, for example, cites 98% in his book, *The Political Mind*, but agrees that it is likely to be higher in his lectures. See: Lakoff, G. (2009). *The political mind: A cognitive scientist's guide to your brain and its politics*; [with a new preface]. Penguin Books. (P. 9); and George Lakoff. (2015, March 14). *George Lakoff: How Brains Think: The Embodiment Hypothesis* [Keynote address]. International Convention of Psychological Science, Amsterdam. https://youtu.be/WuUnMCq-ARQ, respectively. Dr. Jordan Peterson has cited 99%, see: *Sam Harris, Jordan Peterson & Douglas Murray in London—Part 4*. (2018, July 16). Pangburn Philosophy. https://youtu.be/aALsFhZKg-Q (68:20 min. in).

Chapter 2: Source Code

1. Specifically, Dr. Eagleman talks about how far we live in the past we live starting at about fourteen minutes into the lecture. The Flash Lag Effect demonstrates that we live at least 80-100 ms in the past, for vision. But collectively, across all senses, he argues, that we live around a half second in the past. "It turns out in total...on average...we probably live something like a half a second in the past. That's how long it takes for the brain to collect up all this information from our toes and everything else, and put this all together." David Eagleman. (2016, October 4). *The Brain and The Now—David Eagleman* [Keynote address]. The Long Now Member Summit, San Francisco. https://youtu.be/vv_e99qbJ4U (14 min in.)
2. Muckli, L., & Petro, L. S. (2013). Network interactions: Non-geniculate input to V1. *Current Opinion in Neurobiology, 23*(2), 195–201. https://doi.org/10.1016/j.conb.2013.01.020
3. Lee, T. S. (2015). The Visual System's Internal Model of the World. *Proceedings of the IEEE, 103*(8), 1359–1378. https://doi.org/10.1109/JPROC.2015.2434601

Chapter 3: The Movie in Your Mind

1. "Complementary information comes from a consideration of the amount of sensory information made available to the brain. For example, it may surprise some to learn that visual information is significantly compressed as it passes from the eye to the visual cortex.[9] Thus, of the information available from the environment, only about 10^{10} bits/s (i.e., 10 billion bits/s) are deposited in the retina. Yet, only 10^4 bits/s (i.e., 0.001% of that which was deposited on the retina) make it to primary visual cortex. These data make it clear that visual cortex receives a very compressed representation of the world, a subject of more than passing interest to those seeking an understanding of visual information processing." Raichle, M. E. (2019). Creativity and the Brain's Default Mode Network. In M. E. Raichle, *Secrets of Creativity* (p. 107–123). Oxford University Press. https://doi.org/10.1093/oso/9780190462321.003.0006

2. "Published in April 2016 in the *Proceedings of the National Academy of Sciences* (PNAS), the findings show how the drug decreases communication between the brain regions that make up the Default Mode Network (DMN), a collection of hub centres that work together to control and repress consciousness. Like the conductor in an orchestra, the DMN polices the amount of sensory information that enters our sphere of awareness, and has been described as the neural correlate of the 'ego'." From: Beckley Foundation. (2016, April 1). The World's First Images of the Brain on LSD [Scientific Institution]. *Beckley in the Press.* https://www.beckleyfoundation.org/the-brain-on-lsd-revealed-first-scans-show-how-the-drug-affects-the-brain/

3. Naomi Austin. (2010, October 18). Is Seeing Believing? (4 of 15) [Documentary]. In *Horizon.* BBC Two. https://youtu.be/2k8fHR9jKVM

4. Tiippana, K. (2014). What is the McGurk effect? *Frontiers in Psychology, 5.* https://doi.org/10.3389/fpsyg.2014.00725

5. "Why do the sight and sound of a slamming car door suddenly appear unsynchronized if you view it from more than 30 meters away? This seems to occur because the system perceptually synchronizes signals that arrive less than 80 msec apart (past 30 meters, the difference between the speeds of light and sound exceed this window)." From: David Eagleman. (2013, February 12). Time and the Brain (or, What's happening in the Eagleman Lab) [Author's Website]. *The Timing and Perception and the Timing of Neural Signals.* https://eagleman.com/time-and-the-brain-or-what-s-happening-in-the-eagleman-lab/

6. King, A. J. (2005). Multisensory Integration: Strategies for Synchronization. *Current Biology, 15*(9), R339–R341. https://doi.org/10.1016/j.cub.2005.04.022

7. Tully, K., Bolshakov, V.Y. Emotional enhancement of memory: how norepinephrine enables synaptic plasticity. *Mol Brain* **3,** 15 (2010). https://doi.org/10.1186/1756-6606-3-15

8. Tyng, C. M., Amin, H. U., Saad, M. N. M., & Malik, A. S. (2017). The Influences of Emotion on Learning and Memory. *Frontiers in Psychology, 8,* 1454. https://doi.org/10.3389/fpsyg.2017.01454

9. Roediger, H. L., & Pyc, M. A. (2012). Inexpensive techniques to improve education: Applying cognitive psychology to enhance educational practice. *Journal of Applied Research in Memory and Cognition, 1*(4), 242–248. https://doi.org/10.1016/j.jarmac.2012.09.002

10. Taylor, K., & Rohrer, D. (2010). The effects of interleaved practice. *Applied Cognitive Psychology, 24*(6), 837–848. https://doi.org/10.1002/acp.1598. For a good summary of

many studies concerning interleaved practice, see also: Steven C. Pan. (2015, August 4). The Interleaving Effect: Mixing It Up Boosts Learning. *Scientific American.*

11. Landin, D. K., Hebert, E. P., & Fairweather, M. (1993). The Effects of Variable Practice on the Performance of a Basketball Skill. *Research Quarterly for Exercise and Sport, 64*(2), 232–237. https://doi.org/10.1080/02701367.1993.10608803

Chapter 4: Consciousness as Software

1. Budson, A. E., & Solomon, P. R. (2011). *Memory loss a practical guide for clinicians.* Elsevier Saunders (p. 214-219).

2. Cowan, N. (2010). The Magical Mystery Four: How Is Working Memory Capacity Limited, and Why? *Current Directions in Psychological Science, 19*(1), 51–57. https://doi.org/10.1177/0963721409359277

3. Alan Watts. (2019, December 19). 3.4.11 Transformation of Consciousness Part 3 [Lecture/transcript archive]. *Alan Watts Organization.* https://alanwatts.org/3-4-11-transformation-of-consciousness-part-3/

4. After much deliberation, the topic of attention (as it relates to each hemisphere) was deemed to complex to include inside this conversation. However, for the serious student, I'd recommend picking up a copy of Dr. Iain McGilchrist's book about the two hemispheres, because there, the topic is treated with a fine-tooth comb. Here, the topic is only teased. McGilchrist, I. (2019). *The master and his emissary: The divided brain and the making of the Western world* (New expanded edition). Yale University Press. (p. 28-31; 38-50).

5. 1.1.8. - Myth of Myself—Pt. 2. (2019, April 16). *Alan Watts Organization.* https://alanwatts.org/1-1-8-myth-of-myself-pt-2/

6. 86 Billion Neurons - Herculano-Houzel, S. (2012). The remarkable, yet not extraordinary, human brain as a scaled-up primate brain and its associated cost. *Proceedings of the National Academy of Sciences, 109* (Supplement_1), 10661–10668. https://doi.org/10.1073/pnas.1201895109

7. 10,000 Connections - Eagleman, D. (2011). *Incognito: The secret lives of brains.* Pantheon Books (p. 1-2).

8. MICrONs Consortium et al. Functional connectomics spanning multiple areas of mouse visual cortex. bioRxiv 2021.07.28.454025; doi: https://doi.org/10.1101/2021.07.28.454025

Chapter 5: Intro to ASCs

1. Glickstein, M. (1988). The Discovery of the Visual Cortex. *Scientific American, 259*(3), 118–127.

2. Sandrone, S., & Riva, M. (2014). Bartolomeo Panizza (1785–1867). *Journal of Neurology, 261*(6), 1249–1250. https://doi.org/10.1007/s00415-013-7028-6

3. Danziger, S., Levav, J., & Avnaim-Pesso, L. (2011). Extraneous factors in judicial decisions. *Proceedings of the National Academy of Sciences, 108*(17), 6889–6892. https://doi.org/10.1073/pnas.1018033108

4. Eagleman, D. (2020). *Livewired: The inside story of the ever-changing brain.* Pantheon Books. (p. 3-14) But really, this entire book is centered around the idea that the brain is more "livewired" rather than being hardwired, championing the idea that the brain is a living, dynamic, and adaptable system that shapes itself according to the environment.

5. Wheal, J., & Kotler, S. (2017). *Stealing Fire: How silicon valley, the navy SEALs and maverick scientists are revolutionizing the way we live and work.* Dey Street Books.

(p. 27). As the book's subtitle suggests, professionals, Navy SEALs, and top-performers in every field are figuring out that NOSC, like flow states, are the key to accessing peak-levels of performance and staying there. One such example includes Navy SEALs who've combined sensory deprivation tanks with language-training subsequently allowing them to cut down the time it takes to learn a new language from six months down to six weeks.

6. InformedHealth.org [Internet]. (2016). *How does our sense of taste work?* (2006th ed.). Institute for Quality and Efficiency in Health Care (IQWiG). https://www.ncbi.nlm.nih.gov/books/NBK279408/

7. Institute of Medicine (US) Committee on Military Nutrition Research. (2001). *Caffeine for the Sustainment of Mental Task Performance: Formulations for Military Operations.* National Academies Press (US). https://www.ncbi.nlm.nih.gov/books/NBK223808/

8. Chaudhry, S. R., & Gossman, W. (2021). Biochemistry, Endorphin. In *StatPearls*. StatPearls Publishing. http://www.ncbi.nlm.nih.gov/books/NBK470306/

9. Fuss, J., Steinle, J., Bindila, L., Auer, M. K., Kirchherr, H., Lutz, B., & Gass, P. (2015). A runner's high depends on cannabinoid receptors in mice. *Proceedings of the National Academy of Sciences, 112*(42), 13105–13108. https://doi.org/10.1073/pnas.1514996112

10. Dawkins, R. (2016). *Climbing mount improbable.* W W Norton. (p. 12-14).

11. At least anecdotally, there's a mismatch between how we look and how we think we look, and I still report this with complete knowledge about focal length and the magic of using different lenses. For a demonstration of what focal length can do to someone's face, see: https://www.businessinsider.com/cameras-can-make-you-look-fat-2016-7

Chapter 6: Three Cases of Brain Trauma

1. Taylor, J. B. (2016). *My stroke of insight: A brain scientist's personal journey.* Plume. (p. 147).

2. McGilchrist, I. (2019). *The master and his emissary: The divided brain and the making of the Western world* (New expanded edition). Yale University Press. (p. 12).

3. Taylor, J. (2008, February). *My stroke of insight* [Video]. TED Conferences. https://www.ted.com/talks/jill_bolte_taylor_my_stroke_of_insight (10:23 into her talk).

4. Taylor, J. B. (2016). *My stroke of insight: A brain scientist's personal journey.* Plume. (Located in Appendix A of her book) this short list helps provide friends, family, and specialists with a better understanding about what a stroke survivor might be going through. See also: "Forty Things I Needed Most" in Appendix B.

5. Taylor, J. (2008, February). *My stroke of insight* [Video]. TED Conferences. https://www.ted.com/talks/jill_bolte_taylor_my_stroke_of_insight (15:13 into her talk).

6. Harlow, J. M. (1848). Passage of an Iron Rod through the Head. *Boston Medical and Surgical Journal, XXXIX*(20), (p. 389–393).

7. Harlow, J. M. (1868). "Recovery from the passage of an iron bar through the head." *Publications of the Massachusetts Medical Society* 2: 327-47. (Republished in Macmillan, *An Odd Kind of Fame.*)

8. Jane Treays. (2005). *The Man With The Seven Second Memory (Amnesia Documentary)* [Documentary]. https://www.youtube.com/watch?v=k_P7Y0-wgos

9. Maguire, E. A., Gadian, D. G., Johnsrude, I. S., Good, C. D., Ashburner, J., Frackowiak, R. S., & Frith, C. D. (2000). Navigation-related structural change in the hippocampi of

taxi drivers. *Proceedings of the National Academy of Sciences of the United States of America, 97*(8), 4398–4403. https://doi.org/10.1073/pnas.070039597

10. Jane Treays. (2005). *The Man With The Seven Second Memory (Amnesia Documentary)* [Documentary]. https://www.youtube.com/watch?v=k_P7Y0-wgos (To see footage of his diaries, skip to 18 min in).

11. John Dollar. (1986, August 4). Prisoner of Consciousness [Documentary]. In *Equinox*. Channel Four. https://youtu.be/aqiw2nx6gjY (For moments when Clive is playing music, skip to 6 minutes in, 34 minutes in, and 37 minutes in). At 38 min in, when Clive's being shown a tape of himself performing, you can really catch sight the compartmentalized nature of consciousness we're referring to.

12. Moreira-Gonzalez, A., Papay, F. E., & Zins, J. E. (2006). Calvarial Thickness and Its Relation to Cranial Bone Harvest: *Plastic and Reconstructive Surgery, 117*(6), 1964–1971. https://doi.org/10.1097/01.prs.0000209933.78532.a7

13. Kitamura, T., Ogawa, S. K., Roy, D. S., Okuyama, T., Morrissey, M. D., Smith, L. M., Redondo, R. L., & Tonegawa, S. (2017). Engrams and circuits crucial for systems consolidation of a memory. *Science, 356*(6333), 73–78. https://doi.org/10.1126/science.aam6808

14. Taylor, J. B. (2016). *My stroke of insight: A brain scientist's personal journey.* Plume. (p. 50-51).

15. Morrow-Odom, K. L., & Swann, A. B. (2013). Effectiveness of melodic intonation therapy in a case of aphasia following right hemisphere stroke. *Aphasiology, 27*(11), 1322–1338. https://doi.org/10.1080/02687038.2013.817522

Chapter 7: Sensory Awareness

1. For a measurement on the amount of bits streaming in through all our five senses (11 million bits) see: Wiliam, D. (2006). The half-second delay: What follows? *Pedagogy, Culture & Society, 14*(1), 71–81. https://doi.org/10.1080/14681360500487470, and Zimmermann, M. (1989). The Nervous System in the Context of Information Theory. In R. F. Schmidt & G. Thews (Eds.), *Human Physiology* (p. 166–173). Springer Berlin Heidelberg. https://doi.org/10.1007/978-3-642-73831-9_7. Both sources reference the idea of 11 million bits, however the clarification comes in regards to the amount our conscious selves can attend to. The man who's done the most pioneering work on flow, Mihaly Csikszentmihalyi, has estimated that the conscious self can only attend to about 120 bits in: Csikszentmihalyi, M. (2014). *Flow and the Foundations of Positive Psychology.* Springer Netherlands. https://doi.org/10.1007/978-94-017-9088-8 (p. xvi). On the contrary, Bell Labs engineer Robert Lucky has estimated 50 bits per second for the conscious, and billions for the unconscious in: Lucky, R. (1989). *Silicon dreams: Information, man, and machine.* New York, NY: St. Martin's Press. (p. 29). In any case, the common denominator worth focusing on here is the ratio of conscious attention to unconscious attention, because on that point, everyone does seem to agree: *It is the conscious self that fails to compare.* As new data continues to come in, these values are likely to change, but until that time, we have settled upon an estimate of 200 bits for the conscious self, and 11 million bits for the unconscious self.

2. Eagleman, D. (2020). *Livewired: The inside story of the ever-changing brain.* Pantheon Books (p. 168-69). See also: Eagleman, D. (2011). *Incognito: The secret lives of brains.* Pantheon Books (p. 26-27; 48-50). At least as far back as 1956, Donald MacKay was

proposing that the whole purpose of the visual cortex was to generate a model of the exterior world.

3. The three-dimensional representation of reality that we perceive in our minds is actually more like a 2½-D sketch, rather than an actual 3-D model. Remember, the only reason we perceive objects as being "a certain distance away" is because we have two different retinal images to compare against one another. For more on the 2½-D Sketch idea, see: Stevens, K. A. (2012). The Vision of David Marr. *Perception, 41*(9), 1061–1072. https://doi.org/10.1068/p7297

4. Eagleman, D. (2011). *Incognito: The secret lives of brains.* Pantheon Books (p. 26-7; 35).

5. Li, S., Xu, Y., Cong, W., Ma, S., Zhu, M., & Qi, M. (2018). Biologically Inspired Hierarchical Contour Detection with Surround Modulation and Neural Connection. *Sensors, 18*(8), 2559. https://doi.org/10.3390/s18082559

6. Cant, J. S., Large, M.-E., McCall, L., & Goodale, M. A. (2008). Independent Processing of Form, Colour, and Texture in Object Perception. *Perception, 37*(1), 57–78. https://doi.org/10.1068/p5727

7. Most interesting is the movement vision impairment case reported on in: Zihl, J., Von Cramon, D., & Mai, N. (1983). Selective disturbance of movement vision after bilateral brain damage. *Brain, 106*(2), 313–340. https://doi.org/10.1093/brain/106.2.313 and Zihl, J., Von Cramon, D., Mai, N., & Schmid, Ch. (1991). Disturbance of movement vision after bilateral posterior brain damage: Further evidence and follow up observations. *Brain, 114*(5), 2235–2252. https://doi.org/10.1093/brain/114.5.2235

8. O'Reilly and Munakata. (2020, August 13). *6.3: Oriented Edge Detectors in Primary Visual Cortex.* Medicine LibreTexts. https://med.libretexts.org/@go/page/12598

9. Winerman, L. (2012, December). *Neuroscientist brings light to the blind—And to vision research.* American Psychological Association. https://www.apa.org/monitor/2012/12/neuroscientist-sinha

10. Daw, N. W., Fox, K., Sato, H., & Czepita, D. (1992). Critical period for monocular deprivation in the cat visual cortex. *Journal of Neurophysiology, 67*(1), 197–202. https://doi.org/10.1152/jn.1992.67.1.197

11. Wiesel, T. N., & Hubel, D. H. (1965). Extent of recovery from the effects of visual deprivation in kittens. *Journal of Neurophysiology, 28*(6), 1060–1072. https://doi.org/10.1152/jn.1965.28.6.1060

12. Sinha, P. (2009, November). *How brains learn to see* [Video]. TED Conferences. https://www.ted.com/talks/pawan_sinha_how_brains_learn_to_see (9:35 into the video).

13. Gandhi, T. K., Singh, A. K., Swami, P., Ganesh, S., & Sinha, P. (2017). Emergence of categorical face perception after extended early-onset blindness. *Proceedings of the National Academy of Sciences, 114*(23), 6139–6143. https://doi.org/10.1073/pnas.1616050114

14. Ostrovsky, Y. (2010). *Learning to see: The early stages of perceptual organization.* [Massachusetts Institute of Technology, Dept. of Brain and Cognitive Sciences]. http://hdl.handle.net/1721.1/62087 (p. 14).

15. Ganesh, S., Arora, P., Sethi, S., Gandhi, T. K., Kalia, A., Chatterjee, G., & Sinha, P. (2014). Results of late surgical intervention in children with early-onset bilateral cataracts. *British Journal of Ophthalmology, 98*(10), 1424–1428. https://doi.org/10.1136/bjophthalmol-2013-304475

16. Chatterjee, R. (2015). Out of the darkness. *Science, 350*(6259), 372–375. https://doi.org/10.1126/science.350.6259.372

17. Sinha, P., Balas, B. J., Ostrovsky, Y., & Wulff, J. (2009). Visual object discovery. In *Object Categorization: Computer and Human Vision Perspectives*. Cambridge University Press (p. 301-19). See also: Sinha, P. (2009, November). *How brains learn to see* [Video]. TED Conferences. https://www.ted.com/talks/pawan_sinha_how_brains_learn_to_see (9:50 in).

18. Ostrovsky, Y., Andalman, A., & Sinha, P. (2006). Vision Following Extended Congenital Blindness. *Psychological Science, 17*(12), 1009–1014. https://doi.org/10.1111/j.1467-9280.2006.01827.x

19. Like quantifying unconscious processing capabilities, comparing the human eye to pixels can be a bit problematic. For more about this subject, you could read: Jeff Johnson, Kate Finn, Chapter 3 - Vision, Editor(s): Jeff Johnson, Kate Finn, *Designing User Interfaces for an Aging Population*, Morgan Kaufmann, 2017, (p. 27-53). Because there, the general point is made, "It's just how human eyes are: we have high-resolution vision only in a small area in the very center of each eye's visual field—an area called the "fovea" that is about 1% of the total visual field (see box Human Vision Is Mostly Low Resolution). Most of our visual field—the other 99%—has very low resolution: it sees the world as if through frosted glass [Eagleman, 2011]."

20. Stewart, E. E. M., Valsecchi, M., & Schütz, A. C. (2020). A review of interactions between peripheral and foveal vision. *Journal of Vision, 20*(12), 2. https://doi.org/10.1167/jov.20.12.2

21. Rehman, I., Mahabadi, N., Motlagh, M., & Ali, T. (2021). Anatomy, Head and Neck, Eye Fovea. In *StatPearls*. StatPearls Publishing. http://www.ncbi.nlm.nih.gov/books/NBK482301/

22. Eagleman, D. (2011). *Incognito: The secret lives of brains*. Pantheon Books (p. 31-33).

23. Jeff Johnson, Chapter 5 - Our Peripheral Vision is Poor, Editor(s): Jeff Johnson, *Designing with the Mind in Mind* (Second Edition), Morgan Kaufmann, 2014, (p. 109-16).

24. Elizabeth Thomson. (1996, December 19). MIT Research—Brain Processing of Visual Information. *MIT News*. https://news.mit.edu/1996/visualprocessing

25. Jeff Johnson, Chapter 5 - Our Peripheral Vision is Poor, Editor(s): Jeff Johnson, *Designing with the Mind in Mind* (Second Edition), Morgan Kaufmann, 2014, (p. 109-16).

26. Schmid, S., Wilson, D. A., & Rankin, C. H. (2015). Habituation mechanisms and their importance for cognitive function. *Frontiers in Integrative Neuroscience, 8*. https://doi.org/10.3389/fnint.2014.00097

27. See item one from this list.

28. Rankin, C. H., Abrams, T., Barry, R. J., Bhatnagar, S., Clayton, D. F., Colombo, J., Coppola, G., Geyer, M. A., Glanzman, D. L., Marsland, S., McSweeney, F. K., Wilson, D. A., Wu, C.-F., & Thompson, R. F. (2009). Habituation revisited: An updated and revised description of the behavioral characteristics of habituation. *Neurobiology of Learning and Memory, 92*(2), 135–138. https://doi.org/10.1016/j.nlm.2008.09.012

29. First, "During the process of meditation, accumulated stresses are removed, energy is increased, and health is positively affected overall.[7] Research has confirmed a myriad of health benefits associated with the practice of meditation. These include stress reduction,[1,2,17,18,19,20] decreased anxiety,[1,17,19,21,22] decreased depres-

sion,[1,17,18,23,24] reduction in pain (both physical and psychological),[2,25,26] improved memory,[2,27] and increased efficiency.[12,28,29,30]" And also from the same article: "Meditation increases regional cerebral blood flow in the frontal and anterior cingulate regions of the brain,[46,47,48,49,50] increases efficiency in the brain's executive attentional network,[12,28,29,30] and increases electroencephalogram (EEG) coherence.[13,14] A study on the effect of meditation on the executive attentional network found that meditators were faster on all tasks.[12] With aging, the brain cortical thickness (grey matter, which contains neurons) decreases, whereas meditation experience is associated with an increase in grey matter in the brain.[11,26,51,52]. These are direct quotes from: Sharma, H. (2015). Meditation: Process and effects. *AYU (An International Quarterly Journal of Research in Ayurveda), 36*(3), 233. https://doi.org/10.4103/0974-8520.182756

30. "Brain regions associated with attention, interoception and sensory processing were thicker in meditation participants than matched controls..." Lazar, S. W., Kerr, C. E., Wasserman, R. H., Gray, J. R., Greve, D. N., Treadway, M. T., McGarvey, M., Quinn, B. T., Dusek, J. A., Benson, H., Rauch, S. L., Moore, C. I., & Fischl, B. (2005). Meditation experience is associated with increased cortical thickness. *Neuroreport, 16*(17), 1893–1897. https://doi.org/10.1097/01.wnr.0000186598.66243.19

31. "Mindfulness meditation increased thickness in the prefrontal cortex and parietal lobes, both linked to attention control..." Valk, S. L., Bernhardt, B. C., Trautwein, F.-M., Böckler, A., Kanske, P., Guizard, N., Collins, D. L., & Singer, T. (2017). Structural plasticity of the social brain: Differential change after socio-affective and cognitive mental training. *Science Advances, 3*(10), e1700489. https://doi.org/10.1126/sciadv.1700489

32. McAlonan, K. (2006). Attentional Modulation of Thalamic Reticular Neurons. *Journal of Neuroscience, 26*(16), 4444–4450. https://doi.org/10.1523/JNEUROSCI.5602-05.2006

33. BBC Two (Producers), & Austin, Naomi (Director). (2010). *Is Seeing Believing?* [Video File]. Retrieved from https://youtu.be/2k8fHR9jKVM

34. Dr. Iain McGilchrist. (2010, November 22). *The Divided Brain and the Making of the Western World* [RSA Lecture]. https://youtu.be/SbUHxC4wiWk (12:50 in).

35. David Eagleman. (2017, May 31). *David Eagleman: A Brainy Approach to Innovation* [Video]. https://youtu.be/9N00kDGMB5w (10 min in).

36. Zimmermann, M. (1989). The Nervous System in the Context of Information Theory. In R. F. Schmidt & G. Thews (Eds.), *Human Physiology* (p. 166–173). Springer Berlin Heidelberg. https://doi.org/10.1007/978-3-642-73831-9_7

37. Frith, C. D. (2007). *Making up the mind: How the brain creates our mental world.* Wiley Publishing. (p. 111).

Chapter 8: Sensory Input = Consciousness Output*

1. AMNH. (2018, January 12). To Hunt, the Platypus Uses Its Electric Sixth Sense [Museum Website]. *News & Blogs.* https://www.amnh.org/explore/news-blogs/news-posts/to-hunt-the-platypus-uses-its-electric-sixth-sense

2. Merabet, L. B., Swisher, J. D., McMains, S. A., Halko, M. A., Amedi, A., Pascual-Leone, A., & Somers, D. C. (2007). Combined Activation and Deactivation of Visual Cortex During Tactile Sensory Processing. *Journal of Neurophysiology, 97*(2), 1633–1641. https://doi.org/10.1152/jn.00806.2006

3. Merabet, L. B., Hamilton, R., Schlaug, G., Swisher, J. D., Kiriakopoulos, E. T., Pitskel, N. B., Kauffman, T., & Pascual-Leone, A. (2008). Rapid and Reversible Recruitment of Early Visual Cortex for Touch. *PLoS ONE, 3*(8), e3046. https://doi.org/10.1371/journal.pone.0003046

4. San Diego Zoo. (2021, April 13). Polar Bear Ursus maritimus [San Diego Zoo Website]. *Animals.* https://animals.sandiegozoo.org/animals/polar-bear

5. Dalton, Phillip. (2014). *Nature: Snow Bears* [DVD]. BBC Earth. https://www.pbs.org/video/snow-bears-trpvsn/

6. Haldane, J. (2017). *Possible Worlds.* United Kingdom: Taylor & Francis. (p. 260-8)

7. Dawkins, R. (2005, July). *Why the universe seems so strange* [Video]. TED Conferences. https://www.ted.com/talks/richard_dawkins_why_the_universe_seems_so_strange/ (12:10)

8. Ibid. (12:20).

9. *Sam Harris, Jordan Peterson & Douglas Murray in Dublin—Part 3—Presented by Pangburn.* (2008, July 14). [Video File]. Pangburn Philosophy. https://youtu.be/PqpYxD71hJU (70:30).

10. Dr. David Eagleman. (2012). The Umwelt [Author's Website]. *Latest.* Retrieved April 18, 2021, from: https://eagleman.com/latest/umwelt/

11. San Diego Zoo. (2021, April 13). Polar Bear *Ursus maritimus* [San Diego Zoo Website]. *Animals.* https://animals.sandiegozoo.org/animals/polar-bear

12. Brinkløv, S., Fenton, M. B., & Ratcliffe, J. M. (2013). Echolocation in Oilbirds and swiftlets. *Frontiers in Physiology, 4.* https://doi.org/10.3389/fphys.2013.00123

13. Fang, J. (2010). Snake infrared detection unravelled. *Nature,* news.2010.122. https://doi.org/10.1038/news.2010.122

14. Dr. David Eagleman. (2012). The Umwelt [Author's Website]. *Latest.* Retrieved April 18, 2021, from: https://eagleman.com/latest/umwelt/

15. Eagleman, D. (2015, March). *Can we create new senses for humans?* [Video]. TED Conferences. https://www.ted.com/talks/david_eagleman_can_we_create_new_senses_for_humans

16. Ibid. (17:20 in).

17. McGilchrist, I. (2019). *The master and his emissary: The divided brain and the making of the Western world* (New expanded edition). Yale University Press. (p. 28).

Chapter 9: What is Reality?

1. King, B., & Long, J. (2018, February 12). The shocking facts revealed: How sharks and other animals evolved electroreception to find their prey [Independent News Organization]. The Conversation. http://theconversation.com/the-shocking-facts-revealed-how-sharks-and-other-animals-evolved-electroreception-to-find-their-prey-91066

2. Fang, J. (2010). Snake infrared detection unravelled. Nature. https://doi.org/10.1038/news.2010.122

3. Davis, A. (2015, March). New video technology that reveals an object's hidden properties [Video]. TED Conferences. https://www.ted.com/talks/abe_davis_new_video_technology_that_reveals_an_object_s_hidden_properties

4. Wang, P., Liang, J., & Wang, L. V. (2020). Single-shot ultrafast imaging attaining 70 trillion frames per second. Nature Communications, 11(1), 2091. https://doi.org/10.1038/s41467-020-15745-4

5. Baluch, P., & Gonzales, A. (2015, July 1). How Do We See? [ASU School of Life Sciences]. ASU - Ask A Biologist. https://askabiologist.asu.edu/explore/how-do-we-see

6. The three-dimensional representation of reality that we perceive in our minds is actually more like a 2½-D sketch, rather than an actual 3D model. Remember, the only reason we perceive objects as being "a certain distance away" is because we have two different retinal images to compare against one another. For more on the 2½-D Sketch idea, see: Stevens, K. A. (2012). The Vision of David Marr. Perception, 41(9), 1061–1072. https://doi.org/10.1068/p7297

7. Ramachandran, V. S., & Rogers-Ramachandran, D. (2009). Seeing in Stereo. Scientific American Mind, 20(4), 20–22. https://doi.org/10.1038/scientificamericanmind0709-20

8. Gotts, Jason. (2017, November). David Eagleman (neuroscientist)—Your Creative Brain. Retrieved December 01, 2017, from https://podcasts.apple.com/us/podcast/think-again-a-big-think-podcast/id1002073669?i=1000394117931 (26:45 in).

9. New York University. (2011, January 12). How human vision perceives rapid changes: Brain predicts consequences of eye movements based on what we see next [Science Magazine]. ScienceDaily. https://www.sciencedaily.com/releases/2011/01/110110103737.htm

Chapter 10: Emotional Coding

1. Takes statistics - Turns out our brain's take statistics and draw conclusions from the data in the same way a statistician might pull patterns out of noisy data. The authors of this paper suggest, "that the feeling of confidence originates from a mental computation of statistical confidence." Sanders, J. I., Hangya, B., & Kepecs, A. (2016). Signatures of a Statistical Computation in the Human Sense of Confidence. Neuron, 90(3), 499–506. https://doi.org/10.1016/j.neuron.2016.03.025

2. Association learning & the Garcia effect - Jeremy Wolfe. (2004, Fall). Learning: The Power of Association [9.00 Introduction to Psychology]. MIT OpenCourseWare, Massachusetts Institute of Technology. https://ocw.mit.edu/courses/brain-and-cognitive-sciences/9-00-introduction-to-psychology-fall-2004/lecture-notes/3-learning-the-power-of-association/

3. Makes associations/finds connections: Mattson, M. P. (2014). Superior pattern processing is the essence of the evolved human brain. Frontiers in Neuroscience, 8. https://doi.org/10.3389/fnins.2014.00265 The author of this paper argues that, superior pattern processing is "the fundamental basis of most, if not all, unique features of the human brain including intelligence, language, imagination, invention, and the belief in imaginary entities such as ghosts and gods."

4. A more refined version of Maslow's hierarchy of needs is available here: Kenrick, D. T., Griskevicius, V., Neuberg, S. L., & Schaller, M. (2010). Renovating the Pyramid of Needs: Contemporary Extensions Built Upon Ancient Foundations. Perspectives on Psychological Science, 5(3), 292–314. https://doi.org/10.1177/1745691610369469

Chapter 11: Cuckoos Part II

1. The cuckoo's egg must match in size, color, marking variation, and marking dispersion to avoid being chucked out. Davies, N. (2015). Cuckoo: Cheating by nature. Bloomsbury. (p. 125-126).

2. Cutler, J. (2017, December 4). Cheetahs, world's fastest cats, can go from 0 to 60 mph in 3 seconds. USA Today. https://www.usatoday.com/story/news/world/2017/12/04/cheetah-worlds-fastest-cat-national-geographic-big-cats/108067318/

3. Dawkins, R. (2006). *The selfish gene* (30th anniversary ed). Oxford University Press. (p. 250).

4. Davies, N. (2015). *Cuckoo: Cheating by nature*. Bloomsbury. (p. 14)

5. Moksnes, A., & Roskaft, E. (1992). Responses of Some Rare Cuckoo Hosts to Mimetic Model Cuckoo Eggs and to Foreign Conspecific Eggs. *Ornis Scandinavica, 23*(1), 17. https://doi.org/10.2307/3676422

6. Stevens, M. (2013). Bird brood parasitism. *Current Biology, 23*(20), R909–R913. https://doi.org/10.1016/j.cub.2013.08.025

7. Davies, N. (2015). *Cuckoo: Cheating by nature*. Bloomsbury. (p. 110-12).

8. See *Rare-Enemy Effect* (Glossary)

9. Davies, N. B., & Brooke, M. de L. (1988). Cuckoos versus reed warblers: Adaptations and counteradaptations. *Animal Behaviour*, 36(1), 262–284. https://doi.org/10.1016/S0003-3472(88)80269-0

10. "Manipulated" - Dawkins, R. (2006). *The selfish gene* (30th anniversary ed). Oxford University Press. (p. 249).

11. The magnificent begging call of the cuckoo is instrumental in manipulating its foster parents into working hard enough to feed the over-sized cuckoo. See: Davies, N. (2015, May 14) *Cuckoos and their victims: An evolutionary arms race* [Royal Society Lecture]. https://youtu.be/n0O6S4hDDfE (51:40 in).

12. Ibid. (53:00-56:00).

13. In regards to bird's "seeing" an entire brood of reed warblers, I say it's not that far off to suggest since we've already shown how conflicting sense data can cause a tilt in perception. Just as it is in the McGurk effect, where the brain trusts the visual cue of the "F" sound more than what the ears hear, perhaps we're just witnessing the avian version of the McGurk effect. Only it's the begging cry and the gape that causes a tilt in perception, leading the hosts to actually believe they really do have an entire brood of reed warblers on their hands.

14. Dawkins, R. (2006). The selfish gene (30th anniversary ed). Oxford University Press. (p. 248).

15. Darwin, C. (1998). *The origin of species*. Wordsworth. (p. 166-67).

16. Dawkins, R. (2006). The selfish gene (30th anniversary ed). Oxford University Press. (p. 248-252).

17. Dawkins, R. (2016). *Extended phenotype*. Oxford University Press.

18. Dawkins, R. (2015, April 30). *This Is My Vision of "Life"* [Video Interview]. https://www.edge.org/conversation/richard_dawkins-this-is-my-vision-of-life (8 min. in).

19. Dawkins, R. (1997). *Climbing mount improbable*. W.W. Norton & Company. (p. 17-18).

20. Jacklyn, P. M. (1992). "Magnetic" termite mound surfaces are oriented to suit wind and shade conditions. *Oecologia, 91*(3), 385–395. https://doi.org/10.1007/BF00317628

Chapter 12: Non-Ordinary States

1. Sensory deprivation tanks really are interesting ways to explore the Self, and we find the experience to only get better with each progressive use. Reason being, there is a certain amount of relaxation that needs to be achieved, which may not be given the chance to manifest on shorter floats. In regards to the mental benefits from using sensory deprivation tanks, see: Kjellgren, A., & Westman, J. (2014). Beneficial effects of treatment with sensory isolation in flotation-tank as a preventive health-care inter-

vention – a randomized controlled pilot trial. *BMC Complementary and Alternative Medicine, 14*(1), 417. https://doi.org/10.1186/1472-6882-14-417 See also: Feinstein, J. S., Khalsa, S. S., Yeh, H.-W., Wohlrab, C., Simmons, W. K., Stein, M. B., & Paulus, M. P. (2018). Examining the short-term anxiolytic and antidepressant effect of Floatation-REST. *PloS One, 13*(2), e0190292. https://doi.org/10.1371/journal.pone.0190292. For the bit about how Navy SEALs are using them to help acquire languages faster, see: Wheal, J., & Kotler, S. (2017). *Stealing Fire: How silicon valley, the navy SEALs and maverick scientists are revolutionizing the way we live and work.* Dey Street Books. (p. 25-28).

2. Iwata, K., Nakao, M., Yamamoto, M., & Kimura, M. (2001). Quantitative characteristics of alpha and theta EEG activities during sensory deprivation. *Psychiatry and Clinical Neurosciences, 55*(3), 191–192. https://doi.org/10.1046/j.1440-1819.2001.00821.x

3. Braboszcz, C., Cahn, B. R., Levy, J., Fernandez, M., & Delorme, A. (2017). Increased Gamma Brainwave Amplitude Compared to Control in Three Different Meditation Traditions. *PLOS ONE, 12*(1), e0170647. https://doi.org/10.1371/journal.pone.0170647

4. Malik, A. S., & Amin, H. U. (2017). Designing an EEG Experiment. In *Designing EEG Experiments for Studying the Brain* (p. 1–30). Elsevier. https://doi.org/10.1016/B978-0-12-811140-6.00001-1

5. Lilly, J. C. (1997). *The scientist: A metaphysical autobiography.* Ronin Publishing, Inc. http://books.google.com/books?id=0JV9AAAAMAAJ (p. 121).

6. *41.5A: Epinephrine and Norepinephrine.* (2020, August 14). [Open Access Texts]. Biology LibreTexts. https://bio.libretexts.org/@go/page/14081

7. *NIMH » Post-Traumatic Stress Disorder (PTSD).* (n.d.). [Federal Agency Website]. National Institute of Mental Health. Retrieved November 2, 2021, from https://www.nimh.nih.gov/health/statistics/post-traumatic-stress-disorder-ptsd Also here: Harvard Medical School, 2007. National Comorbidity Survey (NCS). (2017, August 21). Retrieved from https://www.hcp.med.harvard.edu/ncs/index.php. Data Table 1: Lifetime prevalence DSM-IV/WMH-CIDI disorders by sex and cohort.

8. Koenigs, M., & Grafman, J. (2009). Posttraumatic Stress Disorder: The Role of Medial Prefrontal Cortex and Amygdala. *The Neuroscientist, 15*(5), 540–548. https://doi.org/10.1177/1073858409333072

9. Attention - There are many different types of attention that the brain can have, and we argue that our conscious experience actually runs on attention. Without attention (i.e., the capacity to attend) there is no experience. As previously mentioned, our organism may see or hear things that we, the conscious selves, don't actually experience. Because again, not every sensation that passes through the Subconscious OS graduates to the level of perception. The philosopher Daniel Dennett has an interesting way of describing this; he calls it "fame in the brain." For a sensation to become a perception, a certain amount of "popularity" needs to be achieved. Or stated in a slightly different way: the sensation has to be worthy of reporting. Take as an example, a child tapping you on the leg while you're riding the metro. Amidst all the other sensations happening around you, you may not notice the tapping until about the sixth or seventh tap. Did your organism register the first tap? Probably, however, it would still take a few more taps to accrue enough 'fame' and be 'worthy of' entering your sphere of awareness. For more on Dennett's "Fame in the Brain" see: Daniel Dennett. (2000). *Are we Explaining Consciousness Yet?* https://ase.tufts.edu/cogstud/dennett/papers/cognition.fin.htm

10. Mishara Aaron L, & Schwartz Michael A. (2012). *Altered States of Consciousness (ASC) as Paradoxically Healing.* https://doi.org/10.13140/RG.2.1.3956.4884

11. The practice of meditation has been a spiritual and healing enterprise for more than the last 5,000 years. See: Ospina, M. B., Bond, K., Karkhaneh, M., Tjosvold, L., Vandermeer, B., Liang, Y., Bialy, L., Hooton, N., Buscemi, N., Dryden, D. M., & Klassen, T. P. (2007). Meditation practices for health: State of the research. *Evidence Report/ Technology Assessment, 155*, 1–263.

12. Flow states, exposure to Nature, and ocean therapy have all been used with success to treat veterans suffering from PTSD. Rogers, C. M., Mallinson, T., & Peppers, D. (2014). High-Intensity Sports for Posttraumatic Stress Disorder and Depression: Feasibility Study of Ocean Therapy With Veterans of Operation Enduring Freedom and Operation Iraqi Freedom. *The American Journal of Occupational Therapy, 68*(4), 395–404. https://doi.org/10.5014/ajot.2014.011221

13. Psychedelics as a treatment for Addiction, see: Nichols, D., Johnson, M., & Nichols, C. (2017). Psychedelics as Medicines: An Emerging New Paradigm. *Clinical Pharmacology & Therapeutics, 101*(2), 209–219. https://doi.org/10.1002/cpt.557

14. "For mood and anxiety disorders, three controlled trials have suggested that psilocybin may decrease symptoms of depression and anxiety in the context of cancer-related psychiatric distress for at least 6 months following a single acute administration." See full article here: Johnson, M. W., & Griffiths, R. R. (2017). Potential Therapeutic Effects of Psilocybin. *Neurotherapeutics*, 14(3), 734–740. https://doi.org/10.1007/s13311-017-0542-y

15. Psilocybin has been found to reduce symptoms of patients with major depressive disorder. See: Davis, A. K., Barrett, F. S., May, D. G., Cosimano, M. P., Sepeda, N. D., Johnson, M. W., Finan, P. H., & Griffiths, R. R. (2021). Effects of Psilocybin-Assisted Therapy on Major Depressive Disorder: A Randomized Clinical Trial. *JAMA Psychiatry, 78*(5), 481. https://doi.org/10.1001/jamapsychiatry.2020.3285

16. "During the 1950s and 1960s, classic psychedelics (also known as serotonergic hallucinogens) such as lysergic acid diethylamide (LSD) and phosphoryloxy- *N,N*-dimethyltryptamine (psilocybin) were extensively investigated in psycholytic (low dose) and psychedelic (low to high dose) substance-assisted psychotherapy, resulting in more than 1,000 scientific papers and reports that included findings from about 40,000 subjects." See: Vollenweider, F. X., & Preller, K. H. (2020). Psychedelic drugs: Neurobiology and potential for treatment of psychiatric disorders. *Nature Reviews Neuroscience, 21*(11), 611–624. https://doi.org/10.1038/s41583-020-0367-2

17. Ayahuasca and its antidepressant effects - Sanches, R. F., de Lima Osório, F., dos Santos, R. G., Macedo, L. R. H., Maia-de-Oliveira, J. P., Wichert-Ana, L., de Araujo, D. B., Riba, J., Crippa, J. A. S., & Hallak, J. E. C. (2016). Antidepressant Effects of a Single Dose of Ayahuasca in Patients With Recurrent Depression: A SPECT Study. *Journal of Clinical Psychopharmacology, 36*(1), 77–81. https://doi.org/10.1097/JCP.0000000000000436. See also: Jiménez-Garrido, D. F., Gómez-Sousa, M., Ona, G., Dos Santos, R. G., Hallak, J. E. C., Alcázar-Córcoles, M. Á., & Bouso, J. C. (2020). Effects of ayahuasca on mental health and quality of life in naïve users: A longitudinal and cross-sectional study combination. *Scientific Reports, 10*(1), 4075. https://doi.org/10.1038/s41598-020-61169-x

18. The entire book (*Stealing Fire*) seems as if it was written in defense of ASCs. However, the second chapter, entitled: "Why it matters" certainly captures the essential

idea. See: Wheal, J., & Kotler, S. (2017). *Stealing Fire: How silicon valley, the navy SEALs and maverick scientists are revolutionizing the way we live and work*. Dey Street Books. (p. 33-50).

19. Jan M Keppel, H. (2018). Kambo and its Multitude of Biological Effects: Adverse Events or Pharmacological Effects? *International Archives of Clinical Pharmacology, 4*(1). https://doi.org/10.23937/2572-3987.1510017

20. de Haro, L., & Pommier, P. (2006). Hallucinatory Fish Poisoning (Ichthyoallyeinotoxism): Two Case Reports From the Western Mediterranean and Literature Review. *Clinical Toxicology, 44*(2), 185–188. https://doi.org/10.1080/15563650500514590

21. Weil, A. T., & Davis, W. (1994). Bufo alvarius: A potent hallucinogen of animal origin. *Journal of Ethnopharmacology, 41*(1–2), 1–8. https://doi.org/10.1016/0378-8741(94)90051-5

22. Extreme rituals - Lee, E. M., Klement, K. R., Ambler, J. K., Loewald, T., Comber, E. M., Hanson, S. A., Pruitt, B., & Sagarin, B. J. (2016). Altered States of Consciousness during an Extreme Ritual. *PLOS ONE, 11*(5), e0153126. https://doi.org/10.1371/journal.pone.0153126

Chapter 13: Unlearning

1. Garrison, K. A., Zeffiro, T. A., Scheinost, D., Constable, R. T., & Brewer, J. A. (2015). Meditation leads to reduced default mode network activity beyond an active task. *Cognitive, Affective, & Behavioral Neuroscience, 15*(3), 712–720. https://doi.org/10.3758/s13415-015-0358-3

2. Csikszentmihalyi, M. (1996, September 1). *Go With The Flow* [Wired Magazine]. https://www.wired.com/1996/09/czik/

3. Wheal, J., & Kotler, S. (2017). *Stealing Fire: How silicon valley, the navy SEALs and maverick scientists are revolutionizing the way we live and work*. Dey Street Books. (p. 42).

4. Jackson, S. A., & Marsh, H. W. (1996). Development and Validation of a Scale to Measure Optimal Experience: The Flow State Scale. *Journal of Sport and Exercise Psychology, 18*(1), 17–35. https://doi.org/10.1123/jsep.18.1.17

5. Mountain biker, Colin Gray, said this on his blog a couple years back, although I can no longer find a record for it. The original blog post was likely removed and replaced.

6. Spreng, R. N., & Grady, C. L. (2010). Patterns of Brain Activity Supporting Autobiographical Memory, Prospection, and Theory of Mind, and Their Relationship to the Default Mode Network. *Journal of Cognitive Neuroscience, 22*(6), 1112–1123. https://doi.org/10.1162/jocn.2009.21282 (About autobiographical thinking being so frequent; it is called the "default mode" of the brain for a reason. Although, if you wanted further evidence of our apparent disconnect with reality, I recommend: Killingsworth, M. A., & Gilbert, D. T. (2010). A Wandering Mind Is an Unhappy Mind. *Science, 330*(6006), 932–932. https://doi.org/10.1126/science.1192439

7. Raichle, M. E. (2010, March 1). The Brain's Dark Energy. *Scientific American, 302*(3), 44–49.

8. Carhart-Harris, R. L., & Friston, K. J. (2010). The default-mode, ego-functions and free-energy: A neurobiological account of Freudian ideas. *Brain, 133*(4), 1265–1283. https://doi.org/10.1093/brain/awq010

9. Carhart-Harris, R. L., Leech, R., Hellyer, P. J., Shanahan, M., Feilding, A., Tagliazucchi, E., Chialvo, D. R., & Nutt, D. (2014). The entropic brain: A theory of conscious

states informed by neuroimaging research with psychedelic drugs. *Frontiers in Human Neuroscience, 8*, 20. https://doi.org/10.3389/fnhum.2014.00020

10. *Dr. Robin Carhart-Harris: Brain Imaging with Psilocybin and MDMA* (No. 17). (2017, September 28). [Podcast]. https://podcasts.apple.com/us/podcast/episode-17-dr-robin-carhart-harris-brain-imaging-with/id1217974024?i=1000392843298 (18 min in).

11. Control center of the brain - David Nutt. (2017, April 26). *David Nutt: Psychedelic Research, From Brain Imaging to Policy Reform.* Psychedelic Science 2017, Oakland, Ca. https://youtu.be/ZzepSK6Gzk8

12. Orchestrator of the self - Raichle, M. E. (2010, March 1). The Brain's Dark Energy. *Scientific American, 302*(3), 44–49.

13. Integration Hub - *Dr. Robin Carhart-Harris: Brain Imaging with Psilocybin and MDMA* (No. 17). (2017, September 28). [Podcast]. from https://podcasts.apple.com/us/podcast/episode-17-dr-robin-carhart-harris-brain-imaging-with/id1217974024?i=1000392843298 (18 min in).

14. DMN candidate for 'the ego' - Ibid. (20:30 in).

15. DMN candidate for 'the ego' - Carhart-Harris, R. L., & Friston, K. J. (2010). The default-mode, ego-functions and free-energy: A neurobiological account of Freudian ideas. *Brain, 133*(4), 1265–1283. https://doi.org/10.1093/brain/awq010

Chapter 14: Pinning it Down

1. Daniel Dennett. (2007, November 2). *Cartesian Theatre—Daniel Dennett* [Video Interview]. https://youtu.be/a3a2FFoRpzQ (2 min. in) Dennett has made this point in other lectures as well, which is where our particular quote has been pulled from, however this lecture has since been taken down.

2. Karl Friston. (2018, June 1). *Embodied Cognition Karl Friston* [Video Interview]. http://serious-science.org/embodied-cognition-9027 (Quote located at the top of the interview).

3. Wolpert, D. (2011, July). *The real reason for brains* [Video]. TED Conferences. https://www.ted.com/talks/daniel_wolpert_the_real_reason_for_brains

4. Charles Choi. (2007, May 24). Strange but True: When Half a Brain Is Better than a Whole One. *Scientific American.* https://www.scientificamerican.com/article/strange-but-true-when-half-brain-better-than-whole/

5. Referencing split-brain operations - McGilchrist, I. (2019). *The master and his emissary: The divided brain and the making of the Western world* (New expanded edition). Yale University Press. (p. 18, 35).

6. Ibid. (p. 18).

7. Taylor, J. B. (2016). *My stroke of insight: A brain scientist's personal journey.* Plume.

8. Singer, Chedd & Angier. (1997, January 22). Severed Corpus Callosum (S07E03) [Video File]. In *Scientific American Frontiers—Piece of Mind.* https://youtu.be/q6ryKGiQh3w

9. McGilchrist, I. (2019). *The master and his emissary: The divided brain and the making of the Western world* (New expanded edition). Yale University Press. (p. 35; 78-82).

10. Ibid. (p. 10).

11. Brewer, J. A., Worhunsky, P. D., Gray, J. R., Tang, Y.-Y., Weber, J., & Kober, H. (2011). Meditation experience is associated with differences in default mode network activity and connectivity. *Proceedings of the National Academy of Sciences, 108*(50), 20254–20259. https://doi.org/10.1073/pnas.1112029108

12. Kang, D.-H., Jo, H. J., Jung, W. H., Kim, S. H., Jung, Y.-H., Choi, C.-H., Lee, U. S., An, S. C., Jang, J. H., & Kwon, J. S. (2013). The effect of meditation on brain structure: Cortical thickness mapping and diffusion tensor imaging. *Social Cognitive and Affective Neuroscience, 8*(1), 27–33. https://doi.org/10.1093/scan/nss056

Chapter 15: Disentangling the Self

1. Carhart-Harris, R. L. (2019). How do psychedelics work? *Current Opinion in Psychiatry, 32*(1), 16–21. https://doi.org/10.1097/YCO.0000000000000467
2. Ibid.
3. Garrison, K. A., Zeffiro, T. A., Scheinost, D., Constable, R. T., & Brewer, J. A. (2015). Meditation leads to reduced default mode network activity beyond an active task. *Cognitive, Affective, & Behavioral Neuroscience, 15*(3), 712–720. https://doi.org/10.3758/s13415-015-0358-3
4. Van der Linden, D., Tops, M., & Bakker, A. B. (2021). The Neuroscience of the Flow State: Involvement of the Locus Coeruleus Norepinephrine System. *Frontiers in Psychology, 12*, 645498. https://doi.org/10.3389/fpsyg.2021.645498
5. Baillargeon, R., & DeVos, J. (1991). Object permanence in young infants: Further evidence. *Child Development, 62*(6), 1227–1246.
6. First, it should be mentioned that theory of mind (ToM) isn't a unitary construct, and that pieces of ToM unpack before the age of five and six. That said, in reference to second-order false belief tasks, mastery does come at about age five or six. Miller, S. A. (2012). *Theory of mind: Beyond the preschool years*. Psychology Press. (p. 63).
7. Fair, D. A., Cohen, A. L., Dosenbach, N. U. F., Church, J. A., Miezin, F. M., Barch, D. M., Raichle, M. E., Petersen, S. E., & Schlaggar, B. L. (2008). The maturing architecture of the brain's default network. *Proceedings of the National Academy of Sciences, 105*(10), 4028–4032. https://doi.org/10.1073/pnas.0800376105. See also: Saxe, R. (2009, July). *How we read each other's minds* [Video]. TED Conferences. https://www.ted.com/talks/rebecca_saxe_how_we_read_each_other_s_minds
8. A wonderful explanation for "infantile amnesia," see: Nelson, K., & Fivush, R. (2004). The Emergence of Autobiographical Memory: A Social Cultural Developmental Theory. *Psychological Review, 111*(2), 486–511. https://doi.org/10.1037/0033-295X.111.2.486
9. Oh, I.-S., Wang, G., & Mount, M. K. (2011). Validity of observer ratings of the five-factor model of personality traits: A meta-analysis. *Journal of Applied Psychology, 96*(4), 762–773. https://doi.org/10.1037/a0021832
10. Borza, D., Danescu, R., Itu, R., & Darabant, A. (2017). High-Speed Video System for Micro-Expression Detection and Recognition. *Sensors, 17*(12), 2913. https://doi.org/10.3390/s17122913
11. Kahneman, D. (2010, February). *The riddle of experience vs. memory* [Video]. TED Conferences. https://www.ted.com/talks/daniel_kahneman_the_riddle_of_experience_vs_memory
12. Freud, S., Strachey, J., & Freud, S. (1989). *The ego and the id*. Norton.
13. Kahneman, D., & Riis, J. (2005). Living, and thinking about it: Two perspectives on life. In F. A. Huppert, N. Baylis, & B. Keverne (Eds.), *The Science of Well-Being* (p. 284–305). Oxford University Press. https://doi.org/10.1093/acprof:oso/9780198567523.003.0011
14. Ibid.

15. Tyng CM, Amin HU, Saad MNM, Malik AS. The Influences of Emotion on Learning and Memory. *Front Psychol.* 2017;8:1454. Published 2017 Aug 24. doi:10.3389/fpsyg.2017.01454

16. "The psychological present is said to be about three seconds long; that means that, you know, the in a life there are about 600 million of them; in a month, there are about 600,000—most of them don't leave a trace. Most of them are completely ignored by the remembering self." - Kahneman, D. (2010, February). *The riddle of experience vs. memory* [Video]. TED Conferences. https://www.ted.com/talks/daniel_kahneman_the_riddle_of_experience_vs_memory (6:53 in).

17. Eleanor Maguire. (2014, March 13). *The Neuroscience of Memory—Eleanor Maguire* [Royal Institution Lecture]. Ri, London, England. https://youtu.be/gdzmNwTLakg (45:30 in).

18. Stevens, A. (1999). *On Jung* (2. ed). Princeton Univ. Pr. (p. 43-46).

19. Anil Seth. (2016, November 2). The real problem [Charity]. *Psychology.* https://aeon.co/essays/the-hard-problem-of-consciousness-is-a-distraction-from-the-real-one

20. The brainstem and the hypothalamus, for example, are two key parts of the brain that help give rise to consciousness and drive conscious behavior, although the goings-on behind both of these areas are completely outside of our awareness. "Voluntary (cognitive) control of behavior requires the cerebral cortex; whereas control of innate (instinctive) behaviors is classically associated with the hypothalamus...Classic lesion experiments have shown that innate behaviors can be performed to some extent without the cerebral cortex, and spinal reflexes without the forebrain and much of the brainstem." - Hahn, J. D., Fink, G., Kruk, M. R., & Stanley, B. G. (2019). Editorial: Current Views of Hypothalamic Contributions to the Control of Motivated Behaviors. *Frontiers in Systems Neuroscience, 13*, 32. https://doi.org/10.3389/fnsys.2019.00032 See also: Brainstem - The brainstem controls numbers vital bodily functions like swallowing, blood pressure regulation, breathing, etc. Damage to the upper brainstem is known to cause coma or persistent vegetative states. Yet, patients with a severely damaged cortex and a relatively spared brainstem typically remain in a vegetative state, suggesting that brainstem activity alone is insufficient to sustain consciousness. Koch, C., Massimini, M., Boly, M., & Tononi, G. (2016). Neural correlates of consciousness: Progress and problems. *Nature Reviews Neuroscience, 17*(5), 307–321. https://doi.org/10.1038/nrn.2016.22

21. Kahneman, D. (2010, February). *The riddle of experience vs. memory* [Video]. TED Conferences. https://www.ted.com/talks/daniel_kahneman_the_riddle_of_experience_vs_memory (11 min in).

22. Kahneman, D., & Riis, J. (2005). Living, and thinking about it: Two perspectives on life. In F. A. Huppert, N. Baylis, & B. Keverne (Eds.), *The Science of Well-Being* (p. 284–305). Oxford University Press. https://doi.org/10.1093/acprof:oso/9780198567523.003.0011 (p.285-86).

23. Anil Seth describes the same experience (loss of sense of time) in his TED talk. Seth, A. (2017, April). *Your brain hallucinates your conscious reality* [Video]. TED Conferences. https://www.ted.com/talks/anil_seth_your_brain_hallucinates_your_conscious_reality

24. Within the anesthetized state, there are many different levels that are possible. See: Bonhomme, V., Staquet, C., Montupil, J., Defresne, A., Kirsch, M., Martial, C., Vanhaudenhuyse, A., Chatelle, C., Larroque, S. K., Raimondo, F., Demertzi, A., Bodart, O., Laureys, S., & Gosseries, O. (2019). General Anesthesia: A Probe to Explore

Consciousness. *Frontiers in Systems Neuroscience, 13*, 36. https://doi.org/10.3389/fnsys.2019.00036

25. Bischoff, P., & Rundshagen, I. (2011). Awareness Under General Anesthesia. *Deutsches Aerzteblatt Online*. https://doi.org/10.3238/arztebl.2011.0001

26. Zajchowski, C. A. B., Schwab, K. A., & Dustin, D. L. (2017). The Experiencing Self and the Remembering Self: Implications for Leisure Science. *Leisure Sciences, 39*(6), 561–568. https://doi.org/10.1080/01490400.2016.1209140

27. George Musser. (2011, September 15). Time on the Brain: How You Are Always Living In the Past, and Other Quirks of Perception [Science Magazine]. *Scientific American: Observations*. https://blogs.scientificamerican.com/observations/time-on-the-brain-how-you-are-always-living-in-the-past-and-other-quirks-of-perception/

28. Suddendorf, T., Addis, D. R., & Corballis, M. C. (2009). Mental time travel and the shaping of the human mind. *Philosophical Transactions of the Royal Society B: Biological Sciences, 364*(1521), 1317–1324. https://doi.org/10.1098/rstb.2008.0301 See also: Klein, S. B., Loftus, J., & Kihlstrom, J. F. (2002). Memory and Temporal Experience: The Effects of Episodic Memory Loss on an Amnesic Patient's Ability to Remember the Past and Imagine the Future. *Social Cognition, 20*(5), 353–379. https://doi.org/10.1521/soco.20.5.353.21125

29. Eleanor Maguire. (2014, March 13). *The Neuroscience of Memory—Eleanor Maguire* [Royal Institution Lecture]. Ri, London, England. https://youtu.be/gdzmNwTLakg See also: Maguire, E. A., & Hassabis, D. (2011). Role of the hippocampus in imagination and future thinking. *Proceedings of the National Academy of Sciences, 108*(11), E39–E39. https://doi.org/10.1073/pnas.1018876108

Chapter 16: Mental Disorders

1. Jung, C. G., & Falzeder, E., Jung, Lorenz, Meyer-Grass, Maria, Woolfson, Tony. (2012). *Children's Dreams: Notes from the Seminar Given in 1936-1940*. https://doi.org/10.1515/9781400843084 (p. 3).

2. Seth, A. (2017, April). *Your brain hallucinates your conscious reality* [Video]. TED Conferences. https://www.ted.com/talks/anil_seth_your_brain_hallucinates_your_conscious_reality (8 min. in).

3. Killingsworth, M. A., & Gilbert, D. T. (2010). A Wandering Mind Is an Unhappy Mind. *Science, 330*(6006), 932–932. https://doi.org/10.1126/science.1192439

4. Duan, L., Van Dam, N. T., Ai, H., & Xu, P. (2020). Intrinsic organization of cortical networks predicts state anxiety: An functional near-infrared spectroscopy (fNIRS) study. *Translational Psychiatry, 10*(1), 402. https://doi.org/10.1038/s41398-020-01088-7

5. Wang, X., Öngür, D., Auerbach, R. P., & Yao, S. (2016). Cognitive Vulnerability to Major Depression: View from the Intrinsic Network and Cross-network Interactions. *Harvard Review of Psychiatry, 24*(3), 188–201. https://doi.org/10.1097/HRP.0000000000000081

6. See Triple network model of major psychopathology. "Deficits in access, engagement and disengagement of largescale neurocognitive networks are shown to play a prominent role in several disorders including schizophrenia, depression, anxiety, dementia and autism." Menon, V. (2011). Large-scale brain networks and psychopathology: A unifying triple network model. *Trends in Cognitive Sciences, 15*(10), 483–506. https://doi.org/10.1016/j.tics.2011.08.003

7.	"Importantly, studies in patient populations have shown that blood oxygen level-de-pendent (BOLD) fMRI can be used to detect altered FC [Functional Connectivity] in individuals suffering from a variety of CNS diseases. Further, patients can often be distinguished from healthy controls with high sensitivity and high specificity. In addi-tion, research has shown connectivity strength to be correlated with severity of disease symptoms, with recovery of connectivity observed following pharmacological treat-ment.[37] These connectivity networks can be disturbed in various psychiatric conditions such as anorexia nervosa,[38] obsessive-compulsive disorder, depression,[40–42] anxiety,[43,44] bipolar disorder and hypomania,[45] trauma,[46] and addiction,[47,48] among others. These citations are by no means exhaustive, but demonstrate that a large number of psychi-atric disorders are associated with disturbances in RSFC." - Nichols, D., Johnson, M., & Nichols, C. (2017). Psychedelics as Medicines: An Emerging New Paradigm. *Clinical Pharmacology & Therapeutics, 101*(2), 209–219. https://doi.org/10.1002/cpt.557

8.	Posner, J., Cha, J., Wang, Z., Talati, A., Warner, V., Gerber, A., Peterson, B. S., & Weissman, M. (2016). Increased Default Mode Network Connectivity in Individuals at High Familial Risk for Depression. *Neuropsychopharmacology, 41*(7), 1759–1767. https://doi.org/10.1038/npp.2015.342

9.	National Institute of Mental Health. (2018, July 1). *Anxiety Disorders* [Federal Agency Website]. NIMH. https://www.nimh.nih.gov/health/topics/anxiety-disorders

10.	Ibid. (See *Panic Disorders*)

11.	Koenigs, M., & Grafman, J. (2009). Posttraumatic Stress Disorder: The Role of Medial Prefrontal Cortex and Amygdala. *The Neuroscientist, 15*(5), 540–548. https://doi.org/10.1177/1073858409333072

12.	Zhou, H.-X., Chen, X., Shen, Y.-Q., Li, L., Chen, N.-X., Zhu, Z.-C., Castellanos, F. X., & Yan, C.-G. (2020). Rumination and the default mode network: Meta-analysis of brain imaging studies and implications for depression. *NeuroImage, 206*, 116287. https://doi.org/10.1016/j.neuroimage.2019.116287

13.	"Depression can be characterized, to some extent, as repetitive, ruminative thinking." - Dr. Anil Seth - (1:18:20) World Science Festival. (2019, April 16). *Revealing the Mind: The Promise of Psychedelics* [Science Festival]. Big Ideas Series, New York. https://youtu.be/Fi66wFfOC-4

14.	W.H.O. (2021, September 13). *Depression* [United Nations Agency Website]. World Health Organization. https://www.who.int/news-room/fact-sheets/detail/depression

15.	Paul Nestadt, M.D. (2020, September 28). *Why Aren't my Antidepressants Working?* [Health & Research]. Johns Hopkins Medicine. https://www.hopkinsmedicine.org/health/wellness-and-prevention/why-arent-my-antidepressants-working

16.	Just taking someone who's suffering from major depressive disorder (MDD) "MDD is found to predict significant decrements in role functioning (e.g., low marital quality low work performance, low earnings). MDD is also associated with elevated risk of inset, persistence, and severity of a wide range of chronic physical disorders and to suicide." Kessler, R. C. (2012). The Costs of Depression. *Psychiatric Clinics of North America, 35*(1), 1–14. https://doi.org/10.1016/j.psc.2011.11.005; See also: Dr. James Pennebaker - "People who are depressed use the word "I" more than people who are not depressed." Then later, "One of the theories of depression is that it's a disease of self-focus, that people are so ruminative and looking inward so much." Dr. James Pennebaker. (2017, March 4). Dialogue: Great U Texas Austin Psych Prof JW Pen-nebaker [Video Interview]. https://youtu.be/hJ4JEypNH2s (38 min. in). Additionally, depression has been characterized by a severe pessimism bias about oneself and future. See: Beevers, C. G., Mullarkey, M. C., Dainer-Best, J., Stewart, R. A., Labrada, J., Allen, J. J. B., McGeary, J. E., & Shumake, J. (2019). Association between negative cognitive bias and depression: A symptom-level approach. *Journal of Abnormal Psychology*, *128*(3), 212–227. https://doi.org/10.1037/abn0000405. After that, grapple with the fact that this pessimism bias appears to be alleviated with just one treatment of psilocybin: Lyons, T., & Carhart-Harris, R. L. (2018). More Realistic Forecasting of Future Life Events After Psilocybin for Treatment-Resistant Depression. *Frontiers in Psychology*, *9*, 1721. https://doi.org/10.3389/fpsyg.2018.01721

17. LSD - Beckley Foundation. (2016, April 1). The World's First Images of the Brain on LSD [Scientific Institution]. *Beckley in the Press*. https://www.beckleyfoundation.org/the-brain-on-lsd-revealed-first-scans-show-how-the-drug-affects-the-brain/

18. Psilocybin (Extraversion and Openness were effected) - Erritzoe, D., Roseman, L., Nour, M. M., MacLean, K., Kaelen, M., Nutt, D. J., & Carhart-Harris, R. L. (2018). Effects of psilocybin therapy on personality structure. *Acta Psychiatrica Scandinavica*, *138*(5), 368–378. https://doi.org/10.1111/acps.12904. See also: MacLean, K. A., Johnson, M. W., & Griffiths, R. R. (2011). Mystical experiences occasioned by the hallucinogen psilocybin lead to increases in the personality domain of openness. *Journal of Psychopharmacology*, *25*(11), 1453–1461. https://doi.org/10.1177/0269881111420188

19. MDMA - Wagner, M. T., Mithoefer, M. C., Mithoefer, A. T., MacAulay, R. K., Jerome, L., Yazar-Klosinski, B., & Doblin, R. (2017). Therapeutic effect of increased openness: Investigating mechanism of action in MDMA-assisted psychotherapy. *Journal of Psychopharmacology*, *31*(8), 967–974. https://doi.org/10.1177/0269881117711712

Chapter 17: MDMA, Psilocybin, and LSD

1. Passie, T. (2018). The early use of MDMA ('Ecstasy') in psychotherapy (1977–1985). *Drug Science, Policy and Law*, *4*, 205032451876744. https://doi.org/10.1177/2050324518767442

2. Benzenhöfer, U., & Passie, T. (2010). Rediscovering MDMA (ecstasy): The role of the American chemist Alexander T. Shulgin: The rediscovery of MDMA by Alexander T. Shulgin. *Addiction*, *105*(8), 1355–1361. https://doi.org/10.1111/j.1360-0443.2010.02948.x

3. Brad Burge. (10/24/2015). *Psychedelics—New Perspectives* (No. 099) [Video]. https://youtu.be/Bq-ewHfJyPs (13 min. in).

4. Ibid. (15 min in).

5. Torsten Passie. (2017, April 26). *Torsten Passie: History of MDMA - An Overview* [Scientific Presentation]. Psychedelic Science 2017, Oakland, Ca]. https://youtu.be/r72hZzJGsNU (24 min in).

6. Doblin, R. (2019, April). *The future of psychedelic-assisted psychotherapy* [Video]. TED Conferences. https://www.ted.com/talks/rick_doblin_the_future_of_psychedelic_assisted_psychotherapy (8:20 in).

7. Feduccia, A. A., Jerome, L., Yazar-Klosinski, B., Emerson, A., Mithoefer, M. C., & Doblin, R. (2019). Breakthrough for Trauma Treatment: Safety and Efficacy of MDMA-Assisted Psychotherapy Compared to Paroxetine and Sertraline. *Frontiers in Psychiatry*, *10*, 650. https://doi.org/10.3389/fpsyt.2019.00650

8. Jerome, L., Feduccia, A. A., Wang, J. B., Hamilton, S., Yazar-Klosinski, B., Emerson, A., Mithoefer, M. C., & Doblin, R. (2020). Long-term follow-up outcomes of MDMA-assisted psychotherapy for treatment of PTSD: A longitudinal pooled analysis of six phase 2 trials. *Psychopharmacology*, *237*(8), 2485–2497. https://doi.org/10.1007/s00213-

020-05548-2 Additionally, it is worth noting that the prevalence of PTSD amongst veterans of OIF and OEF (from 2007-2013) has been estimated to be at 23%! See: Fulton, J. J., Calhoun, P. S., Wagner, H. R., Schry, A. R., Hair, L. P., Feeling, N., Elbogen, E., & Beckham, J. C. (2015). The prevalence of posttraumatic stress disorder in Operation Enduring Freedom/Operation Iraqi Freedom (OEF/OIF) Veterans: A meta-analysis. *Journal of Anxiety Disorders, 31*, 98–107. https://doi.org/10.1016/j.janxdis.2015.02.003

9.   Fuentes, J. J., Fonseca, F., Elices, M., Farré, M., & Torrens, M. (2019). Therapeutic Use of LSD in Psychiatry: A Systematic Review of Randomized-Controlled Clinical Trials. *Frontiers in Psychiatry, 10*, 943. https://doi.org/10.3389/fpsyt.2019.00943

10. Ibid.

11. Brad Burge. (10/24/2015). *Psychedelics—New Perspectives* (No. 099) [Video]. https://youtu.be/Bq-ewHfJyPs (21:20 in).

12. Das, S., Barnwal, P., Ramasamy, A., Sen, S., & Mondal, S. (2016). Lysergic acid diethylamide: A drug of "use"? *Therapeutic Advances in Psychopharmacology, 6*(3), 214–228. https://doi.org/10.1177/2045125316640440

13. Lattin, D. (2012). *Distilled spirits: Getting high, then sober, with a famous writer, a forgotten philosopher, and a hopeless drunk.* University of California Press. (p. 206).

14. The following is a pretty good summary of LSD myths, their origin, and the literature that debunks them: Roberts, T. B., & Council on Spiritual Practices (Eds.). (2001). *Psychoactive sacramentals: Essays on entheogens and religion.* Council on Spiritual Practices. (p. 125-35). However, it is also of note that there were myths that LSD caused chromosome damage, which has also been debunked. See: Grob, C. S., & Grigsby, J. (Eds.). (2021). *Handbook of medical hallucinogens.* The Guilford Press. (p. 18-19).

15. Grof, S. (1980). *LSD psychotherapy.* Hunter House. (p. 297).

16. "80-90% of people rate these experiences in the top 5 of their life," says Dr. Roland Griffiths of Johns Hopkins University when asked to reflect on all his research in the psychedelic domain. See: Jordan Peterson. (March 2, 2021). *The Psychology of Psychedelics—Roland Griffiths* (S4 E20) [Video]. https://youtu.be/NGIP-3Q-p_s (52:50 in).

17. Pamela Caragol. (2009, November 3). Inside LSD [Documentary]. In *National Geographic Explorer.* National Geographic. https://topdocumentaryfilms.com/inside-lsd/ (8 min. in).

18. Beckley Foundation. (2017, April 19). Psychedelic Research Timeline [Research Think Tank]. *Psychedelic Science.* https://www.beckleyfoundation.org/psychedelic-research-timeline-2/

19. Beckley Foundation. (2016, April 1). The World's First Images of the Brain on LSD [Scientific Institution]. *Beckley in the Press.* https://www.beckleyfoundation.org/the-brain-on-lsd-revealed-first-scans-show-how-the-drug-affects-the-brain/

20. Carhart-Harris, R. L., Muthukumaraswamy, S., Roseman, L., Kaelen, M., Droog, W., Murphy, K., Tagliazucchi, E., Schenberg, E. E., Nest, T., Orban, C., Leech, R., Williams, L. T., Williams, T. M., Bolstridge, M., Sessa, B., McGonigle, J., Sereno, M. I., Nichols, D., Hellyer, P. J., ... Nutt, D. J. (2016). Neural correlates of the LSD experience revealed by multimodal neuroimaging. *Proceedings of the National Academy of Sciences, 113*(17), 4853–4858. https://doi.org/10.1073/pnas.1518377113

21. Das, S., Barnwal, P., Ramasamy, A., Sen, S., & Mondal, S. (2016). Lysergic acid diethylamide: A drug of "use"? *Therapeutic Advances in Psychopharmacology, 6*(3), 214–228.

https://doi.org/10.1177/2045125316640440 (This article is a good review about LSD and its uses, however it's worth mentioning that classic psychedelics, like LSD and psilocybin, are both 5-HT2A receptor agonists and do act upon our brains in similar ways. They're both effective agents for generating bona fide "mystical-type" experiences, yet there is far more research being done with psilocybin. And the reason for this seems to be because of its shorter duration time, long history of use, and its "perceived" cultural image, which is typically viewed as being "more natural" than LSD, even though most clinical trials are required to use a synthetic form of psilocybin. All this considered, there would likely be more papers published about the efficacy of LSD, in regards to PTSD, etc., if its duration time were a bit shorter, and its PR campaign a bit stronger.

22. Gasser, P., Kirchner, K., & Passie, T. (2015). LSD-assisted psychotherapy for anxiety associated with a life-threatening disease: A qualitative study of acute and sustained subjective effects. *Journal of Psychopharmacology, 29*(1), 57–68. https://doi.org/10.1177/0269881114555249

23. Muttoni, S., Ardissino, M., & John, C. (2019). Classical psychedelics for the treatment of depression and anxiety: A systematic review. *Journal of Affective Disorders, 258*, 11–24. https://doi.org/10.1016/j.jad.2019.07.076

24. Winkelman, M. (2015). Psychedelics as Medicines for Substance Abuse Rehabilitation: Evaluating Treatments with LSD, Peyote, Ibogaine and Ayahuasca. *Current Drug Abuse Reviews, 7*(2), 101–116. https://doi.org/10.2174/1874473708666150107120011

25. Giorgio Samorini. (2019). The oldest archeological data evidencing the relationship of Homo sapiens with psychoactive plants: A worldwide overview. *Journal of Psychedelic Studies, 3*(2), 63–80. https://doi.org/10.1556/2054.2019.008

26. Peter von Puttkamer. (2009, May 19). *Peyote to LSD: A Psychedelic Odyssey* [Documentary]. History Channel.

27. Davis, W. (1997). *One river: Explorations and discoveries in the Amazon rain forest* (1. Touchstone ed., [Nachdr.]). Simon & Schuster. (p. 240-43).

28. Dasgupta, A. (2019). Abuse of Magic Mushroom, Peyote Cactus, LSD, Khat, and Volatiles. In *Critical Issues in Alcohol and Drugs of Abuse Testing* (pp. 477–494). Elsevier. https://doi.org/10.1016/B978-0-12-815607-0.00033-2

29. Vollenweider, F. X., & Preller, K. H. (2020). Psychedelic drugs: Neurobiology and potential for treatment of psychiatric disorders. *Nature Reviews Neuroscience, 21*(11), 611–624. https://doi.org/10.1038/s41583-020-0367-2

30. Email correspondence with Dr. David Nichols on Dec. 12, 2021. During one of the times Dr. David Nichols met Hofmann, Albert said he had no idea how he got it into his body. Although, Nichols suspects that some of the solution from the column chromatography purification process may have gotten onto his fingers or under his fingernails. Apparently, they didn't wear rubber gloves in those days, Nichols reports, so the solution might've diffused through his skin, or possibly if he rubbed his eyes or mouth.

31. David E. Nichols & Benjamin R. Chemel. (2006). The Neuropharmacology of Religious Experience: Hallucinogens and the Experience of the Divine. In Patrick McNamara (Ed.), *Where God and Science Meet: How Brain and Evolutionary Studies Alter Our Understanding of Religion: Vol. 3 (The Psychology of Religious Experience)* (1st edition). Praeger Publishers. (p. 7).

32. Davis, W. (1997). *One river: Explorations and discoveries in the Amazon rain forest* (1. Touchstone ed., [Nachdr.]). Simon & Schuster. (p. 227-43).

33. Nichols, D., Johnson, M., & Nichols, C. (2017). Psychedelics as Medicines: An Emerging New Paradigm. *Clinical Pharmacology & Therapeutics*, *101*(2), 209–219. https://doi.org/10.1002/cpt.557

34. Andersson, M., Persson, M., & Kjellgren, A. (2017). Psychoactive substances as a last resort-a qualitative study of self-treatment of migraine and cluster headaches. *Harm Reduction Journal*, *14*(1), 60. https://doi.org/10.1186/s12954-017-0186-6

35. Daniel, J., & Haberman, M. (2017). Clinical potential of psilocybin as a treatment for mental health conditions. *The Mental Health Clinician*, *7*(1), 24–28. https://doi.org/10.9740/mhc.2017.01.024

36. Studies using psilocybin as a potential treatment for PTSD are still underway, but there is much support for its potential. Krediet, E., Bostoen, T., Breeksema, J., van Schagen, A., Passie, T., & Vermetten, E. (2020). Reviewing the Potential of Psychedelics for the Treatment of PTSD. *The International Journal of Neuropsychopharmacology*, *23*(6), 385–400. https://doi.org/10.1093/ijnp/pyaa018

37. Bogenschutz, M. P., Forcehimes, A. A., Pommy, J. A., Wilcox, C. E., Barbosa, P., & Strassman, R. J. (2015). Psilocybin-assisted treatment for alcohol dependence: A proof-of-concept study. *Journal of Psychopharmacology*, *29*(3), 289–299. https://doi.org/10.1177/0269881114565144

38. Johnson, M. W., Garcia-Romeu, A., & Griffiths, R. R. (2017). Long-term follow-up of psilocybin-facilitated smoking cessation. *The American Journal of Drug and Alcohol Abuse*, *43*(1), 55–60. https://doi.org/10.3109/00952990.2016.1170135

39. Carhart-Harris, R. L., Muthukumaraswamy, S., Roseman, L., Kaelen, M., Droog, W., Murphy, K., Tagliazucchi, E., Schenberg, E. E., Nest, T., Orban, C., Leech, R., Williams, L. T., Williams, T. M., Bolstridge, M., Sessa, B., McGonigle, J., Sereno, M. I., Nichols, D., Hellyer, P. J., ... Nutt, D. J. (2016). Neural correlates of the LSD experience revealed by multimodal neuroimaging. *Proceedings of the National Academy of Sciences*, *113*(17), 4853–4858. https://doi.org/10.1073/pnas.1518377113

40. "80-90% of people rate these experiences in the top 5 of their life," says Dr. Roland Griffiths of Johns Hopkins University when asked to reflect on all his research in the psychedelic domain. See: Jordan Peterson. (March 2, 2021). *The Psychology of Psychedelics—Roland Griffiths* (S4 E20) [Video]. https://youtu.be/NGIP-3Q-p_s (52:50 in).

41. The locus of action for all classic psychedelics is that they are agonists of the 5-HT2A receptor. Interestingly though, if this receptor is blocked by ketansarin, the subjective effects are also blocked. "Consistent with these animal studies, the administration of the 5-HT2A receptor antagonist ketanserin abolishes virtually all of the subjective effects of psilocybin, LSD and DMT in humans." Vollenweider, F. X., & Preller, K. H. (2020). Psychedelic drugs: Neurobiology and potential for treatment of psychiatric disorders. *Nature Reviews Neuroscience*, *21*(11), 611–624. https://doi.org/10.1038/s41583-020-0367-2

42. Drake Baer. (2015, January 29). *How Steve Jobs' Acid-Fueled Quest For Enlightenment Made Him The Greatest Product Visionary In History* [News Website]. Business Insider. https://www.businessinsider.com/steve-jobs-lsd-meditation-zen-quest-2015-1

43. Adderall and Ritalin are both stimulant medications typically used to treat those diagnosed with attention-deficit hyperactivity disorder. These stimulants typically have a calming or focusing effect on those with ADHD, although for those who have not been diagnosed with ADHD, the effect is similar to that of caffeine, another CNS stimulant, which typically increases wakefulness, increases focus and attention, and

suppresses appetite. See: National Institute of Drug Abuse. (2014, January 1). *Stimulant ADHD Medications: Methylphenidate and Amphetamines* [Government Health Organization]. National Institute on Drug Abuse. https://www.drugabuse.gov/sites/default/files/drugfacts_stimulantadhd_1.pdf

44. Winkelman, M. (2015). Psychedelics as Medicines for Substance Abuse Rehabilitation: Evaluating Treatments with LSD, Peyote, Ibogaine and Ayahuasca. *Current Drug Abuse Reviews*, *7*(2), 101–116. https://doi.org/10.2174/1874473708666150107120011

45. Carhart-Harris, R. L., Leech, R., Hellyer, P. J., Shanahan, M., Feilding, A., Tagliazucchi, E., Chialvo, D. R., & Nutt, D. (2014). The entropic brain: A theory of conscious states informed by neuroimaging research with psychedelic drugs. *Frontiers in Human Neuroscience*, *8*, 20. https://doi.org/10.3389/fnhum.2014.00020

46. Winkelman, M. (2015). Psychedelics as Medicines for Substance Abuse Rehabilitation: Evaluating Treatments with LSD, Peyote, Ibogaine and Ayahuasca. *Current Drug Abuse Reviews*, *7*(2), 101–116. https://doi.org/10.2174/1874473708666150107120011

47. Iboga - Alper, K. R. (2001). Chapter 1 Ibogaine: A review. In *The Alkaloids: Chemistry and Biology* (Vol. 56, p. 1–38). Elsevier. https://doi.org/10.1016/S0099-9598(01)56005-8

48. Psilocybin - Garcia-Romeu, A., Griffiths, R. R., & Johnson, M. W. (2014). Psilocybin-occasioned mystical experiences in the treatment of tobacco addiction. *Current Drug Abuse Reviews*, *7*(3), 157–164. https://doi.org/10.2174/1874473708666150107121331

49. Ayahuasca - Talin, P., & Sanabria, E. (2017). Ayahuasca's entwined efficacy: An ethnographic study of ritual healing from 'addiction.' *International Journal of Drug Policy*, *44*, 23–30. https://doi.org/10.1016/j.drugpo.2017.02.017

50. "We're wired to have these kinds of experiences," says Dr. Roland Griffiths when asked to reflect on twenty years of psychedelic research at Johns Hopkins University. Jordan Peterson. (March 2, 2021). *The Psychology of Psychedelics—Roland Griffiths* (S4 E20) [Video]. https://youtu.be/NGIP-3Q-p_s (67:45 in).

Chapter 18: Piecing it all together

1. Beckley Foundation. (2016, April 1). The World's First Images of the Brain on LSD [Scientific Institution]. *Beckley in the Press*. https://www.beckleyfoundation.org/the-brain-on-lsd-revealed-first-scans-show-how-the-drug-affects-the-brain/

2. Watts, R., Day, C., Krzanowski, J., Nutt, D., & Carhart-Harris, R. (2017). Patients' Accounts of Increased "Connectedness" and "Acceptance" After Psilocybin for Treatment-Resistant Depression. *Journal of Humanistic Psychology*, *57*(5), 520–564. https://doi.org/10.1177/0022167817709585

3. See *Bones of the Experience* (Glossary)

4. Beckley Foundation. (2016, April 1). The World's First Images of the Brain on LSD [Scientific Institution]. *Beckley in the Press*. https://www.beckleyfoundation.org/the-brain-on-lsd-revealed-first-scans-show-how-the-drug-affects-the-brain/

5. Ibid.

6. Carhart-Harris, R. L., & Friston, K. J. (2010). The default-mode, ego-functions and free-energy: A neurobiological account of Freudian ideas. *Brain*, *133*(4), 1265–1283. https://doi.org/10.1093/brain/awq010

7. Spreng, R. N., & Grady, C. L. (2010). Patterns of Brain Activity Supporting Autobiographical Memory, Prospection, and Theory of Mind, and Their Relationship to the Default Mode Network. *Journal of Cognitive Neuroscience*, *22*(6), 1112–1123. https://doi.org/10.1162/jocn.2009.21282

416

8. Raichle, M. E. (2010, March 1). The Brain's Dark Energy. *Scientific American, 302*(3), 44–49.

9. *Dr. Robin Carhart-Harris: Brain Imaging with Psilocybin and MDMA* (No. 17). (2017, September 28). [Podcast]. from https://podcasts.apple.com/us/podcast/episode-17-dr-robin-carhart-harris-brain-imaging-with/id1217974024?i=1000392843298 (20 min. in)

10. Ibid. (18 min. in).

11. Garrison, K. A., Zeffiro, T. A., Scheinost, D., Constable, R. T., & Brewer, J. A. (2015). Meditation leads to reduced default mode network activity beyond an active task. *Cognitive, Affective, & Behavioral Neuroscience, 15*(3), 712–720. https://doi.org/10.3758/s13415-015-0358-3

12. Berman, M. G., Peltier, S., Nee, D. E., Kross, E., Deldin, P. J., & Jonides, J. (2011). Depression, rumination and the default network. *Social Cognitive and Affective Neuroscience*, 6(5), 548–555. https://doi.org/10.1093/scan/nsq080

13. Posner, J., Cha, J., Wang, Z., Talati, A., Warner, V., Gerber, A., Peterson, B. S., & Weissman, M. (2016). Increased Default Mode Network Connectivity in Individuals at High Familial Risk for Depression. *Neuro-psychopharmacology, 41*(7), 1759–1767. https://doi.org/10.1038/npp.2015.342

14. Garrison, K. A., Zeffiro, T. A., Scheinost, D., Constable, R. T., & Brewer, J. A. (2015). Meditation leads to reduced default mode network activity beyond an active task. *Cognitive, Affective, & Behavioral Neuroscience, 15*(3), 712–720. https://doi.org/10.3758/s13415-015-0358-3

15. Beckley Foundation. (2016, April 1). The World's First Images of the Brain on LSD [Scientific Institution]. *Beckley in the Press.* https://www.beckleyfoundation.org/the-brain-on-lsd-revealed-first-scans-show-how-the-drug-affects-the-brain/

16. Carhart-Harris, R. L., Muthukumaraswamy, S., Roseman, L., Kaelen, M., Droog, W., Murphy, K., Tagliazucchi, E., Schenberg, E. E., Nest, T., Orban, C., Leech, R., Williams, L. T., Williams, T. M., Bolstridge, M., Sessa, B., McGonigle, J., Sereno, M. I., Nichols, D., Hellyer, P. J., ... Nutt, D. J. (2016). Neural correlates of the LSD experience revealed by multimodal neuroimaging. *Proceedings of the National Academy of Sciences, 113*(17), 4853–4858. https://doi.org/10.1073/pnas.1518377113

17. Ibid.

18. Ibid.

19. Atasoy, S., Roseman, L., Kaelen, M., Kringelbach, M. L., Deco, G., & Carhart-Harris, R. L. (2017). Connectome-harmonic decomposition of human brain activity reveals dynamical repertoire re-organization under LSD. *Scientific Reports, 7*(1), 17661. https://doi.org/10.1038/s41598-017-17546-0

20. Ibid.

21. Ibid.

22. Carhart-Harris, R. L., Muthukumaraswamy, S., Roseman, L., Kaelen, M., Droog, W., Murphy, K., Tagliazucchi, E., Schenberg, E. E., Nest, T., Orban, C., Leech, R., Williams, L. T., Williams, T. M., Bolstridge, M., Sessa, B., McGonigle, J., Sereno, M. I., Nichols, D., Hellyer, P. J., ... Nutt, D. J. (2016). Neural correlates of the LSD experience revealed by multimodal neuroimaging. *Proceedings of the National Academy of Sciences, 113*(17), 4853–4858. https://doi.org/10.1073/pnas.1518377113

23. Ibid.

24. Beckley Foundation. (2016, April 1). The World's First Images of the Brain on LSD [Scientific Institution]. *Beckley in the Press*. https://www.beckleyfoundation.org/the-brain-on-lsd-revealed-first-scans-show-how-the-drug-affects-the-brain/

25. Nichols, D. E. (2004). Hallucinogens. *Pharmacology & Therapeutics, 101*(2), 131–181. https://doi.org/10.1016/j.pharmthera.2003.11.002

26. J. Krishnamurti. (1979). *Awareness of inattention is attention* [Public Speech]. https://youtu.be/3VrN45mg8gI (6:20 in).

27. Kaiser, R. H., Andrews-Hanna, J. R., Wager, T. D., & Pizzagalli, D. A. (2015). Large-Scale Network Dysfunction in Major Depressive Disorder: A Meta-analysis of Resting-State Functional Connectivity. *JAMA Psychiatry, 72*(6), 603. https://doi.org/10.1001/jamapsychiatry.2015.0071

28. Scult, M. A., Fresco, D. M., Gunning, F. M., Liston, C., Seeley, S. H., García, E., & Mennin, D. S. (2019). Changes in Functional Connectivity Following Treatment With Emotion Regulation Therapy. *Frontiers in Behavioral Neuroscience, 13*, 10. https://doi.org/10.3389/fnbeh.2019.00010

29. Posner, J., Cha, J., Wang, Z., Talati, A., Warner, V., Gerber, A., Peterson, B. S., & Weissman, M. (2016). Increased Default Mode Network Connectivity in Individuals at High Familial Risk for Depression. *Neuropsychopharmacology, 41*(7), 1759–1767. https://doi.org/10.1038/npp.2015.342

30. Csikszentmihalyi, M. (February, 2004). *Flow, the secret to happiness* [Video]. TED Conferences. https://www.ted.com/talks/mihaly_csikszentmihalyi_flow_the_secret_to_happiness

31. Ulrich, M., Keller, J., Hoenig, K., Waller, C., & Grön, G. (2014). Neural correlates of experimentally induced flow experiences. *NeuroImage, 86*, 194–202. https://doi.org/10.1016/j.neuroimage.2013.08.019

32. Shapiro, S., Siegel, R., & Neff, K. D. (2018). Paradoxes of Mindfulness. *Mindfulness, 9*(6), 1693–1701. https://doi.org/10.1007/s12671-018-0957-5

33. Magic mushrooms & Reindeer—Weird Nature—BBC animals. (2009, January 26). [Documentary]. In *Weird Nature*. BBC Four. https://youtu.be/MkCS9ePWuLU

34. Siegel, R. K. (2005). *Intoxication: The universal drive for mind-altering substances*. Park Street Press. (p. 65).

35. Hockings et al. documents chimpanzees using tools made from leaves to scoop out naturally occurring palm wine. However, it is also of note that at least three species of bird have been seen going after this natural sweet beverage. Hockings, K. J., Bryson-Morrison, N., Carvalho, S., Fujisawa, M., Humle, T., McGrew, W. C., Nakamura, M., Ohashi, G., Yamanashi, Y., Yamakoshi, G., & Matsuzawa, T. (2015). Tools to tipple: Ethanol ingestion by wild chimpanzees using leaf-sponges. *Royal Society Open Science, 2*(6), 150150. https://doi.org/10.1098/rsos.150150

36. Birds tapping palm wine, see: Gutiérrez, J. S., Catry, T., & Granadeiro, J. P. (2020). Human facilitation of sap-feeding birds in the Bijagós archipelago, West Africa. *Ibis, 162*(1), 250–254. https://doi.org/10.1111/ibi.12790

37. Janiak, M. C., Pinto, S. L., Duytschaever, G., Carrigan, M. A., & Melin, A. D. (2020). Genetic evidence of widespread variation in ethanol metabolism among mammals: Revisiting the 'myth' of natural intoxication. *Biology Letters, 16*(4), 20200070. https://doi.org/10.1098/rsbl.2020.0070

38. Siegel, R. K. (2005). *Intoxication: The universal drive for mind-altering substances*. Park Street Press. (p.103-5).

418

39. Ibid. (p. 64).
40. Samorini, G. (2002). *Animals and psychedelics: The natural world and the instinct to alter consciousness.* Park Street Press. (p. 76-77).
41. BBC. (2017, February 1). Lemurs get high—Spy in the Wild—BBC (No. 4) [Documentary]. In *Spy in the Wild.* BBC One. https://youtu.be/yYXoCHLqr4o
42. Dolphins Play Catch with a Pufferfish! - Spy In The Wild—BBC Earth (No. 2). (2019, September 19). [Documentary]. In *Spy In The Pod.* BBC One. https://youtu.be/0T5aGLybXEs
43. Siegel, R. K. (2005). *Intoxication: The universal drive for mind-altering substances.* Park Street Press.
44. David Nichols. (2010, August 5). *David E. Nichols Interviewed by Jan Irvin (2010) [5/7]* [Audio]. https://youtu.be/Ih_Yqge2cU0 (9:30 in).

Chapter 19: Experiences.

1. Lebedev, A. V., Kaelen, M., Lövdén, M., Nilsson, J., Feilding, A., Nutt, D. J., & Carhart-Harris, R. L. (2016). LSD-induced entropic brain activity predicts subsequent personality change: LSD-Induced Entropic Brain Activity. *Human Brain Mapping, 37*(9), 3203–3213. https://doi.org/10.1002/hbm.23234
2. Smoking cessation addiction studies - "No significant differences in general intensity of drug effects were found between groups, suggesting that mystical-type subjective effects, rather than overall intensity of drug effects, were responsible for smoking cessation." Garcia-Romeu, A., Griffiths, R. R., & Johnson, M. W. (2014). Psilocybin-occasioned mystical experiences in the treatment of tobacco addiction. *Current Drug Abuse Reviews, 7*(3), 157–164. https://doi.org/10.2174/1874473708666150107121331
3. The mystical-type experience produces "sustained positive changes in attitudes and behavior that were consistent with changes rated by friends and family." - Griffiths, R. R., Richards, W. A., McCann, U., & Jesse, R. (2006). Psilocybin can occasion mystical-type experiences having substantial and sustained personal meaning and spiritual significance. *Psychopharmacology, 187*(3), 268–283. https://doi.org/10.1007/s00213-006-0457-5
4. Anxiety and depression studies - Griffiths, R. R., Johnson, M. W., Carducci, M. A., Umbricht, A., Richards, W. A., Richards, B. D., Cosimano, M. P., & Klinedinst, M. A. (2016). Psilocybin produces substantial and sustained decreases in depression and anxiety in patients with life-threatening cancer: A randomized double-blind trial. *Journal of Psychopharmacology, 30*(12), 1181–1197. https://doi.org/10.1177/0269881116675513
5. Depression studies - "Four separate trials have reported improvements in depressive symptoms after psilocybin-assisted psychotherapy (Griffiths et al. 2016; Ross et al. 2016; Grob et al. 2011; Carhart-Harris et al. 2016), including one in which 'treatment-resistant depression' was the primary criterion for inclusion (Carhart-Harris et al. 2016)." See: Carhart-Harris, R. L., Bolstridge, M., Day, C. M. J., Rucker, J., Watts, R., Erritzoe, D. E., Kaelen, M., Giribaldi, B., Bloomfield, M., Pilling, S., Rickard, J. A., Forbes, B., Feilding, A., Taylor, D., Curran, H. V., & Nutt, D. J. (2018). Psilocybin with psychological support for treatment-resistant depression: Six-month follow-up. *Psychopharmacology, 235*(2), 399–408. https://doi.org/10.1007/s00213-017-4771-x
6. About the mystical-type experience being the key component to the maximization of healing, the authors of this paper were sharp to notice that there are degrees of

mystical experience, and that even 'incomplete' mystical experiences were of benefit. Gasser, P., Kirchner, K., & Passie, T. (2015). LSD-assisted psychotherapy for anxiety associated with a life-threatening disease: A qualitative study of acute and sustained subjective effects. *Journal of Psychopharmacology, 29*(1), 57–68. https://doi.org/10.1177/0269881114555249

7. This entire podcast speaks to the general point, but two sections of note occur at 37 to 39 min in, and at 54 min in. Jordan Peterson. (March 2, 2021). *The Psychology of Psychedelics—Roland Griffiths* (S4 E20) [Video]. https://youtu.be/NGIP-3Q-p_s (54:30 in).

8. Foody, G. M., & Atkinson, P. M. (Eds.). (2002). *Uncertainty in remote sensing and GIS.* J. Wiley. (p. 279).

9. Haden, M. (2017, November). *Psychedelics: Past, present, and future* [Video]. TEDx Conferences. https://youtu.be/JI1dwVsPw2E (6:40 in)

10. This blog post from a ketamine treatment facility in Florida gets at the point of ego death rather succinctly. Xavier Francuski. (2018, November 5). Why We Strive For Ego Death With Psychedelics [Treatment Facility Website]. *Ketamine for Anxiety.* https://revitalizinginfusions.com/why-we-strive-for-ego-death-with-psychedelics/

11. Carhart-Harris, R. L., & Friston, K. J. (2019). REBUS and the Anarchic Brain: Toward a Unified Model of the Brain Action of Psychedelics. *Pharmacological Reviews, 71*(3), 316–344. https://doi.org/10.1124/pr.118.017160

Chapter 20: DEMP Cont'd

1. Wheal, J., & Kotler, S. (2017). *Stealing Fire: How silicon valley, the navy SEALs and maverick scientists are revolutionizing the way we live and work.* Dey Street Books.

2. Dietrich, A. (2011, November). *Surfing the Stream of Consciousness: Tales from the Hallucination Zone* [Video]. TEDxBeirut. https://youtu.be/syfalikXBLA (10:20 min. in)

3. Nichols, D. E. (2016). Psychedelics. *Pharmacological Reviews, 68*(2), 264–355. https://doi.org/10.1124/pr.115.011478

4. Aghajanian, G. K. (1980). Mescaline and LSD facilitate the activation of locus coeruleus neurons by peripheral stimuli. *Brain Research, 186*(2), 492–498. https://doi.org/10.1016/0006-8993(80)90997-X

5. Preller, K. H., Razi, A., Zeidman, P., Stämpfli, P., Friston, K. J., & Vollenweider, F. X. (2019). Effective connectivity changes in LSD-induced altered states of consciousness in humans. *Proceedings of the National Academy of Sciences, 116*(7), 2743–2748. https://doi.org/10.1073/pnas.1815129116

6. *The Master and His Emissary: Conversation with Dr. Iain McGilchrist* (Dr. Jordan Peterson, Interviewer). (2018, February 17). [Video Interview]. https://youtu.be/xtf4F-DlpPZ8 (2:50 in).

7. There are multiple attributions for this quote from Margaret Thatcher to Mahatma Gandhi, although Quote investigator attributes the most modern incarnation to Frank Outlaw. That withstanding, there are versions of this quote going back as far back as 1856. See: Quote Investigator. (2013, January 10). *Watch Your Thoughts, They Become Words; Watch Your Words, They Become Actions* [Quote Tracer]. Quote Investigator. https://quoteinvestigator.com/2013/01/10/watch-your-thoughts/

8. *LSD-induced entropic brain activity predicts subsequent personality change.* (2016). [Research Think Tank]. The Beckley Foundation. https://www.beckleyfoundation.org/

resource/lsd-induced-entropic-brain-activity-predicts-subsequent-personality-change/ See also: Lebedev, A. V., Kaelen, M., Lövdén, M., Nilsson, J., Feilding, A., Nutt, D. J., & Carhart-Harris, R. L. (2016). LSD-induced entropic brain activity predicts subsequent personality change: LSD-Induced Entropic Brain Activity. *Human Brain Mapping*, *37*(9), 3203–3213. https://doi.org/10.1002/hbm.23234

9. Increases in Brain-derived neurotrophic factor, see: Hutten, N. R. P. W., Mason, N. L., Dolder, P. C., Theunissen, E. L., Holze, F., Liechti, M. E., Varghese, N., Eckert, A., Feilding, A., Ramaekers, J. G., & Kuypers, K. P. C. (2021). Low Doses of LSD Acutely Increase BDNF Blood Plasma Levels in Healthy Volunteers. *ACS Pharmacology & Translational Science*, *4*(2), 461–466. https://doi.org/10.1021/acsptsci.0c00099

10. Neurogenesis - Catlow, B. J., Jalloh, A., & Sanchez-Ramos, J. (2016). Hippocampal Neurogenesis. In *Neuropathology of Drug Addictions and Substance Misuse* (pp. 821–831). Elsevier. https://doi.org/10.1016/B978-0-12-800212-4.00077-7

11. Synaptogenesis - Ly, C., Greb, A. C., Cameron, L. P., Wong, J. M., Barragan, E. V., Wilson, P. C., Burbach, K. F., Soltanzadeh Zarandi, S., Sood, A., Paddy, M. R., Duim, W. C., Dennis, M. Y., McAllister, A. K., Ori-McKenney, K. M., Gray, J. A., & Olson, D. E. (2018). Psychedelics Promote Structural and Functional Neural Plasticity. *Cell Reports*, *23*(11), 3170–3182. https://doi.org/10.1016/j.celrep.2018.05.022

12. David Nutt. (2017, April 26). *David Nutt: Psychedelic Research, From Brain Imaging to Policy Reform*. Psychedelic Science 2017, Oakland, Ca. https://youtu.be/ZzepSK-6Gzk8 (21:20 in).

13. Even Carhart-Harris et al., have described it as such with their "entropic brain hypothesis." Carhart-Harris, R. L., Leech, R., Hellyer, P. J., Shanahan, M., Feilding, A., Tagliazucchi, E., Chialvo, D. R., & Nutt, D. (2014). The entropic brain: A theory of conscious states informed by neuroimaging research with psychedelic drugs. *Frontiers in Human Neuroscience*, *8*, 20. https://doi.org/10.3389/fnhum.2014.00020

14. Certainties dissolved - Karl Friston and Dr. Robin Carhart-Harris have co-authored a paper, in which, they argue for psychedelics having the ability to "relax high-level priors" and revise "entrenched pathological priors." See: Carhart-Harris, R. L., & Friston, K. J. (2019). REBUS and the Anarchic Brain: Toward a Unified Model of the Brain Action of Psychedelics. *Pharmacological Reviews*, *71*(3), 316–344. https://doi.org/10.1124/pr.118.017160

Chapter 21: The Chair Incident

1. Nichols, D. E., & Grob, C. S. (2018). Is LSD toxic? *Forensic Science International*, *284*, 141–145. https://doi.org/10.1016/j.forsciint.2018.01.006

2. David Eagleman. (2016, October 4). *The Brain and The Now—David Eagleman* [Keynote address]. The Long Now Member Summit, San Francisco. https://youtu.be/vv_e99qbJ4U (6 min in).

3. Stetson, C., Fiesta, M. P., & Eagleman, D. M. (2007). Does Time Really Slow Down during a Frightening Event? *PLoS ONE*, *2*(12), e1295. https://doi.org/10.1371/journal.pone.0001295

4. David Eagleman. (2016, October 4). *The Brain and The Now—David Eagleman* [Keynote address]. The Long Now Member Summit, San Francisco. https://youtu.be/vv_e99qbJ4U (36 min in).

5. David Eagleman. (2009). Brain Time [Author's Website]. *Latest*. https://eagleman.com/latest/brain-time/

6. Wolpert, D. (2011, July). *The real reason for brains* [Video]. TED Conferences. https://www.ted.com/talks/daniel_wolpert_the_real_reason_for_brains

Chapter 22 : Koreatown Revisited

1. Preller, K. H., Razi, A., Zeidman, P., Stämpfli, P., Friston, K. J., & Vollenweider, F. X. (2019). Effective connectivity changes in LSD-induced altered states of consciousness in humans. *Proceedings of the National Academy of Sciences, 116*(7), 2743–2748. https://doi.org/10.1073/pnas.1815129116

2. The brain can process and identify images seen for as little as 13 milliseconds, roughly equating to us being able to see at about 75fps. See: Anne Trafton. (2014, January 16). In the blink of an eye. *MIT News*. https://news.mit.edu/2014/in-the-blink-of-an-eye-0116

3. Carhart-Harris, R. L., Muthukumaraswamy, S., Roseman, L., Kaelen, M., Droog, W., Murphy, K., Tagliazucchi, E., Schenberg, E. E., Nest, T., Orban, C., Leech, R., Williams, L. T., Williams, T. M., Bolstridge, M., Sessa, B., McGonigle, J., Sereno, M. I., Nichols, D., Hellyer, P. J., ... Nutt, D. J. (2016). Neural correlates of the LSD experience revealed by multimodal neuroimaging. *Proceedings of the National Academy of Sciences, 113*(17), 4853–4858. https://doi.org/10.1073/pnas.1518377113

4. David Biello. (2008, March 20). Self-Experimenters: Psychedelic Chemist Explores the Surreality of Inner Space, One Drug at a Time. *Scientific American*. https://www.scientificamerican.com/article/self-experimenter-chemist-explores-new-psychedelics/

5. Huxley, A. (2009). The doors of perception. Harper Perennial. (p. 27,34).

6. Carhart-Harris, R. L. (2019). How do psychedelics work? *Current Opinion in Psychiatry, 32*(1), 16–21. https://doi.org/10.1097/YCO.0000000000000467

7. "The visual modality is arguably the most developed in the primate and occupies the largest amount of real estate: approximately 50% of cerebral cortex in macaque and 20–30% in humans is devoted to visual processing." In: Sheth, B. R., & Young, R. (2016). Two Visual Pathways in Primates Based on Sampling of Space: Exploitation and Exploration of Visual Information. *Frontiers in Integrative Neuroscience, 10*. https://doi.org/10.3389/fnint.2016.00037. See also: Werner, J. S., & Chalupa, L. M. (Eds.). (2004). *The visual neurosciences* (Vol. 1). MIT Press. (p. 513).

8. David Eagleman. (2016, October 4). *The Brain and The Now—David Eagleman* [Keynote address]. The Long Now Member Summit, San Francisco. https://youtu.be/vv_e99qbJ4U (14 min in.)

9. Dennis, M. Aaron (2021, December 26). *Rodney Brooks. Encyclopedia Britannica.* https://www.britannica.com/biography/Rodney-Allen-Brooks

10. David Eagleman. (2016, October 4). *The Brain and The Now—David Eagleman* [Keynote address]. The Long Now Member Summit, San Francisco. https://youtu.be/vv_e99qbJ4U (15 min. in).

11. Pamela Caragol. (2009, November 3). Inside LSD [Documentary]. In *National Geographic Explorer*. National Geographic. https://topdocumentaryfilms.com/inside-lsd/ (38:50 in).

12. Krebs, R. M., Park, H. R. P., Bombeke, K., & Boehler, C. N. (2018). Modulation of locus coeruleus activity by novel oddball stimuli. *Brain Imaging and Behavior, 12*(2), 577–584. https://doi.org/10.1007/s11682-017-9700-4

13. Tse, P. U., Intriligator, J., Rivest, J., & Cavanagh, P. (2004). Attention and the subjective expansion of time. *Perception & Psychophysics, 66*(7), 1171–1189. https://doi.org/10.3758/BF03196844

14. NatGeo. (2014, June 19). *Brain Games—Time Perception (Oddball Effect)* [Documentary; Video File]. National Geographic. https://youtu.be/W7uLwUHuxRM

15. Eagleman's Texas Tower - Stetson, C., Fiesta, M. P., & Eagleman, D. M. (2007). Does Time Really Slow Down during a Frightening Event? *PLoS ONE, 2*(12), e1295. https://doi.org/10.1371/journal.pone.0001295

16. Zago, L., Fenske, M. J., Aminoff, E., & Bar, M. (2005). The Rise and Fall of Priming: How Visual Exposure Shapes Cortical Representations of Objects. *Cerebral Cortex, 15*(11), 1655–1665. https://doi.org/10.1093/cercor/bhi060

Chapter 23: Synesthesia

1. James Wannerton. (2020). *James Wannerton—Synesthete* [Artist website]. https://jameswannerton.com/about/

2. Sakai, J. (2020). Core Concept: How synaptic pruning shapes neural wiring during development and, possibly, in disease. *Proceedings of the National Academy of Sciences, 117*(28), 16096–16099. https://doi.org/10.1073/pnas.2010281117

3. Dan Clifton, Catherine Gale, Johanna Woolford Gibbon. (2015, October 21). What Makes Me? (No. 2) [Documentary]. In *The Brain with David Eagleman*. PBS. https://www.pbs.org/video/brain-david-eagleman-what-makes-me_ep2/

4. AMNH. (2012, May 21). *Mass Extinction Events* [Museum Website]. American Museum of Natural History. https://www.amnh.org/exhibitions/dinosaurs-ancient-fossils/extinction/mass-extinction

5. Brang, D., & Ramachandran, V. S. (2011). Survival of the Synesthesia Gene: Why Do People Hear Colors and Taste Words? *PLoS Biology, 9*(11), e1001205. https://doi.org/10.1371/journal.pbio.1001205

6. Chrissie Giles. (2017, October 16). What it's like to have synaesthesia: Meet the man who can taste sounds. *Independent*. https://www.independent.co.uk/news/long_reads/synaesthesia-sound-taste-health-science-brain-a7996766.html

7. Harman, W. W., McKim, R. H., Mogar, R. E., Fadiman, J., & Stolaroff, M. J. (1966). Psychedelic Agents in Creative Problem-Solving: A Pilot Study. *Psychological Reports, 19*(1), 211–227. https://doi.org/10.2466/pr0.1966.19.1.211

8. Scott, G., & Carhart-Harris, R. L. (2019). Psychedelics as a treatment for disorders of consciousness. *Neuroscience of Consciousness, 2019*(1). https://doi.org/10.1093/nc/niz003

Chapter 24: Synesthesia Cont'd

1. Grossenbacher, P. G., & Lovelace, C. T. (2001). Mechanisms of synesthesia: Cognitive and physiological constraints. *Trends in Cognitive Sciences, 5*(1), 36–41. https://doi.org/10.1016/S1364-6613(00)01571-0

2. Dana Smith. (2013, December 4). Can Synesthesia in Autism Lead to Savantism? [Magazine Website]. *Scientific American (MIND Guest Blog)*. https://blogs.scientificamerican.com/mind-guest-blog/can-synesthesia-in-autism-lead-to-savantism/

3. Peiffer-Smadja, N., & Cohen, L. (2019). The cerebral bases of the bouba-kiki effect. *NeuroImage, 186*, 679–689. https://doi.org/10.1016/j.neuroimage.2018.11.033

4. David Eagleman. (2009, June). *Synesthesia: Hearing colours, tasting sounds* [Chast Lecture]. https://www.youtube.com/watch?v=nvCw-H8h6E4 (14:30 in).

Chapter 25: Off the Rails

1. Watts, R., Day, C., Krzanowski, J., Nutt, D., & Carhart-Harris, R. (2017). Patients' Accounts of Increased "Connectedness" and "Acceptance" After Psilocybin for Treatment-Resistant Depression. *Journal of Humanistic Psychology, 57*(5), 520–564. https://doi.org/10.1177/0022167817709585

2. Erowid. (2012, September 12). *25I-NBOMe (2C-I-NBOMe) Fatalities / Deaths* [Nonprofit research and educational organization]. Erowid Center. https://erowid.org/chemicals/2ci_nbome/2ci_nbome_death.shtml

3. World Health Organization. (2014). *25I-NBOMe* [Critical Review Report]. W.H.O. https://www.who.int/medicines/areas/quality_safety/4_19_review.pdf

4. Gee, P., Schep, L. J., Jensen, B. P., Moore, G., & Barrington, S. (2016). Case series: Toxicity from 25B-NBOMe – a cluster of N-bomb cases. *Clinical Toxicology, 54*(2), 141–146. https://doi.org/10.3109/15563650.2015.1115056

5. Wood, D. M., Sedefov, R., Cunningham, A., & Dargan, P. I. (2015). Prevalence of use and acute toxicity associated with the use of NBOMe drugs. *Clinical Toxicology, 53*(2), 85–92. https://doi.org/10.3109/15563650.2015.1004179

6. Watts, A. (2017). *Out of your mind: Tricksters, interdependence, and the cosmic game of hide-and-seek.* Sounds True, Inc. (p. 24).

Chapter 26: What Really Happened

1. Nichols, D. E., & Grob, C. S. (2018). Is LSD toxic? *Forensic Science International, 284*, 141–145. https://doi.org/10.1016/j.forsciint.2018.01.006

2. Dan Hooper. (2020, February 10). *What Happened At The Beginning Of Time? - With Dan Hooper* [Royal Institution Lecture]. https://www.youtube.com/watch?v=dB7d89-YHjM

3. Krishnamurti, J. (1987). *The awakening of intelligence* (1st Harper & Row pbk. ed). Harper & Row. (p. 94).

4. "Genes are immortal..." in: Richard Dawkins. (2013, September 25). An Appetite for Wonder: Richard Dawkins in Conversation with Adam Rutherford [Video Interview]. https://www.youtube.com/watch?v=omsUZ3u5TX4 (48 min in).

5. "The genes that survive..." in: Russell Barnes, Dan Hillman. (2008, August 11). Richard Dawkins Presents: The Genius of Charles Darwin (Part 2: The Fifth Ape) [Biology; Documentary]. Channel 4. https://youtu.be/xuCfju7JN_4 (35:20 in).

6. Richard Dawkins. (2013, September 25). An Appetite for Wonder: Richard Dawkins in Conversation with Adam Rutherford [Video Interview]. https://www.youtube.com/watch?v=omsUZ3u5TX4 (49:15 min in).

7. Dawkins, R. (2006). *The selfish gene* (30th anniversary ed). Oxford University Press. (p. 21).

8. The cdc2 gene, which was originally discovered by Sir Paul Nurse, is a gene that controls cell division in fission yeast. Sir Paul Nurse was awarded the Nobel prize for this discovery, because apparently, this cdc2 gene is also found in humans. To put that in perspective, that means the gene for controlling cell division has been conserved for well over a billion years, in light of the fact that humans and yeasts shared a common ancestor between 1.2 and 1.5 billion years ago. Nurse, P. (2021). *What is life? Five great ideas in biology* (First American edition). W.W. Norton & Company. (p. 50-51).

9. While the original "out of Africa" hypothesis of human migration may have been a little too simple (positing one single giant migration occurring about 60,000 years ago) recent archaeological finds are more in support of *multiple* migrations out of Africa. Some of which are believed to have occurred as early as 194-250k years ago. For early migrations, see: Beyer, R. M., Krapp, M., Eriksson, A., & Manica, A. (2021). Climatic windows for human migration out of Africa in the past 300,000 years. *Nature Communications, 12*(1), 4889. https://doi.org/10.1038/s41467-021-24779-1

10. For Multiple dispersal support, see also: Bae, C. J., Douka, K., & Petraglia, M. D. (2017). On the origin of modern humans: Asian perspectives. *Science*, 358(6368), eaai9067. https://doi.org/10.1126/science.aai9067

11. For evidence pointing towards a major dispersal around 65k years ago, see: Malaspinas, A.-S., Westaway, M. C., Muller, C., Sousa, V. C., Lao, O., Alves, I., Bergström, A., Athanasiadis, G., Cheng, J. Y., Crawford, J. E., Heupink, T. H., Macholdt, E., Peischl, S., Rasmussen, S., Schiffels, S., Subramanian, S., Wright, J. L., Albrechtsen, A., Barbieri, C., … Willerslev, E. (2016). A genomic history of Aboriginal Australia. *Nature, 538*(7624), 207–214. https://doi.org/10.1038/nature18299

12. For additional evidence regarding a clear dispersal of modern humans from southern to eastern Africa 60-70k years ago, see: Rito, T., Vieira, D., Silva, M., Conde-Sousa, E., Pereira, L., Mellars, P., Richards, M. B., & Soares, P. (2019). A dispersal of Homo sapiens from southern to eastern Africa immediately preceded the out-of-Africa migration. *Scientific Reports, 9*(1), 4728. https://doi.org/10.1038/s41598-019-41176-3

13. Amos, W., & Hoffman, J. I. (2010). Evidence that two main bottleneck events shaped modern human genetic diversity. *Proceedings of the Royal Society B: Biological Sciences, 277*(1678), 131–137. https://doi.org/10.1098/rspb.2009.1473

14. Geneticist/anthropologist Spencer Wells describes a near-extinction event which happened about 70,000 years ago, where the total number of human beings alive dropped down to as few as 2,000 individuals. Wells, Spencer. (2014, May 5). *The human journey—a genetic odyssey* [Video]. TEDx Talks. https://youtu.be/xnbxrDGZoBQ (7 min in).

15. University of Copenhagen. (2008, January 31). Blue-eyed humans have a single, common ancestor. *ScienceDaily*. Retrieved January 16, 2022 from www.sciencedaily.com/releases/2008/01/080130170343.htm

16. Krista Conrad. (2020). Countries With The Most Blue-Eyed People—World Facts. In *World Atlas*. https://www.worldatlas.com/articles/countries-with-the-most-blue-eyed-people.html

17. National Human Genome Research Institute. (2018, September 7). *Genetics vs. Genomics Fact Sheet* [Research Organization]. Genome.Gov. https://www.genome.gov/about-genomics/fact-sheets/Genetics-vs-Genomics

18. Seaman, J., & Buggs, R. J. A. (2020). FluentDNA: Nucleotide Visualization of Whole Genomes, Annotations, and Alignments. *Frontiers in Genetics, 11*, 292. https://doi.org/10.3389/fgene.2020.00292

19. Suntsova, M. V., & Buzdin, A. A. (2020). Differences between human and chimpanzee genomes and their implications in gene expression, protein functions and biochemical properties of the two species. *BMC Genomics, 21*(S7), 535. https://doi.org/10.1186/s12864-020-06962-8

20. National Human Genome Research Institute. (2010, July 23). *Why Mouse Matters* [Research Organization]. Genome.Gov. https://www.genome.gov/10001345/importance-of-mouse-genome

21. Nour, M. M., Evans, L., Nutt, D., & Carhart-Harris, R. L. (2016). Ego-Dissolution and Psychedelics: Validation of the Ego-Dissolution Inventory (EDI). *Frontiers in Human Neuroscience, 10*. https://doi.org/10.3389/fnhum.2016.00269

22. Pahnke, W. N. (1969). The Psychedelic Mystical Experience in the Human Encounter with Death. *Harvard Theological Review, 62*(1), 1–21. https://doi.org/10.1017/S0017816000027577

23. End Well. (2018, February 28). Transcendence Through Psilocybin | Anthony Bossis, PhD. https://www.youtube.com/watch?v=jCf3h-F7apM

24. Muraresku, B. (2020). The immortality key: The secret history of the religion with no name (First edition). St. Martin's Press. (p. xv, 5-6, and the whole book really).

25. Good Friday Experiment: Pahnke WN. *Thesis presented to the President and Fellows of Harvard University for the Ph.D. in Religion and Society.* 1963. Drugs and mysticism: An analysis of the relationship between psychedelic drugs and the mystical consciousness.

26. Pahnke, W. N. (1967). LSD and religious experience. *LSD man & society. Wesleyan University Press, Middletown, CT*, 60-85.

27. Griffiths, R. R., Richards, W. A., McCann, U., & Jesse, R. (2006). Psilocybin can occasion mystical-type experiences having substantial and sustained personal meaning and spiritual significance. *Psychopharmacology, 187*(3), 268–283. https://doi.org/10.1007/s00213-006-0457-5

28. Griffiths, R., Richards, W., Johnson, M., McCann, U., & Jesse, R. (2008). Mystical-type experiences occasioned by psilocybin mediate the attribution of personal meaning and spiritual significance 14 months later. *Journal of Psychopharmacology, 22*(6), 621–632. https://doi.org/10.1177/0269881108094300

Chapter 27: Gold, Diamonds and CS5

1. UNC-TV Science. (2013). *Make your own Diamond | UNC-TV: Science* [Science Education]. Science.UNCTV.Org. http://science.unctv.org/content/make-your-own-diamond

2. Bahcall, N. A. (2015). Hubble's Law and the expanding universe. *Proceedings of the National Academy of Sciences, 112*(11), 3173–3175. https://doi.org/10.1073/pnas.1424299112

3. deGrasse Tyson, N. & cloudLibrary. (2017). *Astrophysics for people in a hurry.* W. W. Norton & Company. (p. 17).

4. Christian, D. (2011, March). *The history of our world in 18 minutes* [Video]. TED Conferences. https://www.ted.com/talks/david_christian_the_history_of_our_world_in_18_minutes

5. Howell, E., & published, D. D. (2018, August 23). *What is the cosmic microwave background?* [Astronomy News]. Space.Com. https://www.space.com/33892-cosmic-microwave-background.html

6. deGrasse Tyson, N. & cloudLibrary. (2017). *Astrophysics for people in a hurry.* W. W. Norton & Company. (p. 18-27).

7. Christian, D. (2011, March). *The history of our world in 18 minutes* [Video]. TED Conferences. https://www.ted.com/talks/david_christian_the_history_of_our_world_in_18_minutes

8. There is evidence that at least *some portion* of extremely heavy elements like gold, silver, and platinum are forged whenever two dense neutron stars spiral into one

another and merge. In either case, however (rare supernovae explosions or neutron star mergers) the underlying cataclysmic event is still quite rare. *Remember, neutron stars are only formed when massive stars run out of fuel and collapse. For more on this, see: Croswell, K. (2021). News Feature: Tracing gold's cosmic origin story. *Proceedings of the National Academy of Sciences, 118*(4), e2026110118. https://doi.org/10.1073/pnas.2026110118. For a more in-depth review, see: Cowan, J. J., Sneden, C., Lawler, J. E., Aprahamian, A., Wiescher, M., Langanke, K., Martínez-Pinedo, G., & Thielemann, F.-K. (2021). Origin of the heaviest elements: The rapid neutron-capture process. *Reviews of Modern Physics, 93*(1), 015002. https://doi.org/10.1103/RevModPhys.93.015002

9. *Our Solar System.* (2021, August 30). [Space Agency]. Science.Nasa.Gov. https://solarsystem.nasa.gov/solar-system/our-solar-system/in-depth

10. Nature.com. (2014). *Introduction: What is DNA? | Learn Science at Scitable* [Science Magazine]. https://www.nature.com/scitable/topicpage/introduction-what-is-dna-6579978/

11. Gibson, B., Wilson, D. J., Feil, E., & Eyre-Walker, A. (2018). The distribution of bacterial doubling times in the wild. *Proceedings of the Royal Society B: Biological Sciences, 285*(1880), 20180789. https://doi.org/10.1098/rspb.2018.0789

12. *E. coli – the biotech bacterium.* (2014, March 25). [Science Education]. Science Learning Hub. https://www.sciencelearn.org.nz/resources/1899-e-coli-the-biotech-bacterium

13. Christian, D. (2011, March). *The history of our world in 18 minutes* [Video]. https://www.ted.com/talks/david_christian_the_history_of_our_world_in_18_minutes (10 min in).

14. *Mass extinction facts and information from National Geographic.* (2019, September 26). Science. https://www.nationalgeographic.com/science/article/mass-extinction

15. Barnosky et al. only cite 99% but the general point is still the same (i.e., *death* is the ultimate driver of evolution). And without all the evolutionary dead-ends that came before us, we could not be here now. Barnosky, A. D., Matzke, N., Tomiya, S., Wogan, G. O. U., Swartz, B., Quental, T. B., Marshall, C., McGuire, J. L., Lindsey, E. L., Maguire, K. C., Mersey, B., & Ferrer, E. A. (2011). Has the Earth's sixth mass extinction already arrived? *Nature, 471*(7336), 51–57. https://doi.org/10.1038/nature09678

16. Breithaupt, H. (2012). The science of sex. *EMBO Reports, 13*(5), 394–394. https://doi.org/10.1038/embor.2012.45

17. Dawkins, R. (2006). *The selfish gene* (30th anniversary ed). Oxford University Press. (p. 43).

18. Bar-On, Y. M., Phillips, R., & Milo, R. (2018). The biomass distribution on Earth. *Proceedings of the National Academy of Sciences, 115*(25), 6506–6511. https://doi.org/10.1073/pnas.1711842115

19. Pando is still considered to be the largest organism ever discovered (by mass). Although, there is another contender that is of great interest; the gigantic honey mushroom *Armillaria ostoyae* which was discovered in Oregon. See: Krulwich, R. (2014, May 8). A Question Of Biggitude: What's The Largest Creature On Earth? *NPR.* https://www.npr.org/sections/krulwich/2014/05/08/310259300/a-question-of-biggitude-what-s-the-largest-creature-on-earth

20. Katz, B. (2018, October 18). *Pando, One of the World's Largest Organisms, Is Dying.* Smithsonian Magazine. https://www.smithsonianmag.com/smart-news/pano-one-worlds-largest-organisms-dying-180970579/

21. Weiss, M. C., Sousa, F. L., Mrnjavac, N., Neukirchen, S., Roettger, M., Nelson-Sathi, S., & Martin, W. F. (2016). The physiology and habitat of the last universal common ancestor. *Nature Microbiology, 1*(9), 16116. https://doi.org/10.1038/nmicrobiol.2016.116

22. When referencing Mitochondrial Eve (mt-Eve) it's important to clarify that Mitochondrial Eve is not the first female of a species, but rather the most recent common ancestor all humans can point to. For a more in-depth discussion about what mt-Eve means, see: Learn, J. R. (2016, June 28). *No, a Mitochondrial "Eve" Is Not the First Female in a Species.* Smithsonian Magazine. https://www.smithsonianmag.com/science-nature/no-mitochondrial-eve-not-first-female-species-180959593/. For a more in-depth discussion about mt-Eve and the age of her lineage "L0", see: Brandon Spektor. (2019, October 28). *Scientists Think They've Found "Mitochondrial Eve's" First Homeland | Live Science* [Science Magazine]. LiveScience. https://www.livescience.com/mitochondrial-eve-first-human-homeland.html

23. In listening to nearly the entire Alan Watts archive, I have pulled the quote featured, although I cannot seem to find the original lecture where he said it in this one exact way. Regardless, he has made this same point again and again, and in a multitude of different ways, just as he does here: Alan Watts. (2019, December 18). 2.3.1 Introduction to Zen [Lecture/transcript archive]. *Alan Watts Organization.* https://alanwatts.org/2-3-1-introduction-to-zen/

24. Katy Warner/CSU. (2020, October 14). *Echolocation—Bats (U.S. National Park Service).* National Parks Service. https://www.nps.gov/subjects/bats/echolocation.htm

25. St. Petersburg College. (2009, March 26). *Cosmic Quandaries with Dr. Neil deGrasse Tyson* [Panel Discussion]. https://www.youtube.com/watch?v=CAD25s53wmE (1:14:45 in).

26. Chris Bould. (1993). *Bill Hicks: Revelations* [Stand-up special]. (52:20 in).

27. Neusner, J., & Chilton, B. (Eds.). (2008). *The golden rule: The ethics of reciprocity in world religions.* Continuum.

28. The Golden rule isn't unique to one man's culture, but is instead, a universal ethic, found in at least 12 major religions. See: Scaroboro Missions and Paul Scarboro Missions. (n.d.). Understanding the Golden Rule. *Golden Rule.* Retrieved January 26, 2022, from https://www.scarboromissions.ca/golden-rule/understanding-the-golden-rule

Chapter 28: Unity

1. Steven A. Edwards. (2012, November 19). *Isaac Newton and the problem of color.* American Association for the Advancement of Science. https://www.aaas.org/isaac-newton-and-problem-color

2. White, J. (2012). Herschel and the Puzzle of Infrared. *American Scientist, 100*(3), 218. https://doi.org/10.1511/2012.96.218

3. American Physical Society. (2008, July). July 1820: Oersted and electromagnetism. *APS NEWS, 17*(7). http://www.aps.org/publications/apsnews/200807/physicshistory.cfm

4. Three Models of the Universe - Watts, A. (2017). *Out of your mind: Tricksters, interdependence, and the cosmic game of hide-and-seek.* Sounds True, Inc. (p. 3-15).

5. Jones, A. Raymond (2020, May 19). *Ptolemaic system. Encyclopedia Britannica.* https://www.britannica.com/science/Ptolemaic-system

6. Riebeek, H. (2009, July 7). *Planetary Motion: The History of an Idea That Launched the Scientific Revolution* [Text.Article]. Nasa.Gov; NASA Earth Observatory. https://earthobservatory.nasa.gov/features/OrbitsHistory

7. Abbott, A. (2018). Discovery of Galileo's long-lost letter shows he edited his heretical ideas to fool the Inquisition. *Nature, 561*(7724), 441–442. https://doi.org/10.1038/d41586-018-06769-4

8. Although the history books will show that Galileo was not actually tortured, nor did he actually end up spending the rest of his days in prison, it's important to understand some of the reasons as to why Galileo did not suffer this fate. In truth, he did remain under house arrest for the rest of his days, and he did have to abjure his beliefs. However, if Galileo had not been as forthcoming and as apologetic as he was, it's highly conceivable he would have suffered a much more severe punishment. See: Kelly, H. A. (2016). Galileo's Non-Trial (1616), Pre-Trial (1632–1633), and Trial (May 10, 1633): A Review of Procedure, Featuring Routine Violations of the Forum of Conscience. *Church History, 85*(4), 724–761. https://doi.org/10.1017/S0009640716001190

Chapter 29: Concepts of Infinity

1. Baumgartner, R. J., Van Kranendonk, M. J., Wacey, D., Fiorentini, M. L., Saunders, M., Caruso, S., Pages, A., Homann, M., & Guagliardo, P. (2019). Nano−porous pyrite and organic matter in 3.5-billion-year-old stromatolites record primordial life. *Geology, 47*(11), 1039–1043. https://doi.org/10.1130/G46365.1

2. Dawkins, R. (2005, July). *Why the universe seems so strange* [Video]. TED Conferences. https://www.ted.com/talks/richard_dawkins_why_the_universe_seems_so_strange

3. Turner, A. G. R., Michael S. (2008, September 23). The Expanding Universe: From Slowdown to Speed Up. *Scientific American*. https://www.scientificamerican.com/article/expanding-universe-slows-then-speeds/

4. Baker, H. (2021, August 17). *Pi calculated to a record-breaking 62.8 trillion digits* [Science Magazine]. Livescience.Com. https://www.livescience.com/record-number-of-pi-digits.html

5. This was originally a theory from Freud but also had support from Carl Jung. It is also of note that Wegner et al. have found evidence of suppressed thoughts in dreams. See: Wegner, D. M., Wenzlaff, R. M., & Kozak, M. (2004). Dream Rebound: The Return of Suppressed Thoughts in Dreams. *Psychological Science, 15*(4), 232–236. https://doi.org/10.1111/j.0963-7214.2004.00657.x

6. Pythagoras was supposedly one of Anaximander's students. See: Stewart, D. (2015, November 28). *Anaximander—Biography, Facts and Pictures* [Educational Resource]. https://www.famousscientists.org/anaximander/

7. Evans, J. (2020, September 11). *Anaximander. Encyclopedia Britannica.* https://www.britannica.com/biography/Anaximander

8. World Science Festival. (2013, May). *Infinity: The Science of Endless* [Panel Discussion]. https://www.youtube.com/watch?v=KDCJZ81PwVM (11 min. in).

9. Westerhoff, J. C. (2021). Nāgārjuna. In E. N. Zalta (Ed.), *The Stanford Encyclopedia of Philosophy* (Fall 2021). Metaphysics Research Lab, Stanford University. https://plato.stanford.edu/archives/fall2021/entries/nagarjuna/

10. To my knowledge, this quote was first found in *The Book of the Twenty-four Philosophers* (circa 12th century CE), but has been repeated and championed by various philosophers including Nicholas of Cusa. One of my heroes, Joseph Campbell has

added a little piece to the quote, and so I am quoting his quotation. See: Campbell, J., & Moyers, B. D. (2011). *The power of myth*. Broadway Books. (p. 191).

11. Nadler, S. (2020). Baruch Spinoza. In E. N. Zalta (Ed.), *The Stanford Encyclopedia of Philosophy* (Summer 2020). Metaphysics Research Lab, Stanford University. https://plato.stanford.edu/archives/sum2020/entries/spinoza/

12. World Science Festival. (2013, May). *Infinity: The Science of Endless* [Panel Discussion]. https://www.youtube.com/watch?v=KDCJZ81PwVM (12:30 in).

13. Perkowitz, S. (2021, July 22). *E = mc2. Encyclopedia Britannica*. https://www.britannica.com/science/E-mc2-equation

14. Kauffman, S. (2010, March 8). The Philosophy of Mind, 1. *NPR*. https://www.npr.org/sections/13.7/2010/03/the_philosophy_of_mind.html

15. "The big bang tells us how the universe evolved from a split second after whatever brought it into existence but the big bang theory, many people don't realize is completely silent on what happened at time zero itself, the very beginning. And when we try to fill in that gap as we have been for a number of decades now we find that there is a good chance that there wasn't a single big bang event, that there were possibly many big bang events at various and far flung places throughout a larger cosmos giving rise to universe upon universe upon universe, our universe being the aftermath of one of those bangs. There are other universes which are the aftermaths of the other bangs."
- Brian Greene From: Transcript for Brian Greene on Parallel Universes. (2011, May 1). [Interview]. In *To the best of our Knowledge*. PRX. http://archive.ttbook.org/book/transcript/transcript-brian-greene-parallel-universes

Chapter 30: Piecing it all Together II

1. Watts, A. (2002). *The Tao of philosophy: The edited transcripts*. (p. 34)

2. The question: "What is life?" has been asked again and again, but there is still no great consensus or agreement. Viruses, for example, exist in some kind of bizarre grey area between living and non-living, which is still a heated debate to this day. Having said that, Sir Paul Nurse has done a fine job of synthesizing a handful of principles in his latest book, *What is Life?* He even considers viruses as being alive. See: Nurse, P. (2021). *What is life? Five great ideas in biology* (First American edition). W.W. Norton & Company.

3. Carl Zimmer also takes on the question "Are Viruses Alive?" in his Royal Institution lecture: The Royal Institution. (2021, November 25). *Are Viruses Alive? - With Carl Zimmer* [Ri Lecture]. https://www.youtube.com/watch?v=Tryg5UCp6fI

4. NASA defines life as follows: "Life is a self-sustaining chemical system capable of Darwinian evolution" in: *NASA Astrobiology*. (2022, January). [Government Space Agency]. Nasa.Gov. https://astrobiology.nasa.gov/research/life-detection/about/

5. Vaughan, D. (n.d.). *What Is the Most Widely Practiced Religion in the World?. Encyclopedia Britannica*. https://www.britannica.com/story/what-is-the-most-widely-practiced-religion-in-the-world

6. Majumdar, S. (2018, June 29). 5 facts about religion in India [Think Tank]. *Pew Research Center*. https://www.pewresearch.org/fact-tank/2018/06/29/5-facts-about-religion-in-india/

7. Watts, A. (2017). *Out of your mind: Tricksters, interdependence, and the cosmic game of hide-and-seek*. Sounds True, Inc. (p. 15-29; 121-34).

8. Eknath, E., & Nagler, M. N. (Eds.). (2007). *The Upanishads* (2nd ed). Nilgiri Press. (p. 89).

9. Olivelle, P. (Ed.). (1998). *The early Upanisads: Annotated text and translation.* Oxford University Press. (p. 12-13).

10. Jordan Peterson. (March 2, 2021). *The Psychology of Psychedelics—Roland Griffiths* (S4 E20) [Video]. https://youtu.be/NGIP-3Q-p_s (1:07:45 in).

Chapter 31: Speciation Events

1. Van Wyhe, J. (2013). Dispelling the darkness: Voyage in the Malay Archipelago and the discovery of evolution by Wallace and Darwin. World Scientific.

2. Marshall, E. (2018, July). *160th anniversary of the presentation of "On the tendency of Species...* [Natural History Society]. The Linnean Society. https://www.linnean. org/news/2018/07/01/1st-july-2018-160th-anniversary-of-the-presentation-of-on-the-tendency-of-species-to-form-varieties

3. Wulf, A. (2016). *The Invention of Nature: Alexander von Humboldt's New World* (1st Vintage Books edition). Vintage Books. (p. 327-353).

4. Ibid., (p. 345-47).

5. Quote from Darwin, C. (2009). *The Voyage of the Beagle* (C. W. Elliot, Ed.). P.F. Collier & Son Company. https://www.google.com/books/edition/The_Voyage_of_the_Beagle/ (p. 384).

6. Dawkins, R. (2009). *The Greatest Show on Earth: The Evidence for Evolution* United States: Free Press. (p. 33).

7. Reznick, D. N. (2011). *The "Origin" Then and Now: An Interpretive Guide to the "Origin of Species."* Princeton University Press. https://doi.org/10.1515/9781400833573 (p. 205-16).

8. Byrne, K., & Nichols, R. A. (1999). Culex pipiens in London Underground tunnels: Differentiation between surface and subterranean populations. *Heredity, 82*(1), 7–15. https://doi.org/10.1038/sj.hdy.6884120

9. Morell, V. (2015). From Wolf to Dog. *Scientific American, 313*(1), 60–67. https://doi. org/10.1038/scientificamerican0715-60

10. Cirilli, O., Pandolfi, L., Rook, L., & Bernor, R. L. (2021). Evolution of Old World Equus and origin of the zebra-ass clade. *Scientific Reports, 11*(1), 10156. https://doi. org/10.1038/s41598-021-89440-9

11. Čirjak, A. (2020, February). *What Is A Hinny?* WorldAtlas. https://www.worldatlas. com/what-is-a-hinny.html

12. Human-Chimpanzee split dated to be approximately 7-8 million years ago, in: Langergraber, K. E., Prufer, K., Rowney, C., Boesch, C., Crockford, C., Fawcett, K., Inoue, E., Inoue-Muruyama, M., Mitani, J. C., Muller, M. N., Robbins, M. M., Schubert, G., Stoinski, T. S., Viola, B., Watts, D., Wittig, R. M., Wrangham, R. W., Zuberbuhler, K., Paabo, S., & Vigilant, L. (2012). Generation times in wild chimpanzees and gorillas suggest earlier divergence times in great ape and human evolution. *Proceedings of the National Academy of Sciences, 109*(39), 15716–15721. https://doi.org/10.1073/ pnas.1211740109

13. Human-Chimpanzee split dated to be approximately 6-7 million years ago, in: Young, N. M., Capellini, T. D., Roach, N. T., & Alemseged, Z. (2015). Fossil hominin shoulders support an African ape-like last common ancestor of humans and chimpanzees.

Proceedings of the National Academy of Sciences, 112(38), 11829–11834. https://doi.org/10.1073/pnas.1511220112

14. Human-Chimpanzee split may have occurred as recently as 6.6 million years ago, in: Amster, G., & Sella, G. (2016). Life history effects on the molecular clock of autosomes and sex chromosomes. *Proceedings of the National Academy of Sciences, 113*(6), 1588–1593. https://doi.org/10.1073/pnas.1515798113

15. Theobald, D. L. (2010). A formal test of the theory of universal common ancestry. *Nature, 465*(7295), 219–222. https://doi.org/10.1038/nature09014

16. Weiss, M. C., Sousa, F. L., Mrnjavac, N., Neukirchen, S., Roettger, M., Nelson-Sathi, S., & Martin, W. F. (2016). The physiology and habitat of the last universal common ancestor. *Nature Microbiology, 1*(9), 16116. https://doi.org/10.1038/nmicrobiol.2016.116

17. "The first signs of life appear as carbon isotope signatures in rocks 3.95 billion years of age," cites the following paper: Weiss, M. C., Preiner, M., Xavier, J. C., Zimorski, V., & Martin, W. F. (2018). The last universal common ancestor between ancient Earth chemistry and the onset of genetics. *PLoS Genetics, 14*(8), e1007518. https://doi.org/10.1371/journal.pgen.1007518

18. Botigué, L. R., Song, S., Scheu, A., Gopalan, S., Pendleton, A. L., Oetjens, M., Taravella, A. M., Seregély, T., Zeeb-Lanz, A., Arbogast, R.-M., Bobo, D., Daly, K., Unterländer, M., Burger, J., Kidd, J. M., & Veeramah, K. R. (2017). Ancient European dog genomes reveal continuity since the Early Neolithic. *Nature Communications, 8*(1), 16082. https://doi.org/10.1038/ncomms16082

19. There are at least 354 distinctive dog breeds currently recognized by the World Canine Organization (aka the FCI). See: *Presentation of our organisation.* (2021, June). [World Canine Organisation]. Fédération Cynologique Internationale. http://www.fci.be/en/Presentation-of-our-organisation-4.html

20. Dawkins, R. (2009). *The Greatest Show on Earth: The Evidence for Evolution* United States: Free Press. (p. 27).

21. Darwin, C. (1998). The origin of species. Wordsworth. (p. 3-101).

Additional References for Chapter 31:

The Great Tree of Life. © Len Eisenberg (2008, 2017) is also available as a video here: https://youtu.be/f67Pem71tXM (www.evogeneao.com)

Chapter 32: Interdependence

1. Baumgartner, R. J., Van Kranendonk, M. J., Wacey, D., Fiorentini, M. L., Saunders, M., Caruso, S., Pages, A., Homann, M., & Guagliardo, P. (2019). Nano–porous pyrite and organic matter in 3.5-billion-year-old stromatolites record primordial life. *Geology, 47*(11), 1039–1043. https://doi.org/10.1130/G46365.1

2. Anscombe, G. E. M. (2000). *An introduction to Wittgenstein's Tractatus.* St. Augustine's Press. (p. 151).

3. University of Alaska Fairbanks. (2017, October 11). How rabies can induce frenzied behavior: Researchers better understand the disease that kills 59,000 people annually. *ScienceDaily.* Retrieved February 3, 2022 from www.sciencedaily.com/releases/2017/10/171011091847.htm

4. Oxford Martin School. (2020, March 6). *What is Life? Sir Paul Nurse - 2020 James Martin Memorial Lecture* [Video]. https://youtu.be/92oMfkuOIlA (58:35 in).

5. Darwin, C. (1998). *The Origin of Species*. Wordsworth. (p. 50-51; 368-69)
6. Watts, A. (2009, April 16). 1.2.9. - Taoist Way—Pt. 1 [Lecture/transcript archive]. *Alan Watts Organization*. https://alanwatts.org/1-2-9-taoist-way-pt-1/
7. Watts, A. (n.d.). *The Tao of Philosophy 3: Coincidence of Opposites*. The Library. Retrieved February 2, 2022, from https://www.organism.earth/library/document/tao-of-philosophy-3
8. Dawkins, R. (1997). *Climbing Mount Improbable*. W.W. Norton & Company. (p. 139).
9. Dawkins, R., & Wong, Y. (2016). *The Ancestor's Tale: A pilgrimage to the Dawn of Evolution*. Houghton Mifflin Harcourt. (p. 674).
10. Fang, J. (2010). Snake infrared detection unravelled. *Nature*. https://doi.org/10.1038/news.2010.122
11. Dawkins, R. (1991). *Growing up in the Universe: Climbing Mount Improbable* [Video]. https://www.rigb.org/christmas-lectures/watch/1991/growing-up-in-the-universe/climbing-mount-improbable (3:45 in).
12. Segrest, T. (2008, December 3). *Wandering Toward Wonder: The Incendiary Trail of Werner Herzog's Fever Dreams*. International Documentary Association. https://www.documentary.org/feature/wandering-toward-wonder-incendiary-trail-werner-herzogs-fever-dreams
13. Boeree, Dr. C. G. (1999). *Metaphysics*. https://webspace.ship.edu/cgboer/meta.html
14. Lee C, A. (2006). *Baruch Spinoza, "Human Beings are Determined."* Philosophical Ethics. https://philosophy.lander.edu/intro/spinoza.shtml
15. Campbell, J., & Moyers, B. D. (2011). *The Power of Myth*. Broadway Books. (p. 83).

Chapter 33: Cuckoos III & Story B

1. Basham, A. Llewellyn , Narayanan, . Vasudha , Dimock, . Edward C. , Doniger, . Wendy , Smith, . Brian K. , Gold, . Ann G. and Buitenen, . J.A.B. van (2022, January 26). Hinduism. Encyclopedia Britannica. https://www.britannica.com/topic/Hinduism
2. Shashkevich, A. (2018, August 20). Buddhism and its origins. *Stanford News*. https://news.stanford.edu/2018/08/20/stanford-scholar-discusses-buddhism-origins/
3. Society, N. G. (2020). Taoism. In *National Geographic Society*. http://www.nationalgeographic.org/encyclopedia/taoism/
4. Society, N. G. (2019). Chinese Religions and Philosophies. In *National Geographic Society*. http://www.nationalgeographic.org/article/chinese-religions-and-philosophies/
5. Clark, L. (2018, July). Dharma and the Tao: How Buddhism and Daoism have influenced each other; Why Zen and Taoism can be complementary. *Buddha Weekly*. https://buddhaweekly.com/dharma-and-the-tao-how-buddhism-and-daoism-have-influenced-each-other-why-zen-and-taoism-can-be-compliementary/
6. Westerhoff, J. C. (2021). Nāgārjuna. In E. N. Zalta (Ed.), *The Stanford Encyclopedia of Philosophy* (Fall 2021). Metaphysics Research Lab, Stanford University. https://plato.stanford.edu/archives/fall2021/entries/nagarjuna/
7. Watts, A. (2017). *Out of your mind: Tricksters, interdependence, and the cosmic game of hide-and-seek*. Sounds True, Inc. (p. 161).
8. Graham, D. W. "Heraclitus—The Doctrine of Flux and the Unity of Opposites." In *Internet Encyclopedia of Philosophy*. Retrieved February 4, 2022, from https://iep.utm.edu/heraclit/
9. Wulf, A. (2016). *The Invention of Nature: Alexander von Humboldt's New World* (1st Vintage Books edition). Vintage Books. (p. 347, 367).

10. Davies, N. (2015, May 14). *Cuckoos and their victims: An evolutionary arms race* [Royal Society Lecture]. https://youtu.be/nOO6S4hDDfE (29 min in).

11. Spottiswoode, C. N., & Stevens, M. (2012). Host-Parasite Arms Races and Rapid Changes in Bird Egg Appearance. *The American Naturalist, 179*(5), 633–648. https://doi.org/10.1086/665031

12. Nuwer, R. (2013, September 24). Parasitic Cuckoo Finches Use an Egg Overload to Evade Host Defenses. *Smithsonian Magazine*. https://www.smithsonianmag.com/science-nature/parasitic-cuckoo-finches-use-an-egg-overload-to-evade-host-defenses-39218/

13. Langmore, N. E., Stevens, M., Maurer, G., Heinsohn, R., Hall, M. L., Peters, A., & Kilner, R. M. (2011). Visual mimicry of host nestlings by cuckoos. *Proceedings of the Royal Society B: Biological Sciences, 278*(1717), 2455–2463. https://doi.org/10.1098/rspb.2010.2391

14. Davies, N. (2015, May 14). *Cuckoos and their victims: An evolutionary arms race* [Royal Society Lecture]. https://youtu.be/nOO6S4hDDfE (50 min in).

Chapter 34: A Parting Note on Opposition

1. Stewart, D. (2015, November 28). *Anaximander—Biography, Facts and Pictures* [Educational Resource]. https://www.famousscientists.org/anaximander/

2. KOČANDRLE, R., & KLEISNER, K. (2013). Evolution Born of Moisture: Analogies and Parallels Between Anaximander's Ideas on Origin of Life and Man and Later Pre-Darwinian and Darwinian Evolutionary Concepts. *Journal of the History of Biology, 46*(1), 103–124. http://www.jstor.org/stable/42628763

3. Greek Philosophers: Aristarchus of Samos, Philolaus, and Hicetas. See: Britannica, T. Editors of Encyclopaedia (2019, April 1). *heliocentrism. Encyclopedia Britannica.* https://www.britannica.com/science/heliocentrism

4. Riebeek, H. (2009, July 7). *Planetary Motion: The History of an Idea That Launched the Scientific Revolution* [Text.Article]. Nasa.Gov; NASA Earth Observatory. https://earthobservatory.nasa.gov/features/OrbitsHistory

5. Wulf, A. (2016). *The Invention of Nature: Alexander von Humboldt's New World* (1st Vintage Books edition). Vintage Books. (p. 65-68).

6. Interconnected Web - Ibid., (p. 23). But really, the entire book speaks of this conception.

7. 'Pre-Darwininan Darwinist' - Ibid., (p. 366).

8. Darwin's praise for Humboldt and Humboldt's effect on Darwin Ibid., (p. 23, 327-53).

9. "Greatest Naturalist in the world" see: Humboldt, A., & Wulf, A. (2018). *Selected Writings*. Alfred A. Knopf. (p. vii, x). See also: BBC. (2000, June). Humboldt: Natural Traveler (No. 2) [Documentary]. In *Wilderness Men*. BBC. https://www.youtube.com/watch?v=pgvX0QdYI6M (47:30 in).

10. Wulf, A. (2016). *The Invention of Nature: Alexander von Humboldt's New World* (1st Vintage Books edition). Vintage Books. (p. 142, 137-49).

11. Humboldt, A., & Wulf, A. (2018). *Selected Writings*. Alfred A. Knopf. (p. x).

12. Wasmuth, C. (2019). A name to conjure with [Alexander von Humboldt Foundation]. *Explore*. https://www.humboldt-foundation.de/en/explore/alexander-von-humboldt/a-name-to-conjure-with

13. Campbell, J. (1988). *Myths To Live By*. Bantam Books. (p. 174-77).

14. Watts, A. (2017). *Out of your mind: Tricksters, interdependence, and the cosmic game of hide-and-seek*. Sounds True, Inc. (p. 31).

15. Ibid., (p. 8-9).
16. TEDx Talks. (2014, June). *Consciousness is a mathematical pattern: Max Tegmark at TEDxCambridge 2014* [TEDx Talk]. https://www.youtube.com/watch?v=GzCvl-FRISIM (10:35 min in).
17. Roughly, there are about twelve major religions in the world: Christianity, Islam, Hinduism, Buddhism, Sikhism, Taoism, Judaism, Confucianism, Bahá'í, Shinto, Jainism, and Zoroastrianism. For a summary of the twelve, see: Boyett, J. (2016). *12 major world religions: The beliefs, rituals, and traditions of humanity's most influential faiths* (1st edition). Zephyros Press.
18. Right View, or first step of the Noble Eightfold Path, see: Watts, A. (n.d.). *Out Of Your Mind 11: The World as Emptiness (Part 1)*. The Library. Retrieved February 10, 2022, from https://www.organism.earth/library/document/out-of-your-mind-11 (24 min. in).
19. Malone, A. (1980, November 30). The Edge of Forever (No. 10) [Documentary]. In *Cosmos*.

Image Attributions

(Scan QR Code for color images of all photos and figures)

Chapter 2. Courtesy of the Authors. *Source Code.*

Chapter 3. Kacpura. Adobe Stock #98146562. *West Coast Totem Pole.*

Chapter 4. MICrONs Consortium et al. *Functional connectomics spanning multiple areas of mouse visual cortex.* bioRxiv 2021.07.28.454025; doi: https://doi.org/10.1101/2021.07.28.454025.

Chapter 5. Courtesy of the Authors. *An Incomplete Spectrum of Altered States.*

Chapter 5. Courtesy of the Authors. *Synaptic Gap.*

Chapter 6. Blausen.com staff (2014). *Medical gallery of Blausen Medical 2014.* WikiJournal of Medicine 1 (2). DOI:10.15347/wjm/2014.010. ISSN 2002-4436. Marked with CC by 3.0.

Chapter 7. Mazuryk, Mykola. Shutterstock #1888869910. *Geometric optical illusion. white and black circle psychedelic pattern.*

Chapter 7. Adelson, Edward H. *Checker-Shadow Illusion.*

Chapter 7. Courtesy of the Authors. *Square/Circle Illusion.*

Chapter 7. Ninio, J. and Stevens, K. A. (2000) *Variations on the Hermann grid: an extinction illusion.* Perception, 29, 1209-1217.

Chapter 7. Courtesy of the Authors. *Bits.*

Chapter 7. Courtesy of the Authors. *Thalamic gating.*

Chapter 11. Courtesy of the Authors. *Gene pool.*

Chapter 11. Spottiswoode, Claire & Stevens, Martin (2012). *Host-Parasite Arms Races and Rapid Changes in Bird Egg Appearance*. The American naturalist. 179. 633-48. 10.1086/665031.

Chapter 11. Davies, Nick. *Races of the Common Cuckoo*.

Chapter 12. Courtesy of the Authors. *Brainwaves*.

Chapter 13. Courtesy of the Authors. *Kinesphere*.

Chapter 14. Andreashorn. *Default Mode Network Connectivity* [Graphic]. Wikipedia. en.wikipedia.org/wiki/Default_mode_network#/media/File:Default_Mode_Network_Connectivity.png. Marked with CC by 4.0.

Chapter 15. Courtesy of the Authors. *Multiplicity of Selves (Simple)*.

Chapter 15. Courtesy of the Authors. *Multiplicity of Selves (Detailed)*.

Chapter 16. Courtesy of the Authors. *Necker Cube*.

Chapter 17. Beckley Imperial Research Program. *LSD Revealed: the World's First Images of the Brain on LSD*.

Chapter 17. Courtesy of the Authors. *Three molecules*.

Chapter 19. Courtesy of the Authors. *Spectrum of Alignment*.

Chapter 22. Courtesy of the Authors. *Bayesian inference*.

Chapter 23. Courtesy of the Authors. *Bouba Kiki*.

Chapter 24. Li, Xuejun. Adobe Stock #47181504. *Green sound waveform*.

Chapter 26. Adobe Stock #214282651. *Three times figure-ground perception, face and vase. Figure-ground organization. Perceptual grouping. In Gestalt Psychology known as identifying figure from background. Illustration over white*. [Vector].

Chapter 27. Courtesy of the Authors. *The Singularity*.

Chapter 27. European Space Agency (2021). *Planck satellite cmb.jpg* [Graphic]. *ESA/Planck Satellite Collaboration*. 2013. Wikimedia. commons.wikimedia.org/wiki/File:Planck_satellite_cmb.jpg. Marked with CC by 4.0.

Chapter 27. Georghiou, Christos. Shutterstock #221744677. *An illustration of the planets of our solar system*.

Chapter 27. Ball, Madeleine Price. commons.wikimedia.org/wiki/File:Simplified_tree.png.

Chapter 27. Maryia, Kazakova. Shutterstock #1303947361. *Parts of plant. Morphology of raspberry shrub with berries, green leaves, root system isolated on white background*.

Chapter 27. Intermountain Forest Service, USDA Region 4 (2012). *121003-FS-Fishlake-JZ-002(30469976897)* [Photograph]. Wikimedia Commons. flickr.com/photos/107640324@N05/30469976897. Marked with CC by 1.0 Universal.

Chapter 27. Fisher, J.C (2013). *Fall Aspen in Hope Valley, California* [Photograph]. Flickr. https://www.flickr.com/photos/15729043@N00/10388721565. (CC by 2.0)

Chapter 27. Suren Manvelyan/Hubble. *Nebuleyes* [Photograph].

Chapter 28. Courtesy of the Authors. *Visible light spectrum*.

Chapter 30. Courtesy of the Authors. *Photoshop layers*.

Chapter 31. Courtesy of the Authors. *Types of finches*.

Chapter 31. Courtesy of the Authors. *Finch beaks*.

Chapter 31. Courtesy of the Authors. *Phylogenetic tree*.

Chapter 31. Courtesy of the Authors. *Relations chart*.

Chapter 31. The Great Tree of Life. © Len Eisenberg 2008, 2017 (evogeneao.com)

Chapter 33. Courtesy of the Authors. *Yin Yang*.

Chapter 33. Davies, Nick. *Cuckoo egg in dunnock nest* [Photograph].

Chapter 33. Spottiswoode, C. (2011). *Anomalospiza egg mimicry* [Photograph]. Wikimedia.
commons.wikimedia.org/wiki/File:Anomalospiza_egg_mimicry.jpg. Marked with CC by 4.0.

Chapter 33. Kilner, Rebeca and Naomi Langmore. *Australian bronze-cuckoos* [Photograph].

Chapter 34. Courtesy of the Author. *Yin Yang Fractal*.

Chapter 34. K3Star. Adobe Stock #92389535. *Snake curled in infinity ring. Ouroboros
devouring its own tail*.

Chapter 34. Courtesy of the Authors. *Nested Yin Yang*.

Chapter 34. Sally, Sunny. Shutterstock #1920454910. *Yin Yang Symbol - Vector Sign*.

Chapter 34. Courtesy of the Authors. *Necker cube shadow* [Photograph].

Glossary

Alice in Wonderland Syndrome (AIWS) - is a perceptual disorder featuring distortions in body schema, visual perception, and the experience of time. Aptly named after Lewis Carrol's famous story, those who experience AIWS tend to report experiencing stationary objects as changing in size, position, and distance. For a more complete review, see: Blom, J. D. (2016). Alice in Wonderland syndrome: A systematic review. *Neurology: Clinical Practice, 6*(3), 259–270. https://doi.org/10.1212/CPJ.0000000000000251

Altered States of Consciousness (ASC) - an ASC is any kind of deviation or change from the normal, waking, baseline state of awareness we're calling Consciousness 5.0 or CS5. ASCs are typically brought about as a result of a certain activity, environment, or pharmacological agent and are sometimes also referred to as non-ordinary state of consciousness (or NOSC).

America - North America's "discovery" is largely considered to be an accident as Christopher Columbus was searching for a more direct route from Europe to Asia.

Anesthesia awareness - under general anesthesia, it is rare for patients to ever remember or recall pain or pressure during their surgery. However, in some cases, patients have achieved varying levels of awareness and recall during these operations. In the worst of cases, patients are immobilized but are still able to hear and feel the doctor's operating.

Atman - (Sanskrit: "self," "breath") sometimes it is translated as soul, but it is better understood to be the immortal aspect of our mortal experience, or the eternal Supreme Self hiding behind all things.

Attention - the focussing of perceptual and cognitive awareness on some sort of stimuli at the expense of something else. There may be many different kinds of attention, but at the center of all of them is the idea that the brain has to prioritize certain elements over others, and that attention is limited in both capacity and duration.

Bayesian Brain - Bayes' theorem or Bayesian inference, is an idea first formulated by Reverend Thomas Bayes which helps us describe how the brain works. Sitting inside the skull (being ensconced in silence and darkness), the brain is charged with trying to determine the external causes for all these incoming sensory signals, and Bayesian inference appears to be it's method of accomplishing that. By consistently updating the brain's model as new data comes in, raw sense data is combined with prior expectations/beliefs in an effort to determine the exact causes of those sensations. So, from the top-down, we have a brain that is producing a "generative model" (inclusive of all its prior beliefs concerning the causes of this sense data) and then from the bottom-up we have sense organs sending up "prediction errors" to help refine that model.

Bones of the Experience - after researching the topic of mystical experiences, one comes to notice a few recurring themes amongst the reports, and this pattern is reflected in the questionnaires researchers use, like the OAV and the 5D-ASC. Ego dissolution, time dilation, and feelings of oceanic boundlessness are among the three most essential themes to focus on here, however, it is also best to understand Arnie Dietrich's theory of transient hypofrontality.

Brahman - (concept derived from ancient Hindu texts) it is the force underlying all things; the Ultimate Reality of the cosmos. The unchanging universal spirit of Being. According to Hindu theology the Atman is reincarnated again and again, until it realizes that, "Atman is Brahman," and is consequently, one with all of creation.

Brain-derived neurotrophic factor (BDNF) - a key molecule that regulates plastic changes in brain structures relating to learning and memory.

BrainPort - BrainPort is another example of sensory substitution developed by Paul Bach-y-Rita, and was originally intended to help stroke victims regain their sense of balance. With BrainPort, there's an electrode that sits on the tongue and is fed a particular stream of data, like vision. Eventually, users of the BrainPort get better at "seeing" with their tongue; and believe it or not, but there's at least one blind individual who once used it to climb mountains for years.

Central Executive Network (CEN) - responsible for decision-making and problem-solving during goal-directed behavior. A large-scale brain network involved in the control of attention, working memory, and planning. Largely involved in externally directed cognition vs. the internally directed cognition of the default mode network.

Checker Shadow Illusion - an optical illusion that displays a cylinder casting a shadow over a checkerboard with light and dark squares. Because of the cylinder's shadow, we expect the two labeled squares to be different colors, however we are quickly shown that they are in fact the same identical color.

Classic psychedelics - drugs that are serotonin 2A receptor agonists, i.e. LSD, mescaline, psilocybin, and DMT.

Coevolution - the evolution of one or more species in a mutually dependent manner. A popular example is the coevolution of pollinators and flowers. Another example might be the coevolution between wolves and man. The wolves that figured out how to communicate best and work effectively with humans were often looked after and got more food. While the wolves that didn't embark on this coevolutionary journey, subsequently stayed wolves and never enjoyed the fruits of interspecies friendship.

Cognitive flexibility - roughly, it is our ability to adapt and conform to new situations. Our ability to change our current mindset or state of mind, and adopt new ways of thinking that might best serve the problem. Remember, Dustin Hoffman's character from *Rain Man* exuded very high amounts of cognitive rigidity, and very low amounts of cognitive flexibility.

Cognitive rigidity - the opposite of cognitive flexibility. Those who possess cognitive rigidity have very fixed ways of doing things, thinking about things, and operating in the world. Generally, they are unwilling to change their internal states to match the demands of the environment.

Consciousness 5.0 (CS5) - if the brain can be likened to being *like* the hardware of a computer, than consciousness is like the software running inside it. Supposing that's the case, there is a baseline state of awareness and control we're given over this user-illusion software; and under unaltered conditions, we call this software Consciousness 5.0 or CS5. The second we start ingesting caffeine, alcohol, or go for a 20 minute run, we've just knocked ourselves out of our baseline, conscious awareness (CS5), and into some other type of NOSC, which likely has effects on the way we think and perceive.

Comas - states of prolonged unconsciousness that can be brought about in a number of different ways (trauma, illness, intoxication); and last from weeks to decades.

Cdc2 gene for cell division - a gene that controls cell division in fission yeast. Originally discovered by Sir Paul Nurse in the 1970s, this cdc2 gene is also found in humans. To put that in perspective, that means the gene for controlling cell division has been conserved, down the evolutionary line, for well over a billion years! Humans and yeasts shared a common ancestor between 1.2 and 1.5 billion years ago. As Sir Paul points out, the last dinosaurs went extinct a 'mere' 65 million years ago.

CRISPR - Clusters of Regularly Interspaced Short Palindromic Repeats. CRISPR is a technology that can be used to alter DNA sequences and edit genes.

Default Mode Network (DMN) - a large-scale brain network responsible for your sense of self, and is hypothesized to be the neural correlates of 'the ego'. Originally discovered by Dr. Marcus Raichle, the default mode network consists of regions like the posterior cingulate cortex, medial prefrontal cortex, precuneus, and angular gyrus. Largely, the DMN is responsible for internally directed cognition.

D.E.M.P. - a mnemonic device for remembering the "four mental rites of passage" one must endure when confronted with a transcendent or "mystical-type" experience. Dissolvability of established brain networks, Elasticity of attention, Malleability of the internal experience, and Possibility for change.

Dualism - in general, dualism refers to the idea that there are two fundamentally different kinds of things or principles. In philosophy of mind, it's the idea that the mind and body are two radically different things and exist as separate entities.

Ecology - a multidisciplinary science that focuses on the relationships of organisms to their environment, and between organisms and other organisms.

Ecological Awareness - the recognition of our immense dependency on the environment for survival, its fragility, and the importance of its protection. Alan Watts refers to a state of mind in which a person ceases to feel separate from his or her environment.

Ego [see also : Narrative Self] - the component of one's personality that deals with the external demands of the world and is self-conscious. The ego is what a person is aware of when they think about themselves, and is typically what they try to project to the outside world. Most importantly, however, is the idea that the ego is not a full representation of one's psyche or mental processes.

Ethology - the study of animal behavior, generally with a focus on behavior under natural conditions.

Evolution by Natural Selection - evolution by natural selection is only possible because there are errors in the DNA replication process. Once in a billion rungs of DNA there tends to be an error, and this error is the fundamental building block of evolution. If heredity were perfect, there would be no evolution, plain and simple.

Evolutionary Arms Race - in the long, evolutionary battle between predators and their prey, a kind of arms race has developed between the two, in which predators are constantly evolving new tricks of attack and prey are constantly evolving new means of defense.

Extended Phenotype - (examples of phenotypes include: eye color, fur color, beak size, length of tail) they are physical manifestations of a certain gene or genes. Extended phenotypes, on the other hand, are manifestations of a particular gene or genes which have effects that reach far beyond the body in which those genes are found. Some examples of extended phenotype include, the nest-building behavior in birds, the mound-building behavior in termites, and the dam-building behavior in beavers.

Flow States - are a NOSC capable of occurring during almost any creative or athletic endeavor. Specifically, they refer to those "in the zone" moments where we are entirely focused on the task at hand and are in an optimal state of performance. The pioneer behind much of this research has defined it as, "a state in which people are so involved in an activity that nothing else seems to matter." Time dilates, our sense of self disappears, and usually, they only come on once we've sufficiently overloaded ourselves (i.e., our Challenge/Skills Ratio is met). Flow states, quite simply, are the keys to happiness.

Gene - the basic unit of inheritance. A unit of heredity passed from parent to offspring. Genes are made up of sequences of DNA and are located at specific locations on chromosomes.

Genotype - an animal's genotype is the collection of genes responsible for the various genetic traits (i.e., its genetic makeup).

Gnosticism - a religious collective emerging around the second century CE which emphasized personal spiritual knowledge above traditional religious practices. Gnostics believe in a spark or piece of God being in every human being, and that salvation could be attained by awakening to this knowledge of our true divine identity.

God-of-the-gaps fallacy - the repeated claim that certain features of this world must be supernatural because there is currently no plausible, scientific explanation.

Hollow-Mask Illusion - a rotating mask clues us into how the brain is hardwired to perceive faces. Like a traditional mask, one side is convex and the other side is concave. Therefore, it is entirely normal to perceive the outside of a mask as a face. However, when the mask is rotated, such that the observer can now see the concave side of the mask, the inside appears to look convex, like a face. Interestingly enough, this illusion doesn't seem to work so well on those suffering from schizophrenia, or those who have recently taken LSD. See illusion here: https://youtu.be/pH9dAbPOR6M

Lempel-Ziv Complexity (LZC) - the LZC is a frequently used measure of signal complexity in the study of consciousness. Specifically, the LZC refers to a measure of compressibility for a given data set. The Lempel-Ziv algorithm counts the number of unique patterns in certain sequence, and is used everyday in compressing large computer files.

Locked-in Syndrome (LIS) - kind of like a pseudocoma, LIS patients are awake and having conscious experiences, only they have no capability of producing speech, limb, or facial movements. Essentially they are conscious entities trapped in an immobile body, and they are believed to be in a coma.

Lysergic Acid Diethylamide (LSD) - a derivative of a fungus discovered by accident in 1943 and studied extensively until its criminalization in 1970. One study showed physiological and psychological effects with as little as 7 millionths of a gram.

McGurk Effect - a multisensory illusion that occurs when there is a mis-alignment between what is seen and what is heard (i.e., the individual's lip movements do not match up with what they're saying).

MDMA - induces feelings of pleasure, warmth, connection, and energy. MDMA is supposed to be the most prevalent ingredient in the drug ecstacy, however, clinical trials only use MDMA in its pure form.

Memories - human memory is extremely fallible and prone to many different remembering biases.

Microdosing - microdosing psychedelics has become an extremely popular practice amongst coders in Silicon Valley, billionaire CEOs, and artists/creatives. As an example, one might only take 5 to 20 μg of LSD (a sub-threshold dose) as opposed to a full dose which might be 100 μg. James Fadiman and Paul Stamets have both spoken about their different methods of approach, however Ayelet Waldman has also had some interesting words to say, in: Jason Gots. (10/28/17). *Ayelet Waldman (Author)— Yourself, Only Better* (No. 86) [Audio File]. https://podcasts.apple.com/us/podcast/think-again-a-big-think-podcast/id1002073669?i=1000381375767

Microwaves - microwaves were discovered on accident by Percy Spencer, in 1945.

Monism - is the belief that there is only one kind of ultimate substance and everything in the universe can ultimately be reduced to that reality. For the monist, there is no difference between self and the supreme creator because the fundamental character of the universe is unity. However, for the dualist, there is a difference between the individual self and the supreme creator.

Multi-stable stimulus - an image that can be interpreted at least two different ways, like the Necker Cube, the Rabbit-duck illusion, or Rubin's vase.

Narrative Self [see also : Ego] - a low resolution snapshot the brain uses to experience itself, which is neither complete nor authoritative. Other names include inner-witness, inner-narrator, or the false self.

Nature's Microscope (aka Nature's Key to the unconscious) - a personal coinage representing the fundamental underlying molecular mechanism of action for classic psychedelics. Although each molecule has a slightly different expression of it— psilocybin vs. LSA, for example—the underlying shape is highly similar. LSD just happens to be more "sticky" and more potent. Therefore, if LSA and psilocybin are Nature's "natural" keys to the unconscious, then LSD is just the most refined version.

Necker Cube - an optical illusion produced by drawing a three-dimensional cube on a two-dimensional sheet of paper. Because the sides of the cube are drawn ambiguously, with no mention of the cube's true orientation, the brain wavers back and forth between the two interpretations that are possible.

Neurogenesis - the process by which neural stem cells generate new neurons in the brain. Critical for brain development and maintaining neural highways. It was once believed that neurogenesis, in humans, could only take place during embryonic development, but this has finally been overturned.

Neurotransmitters - chemical messengers released when nerve impulses arrive at synapses. In essence, they are signaling chemicals used by neurons to communicate specific messages.

Neuropsychopharmacology - the study of how drugs effect the mind and behavior. An interdisciplinary science which combines the tools of neuroscience and psychopharmacology.

Nirvana - a place of ultimate centeredness and stillness. Not a physical place, but rather, a state of mind free from wishes, wants, desires, and clinging. In this state, we are no longer being compelled by fear, greed, and delusion, because in nirvana we've finally tapped into a dimension of *selflessness*, compassion, and understanding. (Other names for this state of mind are satori, moksha, liberation, enlightenment, oceanic boundlessness, heaven, the Possibility Space, the beatific vision, etc.)

Non-ordinary states of consciousness (NOSC) - non-ordinary states of consciousness are also sometimes referred to as altered states of consciousness or ASCs. In its most simple encapsulation, there is some kind of baseline, ordinary, conscious state (CS5). And when we knock ourselves out of this baseline state of awareness, we're entering into NOSC territory.

Oddball effect - a perceptual phenomenon in which a novel or unexpected stimuli causes the brain to devote more energy and attentional resources into perceiving it, thus resulting in temporal distortions.

Optical Illusions - tools that deceive the eye by appearing to be something other than what they actually are. By repeatedly exposing the conscious self to certain holes in perception, we tend to gain insight about the way in which our visual system operates (i.e., vision is an active process, not passive).

Persistent Vegetative State - a chronic state of brain dysfunction whereby the patient displays no signs of awareness or higher brain function.

Paul Bach-y-Rita - a scientist named Paul Bach-y-Rita first demonstrated that sensory substitution was possible, back in 1969, using blind people and a modified dental chair that 'drew' images onto their back. See: Bach-Y-Rita, P., Collins, C. C., Saunders, F. A., White, B., & Scadden, L. (1969). Vision Substitution by Tactile Image Projection. *Nature*, 221(5184), 963–964. https://doi.org/10.1038/221963a0

Panic Disorder - an anxiety disorder typically triggered by a perceived threat of danger, whether that be physical or psychological. Panic disorders can be accompanied by severe physical symptoms like heart palpitations, chest pain, dizziness, and shortness of breath.

Persona - the social mask you present to the world. Originally, the Latin word "persona" was used in reference to the theatrical masks worn by actors in ancient times. However, in psychology, the persona refers to the version of yourself you try and project to the outside world.

Perturbational-Complexity Index (PCI) - is measure calculated by blasting the cortex with transcranial magnetic stimulation and then registering the neuronal responses under different conditions, such as in deep sleep or anesthesia, etc. Accordingly, PCI has a potential for use determining levels of consciousness in difficult to determine situations.

Phantom Limb - it's quite common for someone who's lost a limb, or had some kind of amputation to experience "phantom limb pain" in the region of the body that's been removed. Whether it be cramping, pins and needles, throbbing, shooting or burning pain, the individuals feel this sensation as if their hand, arm, or leg, were literally still attached.

Phase Transition - given enough energy, temperature, and pressure, matter is able to transform from one state of being into another. For example, ice that has just melted on the concrete has just made a phase transition from solid to liquid. With phase changes or phase transitions, the constituent parts still stay the same but the molecules inside recombine to create something entirely new.

Phenotype - while an animal's genotype refers only to the genetic information passed between generations, an animal's phenotype refers to the observable expressions of those genes or of that genotype. Examples of phenotype include: eye color, fur color, beak size, length of tail, sound of voice, etc. *Phenotypic expression is also influenced by environmental and epigenetic factors.

Procedural memory - is a type of long-term memory which is stored in the cerebellum, and has to deal with motor skills, like tying your shoes, typing on a keyboard, riding a bike, ice skating, etc. Sometimes also called implicit memories, procedural memories are often unconscious and difficult to articulate.

Psilocybin - a naturally occurring psychoactive chemical found in many species of mushroom with use recorded as early as 7000 BCE in cave paintings. Like LSD, psilocybin decreases communication between the regions that make up the default mode network, allowing for greater communication between disparate regions in the brain. Psilocybin has also been shown as a promising new treatment for end-of-life anxiety, OCD, PTSD, treatment-resistant depression, and addiction.

Psyche - the totality one's mind and personality encompassing all of one's conscious and unconscious behaviors, attitudes, feelings, emotions, and sub-personalities. Means "mind" or "soul" in Greek.

Psychedelics as a Treatment for Disorders of Consciousness - "Psychedelics increase brain complexity above normal levels." See: Scott, G., & Carhart-Harris, R. L. (2019). Psychedelics as a treatment for disorders of consciousness. *Neuroscience of Consciousness, 2019*(1). https://doi.org/10.1093/nc/niz003

Psychoactive - a psychoactive drug or psychotropic is any chemical substance that alters brain function and causes temporary alterations in one's mood, perception, consciousness, behavior, or cognition. Examples include: caffeine, alcohol, aspirin, cannabis, cacao, lithium, cocaine, adderall, LSD, antidepressants, ambien, etc.

Psychonaut - someone who uses altered states of consciousness to perturb and explore different ways of experiencing time (i.e., different modes of conscious experience). Specifically, psychonauts describe and explain the subjective effects of certain compounds or practices in an effort to learn something about the human condition.

Rare-enemy effect - when predators are quite rare to a particular prey species, it is typical for prey species to have little or no defenses. For example, angler fish are quite rare. It would be most unfortunate for an individual prey fish to fall victim to an angler fish's luminescent fin protrusion. However, evolving the sorts of sophisticated visual machinery to discriminate between an actual worm and the clever "worm-looking" protrusion is a costly endeavor. So when predation is quite rare, there is, in effect, not enough selection pressure to warrant the cost of evolving a proper means of defense.

Salience Network (SN) - primarily composed of the anterior insula and dorsal anterior cingulate cortex, the SN is a collection of brain regions that select which stimuli receive our attention. It also plays an important role in emotional information processing.

Sensory Deprivation Tanks - float tanks are great non-invasive, safe, methods for off-lining the self and understanding the mind. There is a lot that can be learned from simply turning inward, and sensory deprivation help us accomplish that by off-lining the Bodily Self and the Narrative Self.

Stealing Fire - there is no precursor for this book, although *Stealing Fire* by Steven Kotler and Jamie Wheal would certainly be a good start, seeing as the authors really do make a powerful argument for ASCs being the key to unlocking happiness, peak performance, and life fulfillment. In sum, altered states lead to altered traits, and that's what leads to long-term happiness and fulfillment.

Sonic Glasses - as early as 1970, Leslie Kay was developing ultrasonic eyeglasses for the blind, and by virtue of the fact that you were seeing with sound, they called it "Sonocular perception." So, if you can imagine a pair of glasses with three sensors mounted on top (one radiating ultrasonic waves, and the other two receiving them), the entire visual landscape would be turned into a sonic experience. Kay, L. (1973).

448

The Sonic Glasses Evaluated. *Journal of Visual Impairment & Blindness, 67*(1), 7–11. https://doi.org/10.1177/0145482X7306700102

Subconscious OS (Background OS) - the supposed background operating system that our baseline wakeful conscious state (CS5) runs on top of. Similar to the way Windows used to run on top of a background OS called MS-DOS, the Subconscious OS is the underlying preconscious substrate the VMs of self need to exist properly.

Synaptogenesis - the process of synapse formation and synapse maintenance between neurons in the nervous system. Interestingly enough, psychedelics actually promote synaptogenesis.

Synesthesia - a perceptual condition, which typically runs in families, whereby stimulation of one sensory modality unconsciously excites stimulation of another sensory modality. So, for example, the sight of black numbers written on a page may unconsciously elicit the sensation of color, causing the synesthete in question to perceive each number or letter as having its own specific color. This would be the most common form of synesthesia but there are many different types.

Theory of Mind (aka mentalizing) - theory of mind refers to the process of inferring what people are up to based on their actions. Seemingly, we cannot read other people's minds. However, we can watch the way someone acts and uses their body and predict what the other person might be thinking.

Umwelt (pronounced oŏmvelt) - originally coined by the German biologist Jakob von Uexküll, this term is meant to describe the idiosyncratic, perceptual world each creature is likely to inhabit, given that not all creatures have the same sensory inputs. For example, a bat's perceptual world is likely to be different than ours, seeing how their main sensory input is sound and not sight. By the same token, a dog's umwelt is likely to be different than a bat's on account of their main sensory modality being smell and not sound.

Virtual Machine (VM) - VMs are virtual computers that are capable of running inside physical computers. Typically, they run on an isolated partition of their host computer and are used by programmers to test certain apps or programs inside of a safe container.

About the Authors

JAY NELSON is a writer, researcher, and award-winning screenwriter obsessed with three questions: (i) What is consciousness? (ii) Is there an edge of experience? And (iii) How best to explain ecological awareness? After originally getting his start in acting, Jay soon found a love for psychology and explaining the human mind. But after a near-death-experience realigned all of his perceptions in 2013, he started moonlighting in neuroscience, philosophy, and evolutionary biology. He has been called a philosopher, a shaman, and a man of science. But, by his own description, he's a "moderately well-read person who wants to test the limits of human perception." You can find him online at www.jaynelson.com.

LINDY NELSON is a lifelong student of what makes a human a human. She has always believed that we are more alike than we are different, and is forever on the journey towards a better understanding of that knowledge and vocabulary for sharing it with the world. From Managing escape rooms to restaurants, film sets to corporate offices, she is adept at seeing all of the pieces that make up the whole. She leads with patience and the mantra that we should each use what we have to improve where we can.

Lightning Source UK Ltd.
Milton Keynes UK
UKHW040716190123
415612UK00004B/42